QUÉBEC STUDIES IN THE PHILOSOPHY OF SCIENCE

PART II

BOSTON STUDIES IN THE PHILOSOPHY OF SCIENCE

VOLUME 178

HUGUES LEBLANC
Courtesy of Suzanne S. Leblanc

QUÉBEC STUDIES IN THE PHILOSOPHY OF SCIENCE

Part II: Biology, Psychology, Cognitive Science and Economics

Essays in Honor of Hugues Leblanc

Edited by

MATHIEU MARION
University of Ottawa

and

ROBERT S. COHEN
Boston University

KLUWER ACADEMIC PUBLISHERS
DORDRECHT / BOSTON / LONDON

Library of Congress Cataloging-in-Publication Data

A C.I.P. Catalogue record for this book is available from the Library of Congress.

ISBN-13: 978-94-010-6537-5 e-ISBN-13: 978-94-009-0113-1
DOI: 10.1007/978-94-009-0113-1

Published by Kluwer Academic Publishers,
P.O. Box 17, 3300 AA Dordrecht, The Netherlands.

Kluwer Academic Publishers incorporates
the publishing programmes of
D. Reidel, Martinus Nijhoff, Dr W. Junk and MTP Press.

Sold and distributed in the U.S.A. and Canada
by Kluwer Academic Publishers,
101 Philip Drive, Norwell, MA 02061, U.S.A.

In all other countries, sold and distributed
by Kluwer Academic Publishers Group,
P.O. Box 322, 3300 AH Dordrecht, The Netherlands.

Printed on acid-free paper

TABLE OF CONTENTS

PHILOSOPHY OF BIOLOGY

PHILOSOPHY OF PSYCHOLOGY AND COGNITIVE SCIENCE

DECISION THEORY AND PHILOSOPHY OF ECONOMICS

EPISTEMOLOGICAL STUDIES

EDITORIAL PREFACE

By North-American standards, philosophy is not new in Québec: the first mention of philosophy lectures given by a Jesuit in the *Collège de Québec* (founded 1635) dates from 1665, and the oldest logic manuscript dates from 1679. In English-speaking universities such as McGill (founded 1829), philosophy began to be taught later, during the second half of the 19th century. The major influence on English-speaking philosophers was, at least initially, that of Scottish Empiricism. On the other hand, the strong influence of the Catholic Church on French-Canadian society meant that the staff of the *facultés* of the French-speaking universities consisted, until recently, almost entirely of Thomist philosophers. There was accordingly little or no work in modern Formal Logic and Philosophy of Science and precious few contacts between the philosophical communities. In the late forties, Hugues Leblanc was a young student wanting to learn Formal Logic. He could not find anyone in Québec to teach him and he went to study at Harvard University under the supervision of W. V. Quine. His best friend Maurice L'Abbé had left, a year earlier, for Princeton to study with Alonzo Church.

After receiving his Ph.D from Harvard in 1948, Leblanc started his professional career at Bryn Mawr College, where he stayed until 1967. He then went to Temple University, where he taught until his retirement in 1992, serving as Chair of the Department of Philosophy from 1973 until 1979. His achievements as a logician include seminal contributions to the development of Free Logic, in particular with the ground breaking paper, written jointly with Theodore Hailperin, 'Nondesignating Singular Terms' (*Philosophical Review* **68** (1959), pp. 239–43). After initial results by Bas van Fraassen, using supervaluation, Hugues Leblanc and Richmond Thomason obtained completeness results in 'Completeness Theorems for Some Presupposition-Free Logic' (*Fundamenta Mathematicae* **62** (1968), pp. 125–64). More recently, Leblanc also made seminal contributions to Truth-Value Semantics (cf. his *Truth-Value Semantics*, Amsterdam, North-Holland, 1976) and, inspired by appendices to Karl Popper's *Logic of Scientific Discovery*, to Probability Semantics and Probability Theory, in his paper 'Probabilistic Semantics for First-Order Logic' (*Zeitschrift für mathematische Logik und Grundlagen der Mathematik* **25** (1979), pp. 498–509). In all, Leblanc has written more than one hundred scientific papers, the more recent of them in collaboration with Peter Roeper (Australian National University), and four books, he collaborated on two books and edited or co-edited four. Many logic students will remember learning the subject from his classic textbook, written with William A. Wisdom, *Deductive Logic* (3rd edn., Englewood Cliffs, Prentice Hall, 1993).

After a long and fruitful career in the United States, Hugues Leblanc is now back in Québec, where the philosophical milieu has changed beyond recognition since his student days. He came back to find studies in logic and in all aspects of philosophy of science in a flourishing state. As a result of the *révolution tranquille* which took place among the French-speaking society in the sixties, philosophy in Québec opened up to external influences such as, initially, phenomenology and Marxism and, increasingly in the past twenty years, Anglo-American analytic philosophy. As a result, there is now a growing number of French-speaking logicians and philosophers of science – although not all of them work from the point of view of analytical philosophy. Conditions were set for fruitful exchanges with the English-speaking philosophical community. (But we should add here that the essential role of immigrants in the evolution of the philosophical life in Québec should not be overlooked. Contributors to the present volumes come not only from other parts of Canada, but also from Argentina, Australia, Belgium, Germany, Ireland, Switzerland, the United Kingdom and the United States).

Such exchanges have led recently to the creation of research groups across Québec. These are now joined together under the name of *Groupe de recherche sur la représentation, l'action et le langage* or GRRAL. Our two volumes of *Québec Studies in the Philosophy of Science* comprise the first full-scale collection of studies in the philosophy and history of science from French- and English-speaking philosophers of Québec to appear in English; they include in particular most members of the GRRAL. As editors, we are happy to join the contributors in dedicating these volumes to Hugues Leblanc, who is, among philosophers, the first *logicien québecois*.

In our second volume, we have collected papers in philosophy of biology, philosophy of psychology and cognitive sciences, philosophy of economics and decision theory, and relevant epistemological studies (while papers in logic, philosophy of mathematics, philosophy of physics and in general philosophy and history of science were collected in volume one). This volume includes members of two of the research groups forming the GRRAL, the group *Naturalisation de l'intentionalité* (D. Laurier, C. Panaccio, M. Seymour, A. Voizard) and a group located at McGill, working on the possibility of a scientific theory of meaning (S. Dwyer, J. McGilvray; the contributions of P. Pietroski and D. Davies are in volume I). Another group, *Rationalité et sciences humaines*, not affiliated to the GRRAL is also represented here (P. Dumouchel, M. Lagueux, R. Nadeau).

In the section on philosophy of biology, François Duchesneau explores the use of teleological arguments in biology, while Paul Dumouchel discusses selection type theories. In papers oriented towards the program of teleosemantics, Daniel Laurier examines the notion of natural function and Murray Clarke discusses the issue of the selection of true beliefs. The section on the philosophy of psychology and cognitive science opens with a paper at the border between theoretical biology and cognitive science, Evan Thompson discusses the notion of syntactic interpretability in relation to the 'symbol-matter' problem. Jim McGilvray looks at bearers of color properties from a subjectivist point of view. Paul Bernier argues against Burge's interpretation of Marr's theory of vision as anti-individualist, while Michel Seymour reexamines thought experiments (by Putnam, Burge & Kripke) which are used to support the later. Finally, Denis Fisette looks at the support given to Davidson's anomalous monism by the normative character of rationality.

The next section opens with papers on the foundations of economics by Robert Nadeau, who argues that intentionality is a phenomenon constitutive of the domain of economic theory, and by Maurice Lagueux, who challenges the possibility of a characterization of irrationality free of value judgements. Then follow three papers concerned with issues in decision theory: Alain Voizard argues for Bayesian decision theory by minimizing the impact of Newcomb's problem. Nicolas Kaufmann argues for the incorporation of prospective intentions in the model of expected utility maximization and Jocelyne Couture attacks individualistic interpretations of decision theory.

In the last section, we collected papers dealing with wider philosophical issues, which are nevertheless relevant and complementary to some of the issues raised in this volume: Susan Dwyer argues for a dispositional explanation of linguistic competence, while Claude Panaccio present his own nominalistic interpretation of belief sentences and Martin Montminy raises objections to Dummett's molecularist conception of meaning.

We would like to thank Alain Voizard for his help in writing this preface. Once again we express our gratitude to Annie Kuipers for her assistance, encouragement and patience.

Boston and Montreal   MATHIEU MARION & ROBERT S. COHEN
April 1995

FRANÇOIS DUCHESNEAU

TELEOLOGICAL ARGUMENTS FROM A METHODOLOGICAL VIEWPOINT

In Michael Ruse's *Philosophy of Biology Today*, the chapter devoted to teleology starts with a paradoxical statement:

A major philosophical puzzle about biology stems from the way in which its practitioners, unlike physicists and chemists, seem generally quite prepared to use teleological or functional language (Woodfield, 1976; Wright, 1976; Rescher, 1986; O'Grady, 1986). They talk about "ends" or "functions" or "purposes" – as in "the sail on the back of the dimetrodon exists *in order to* regulate body temperature" or in "heartbeats serve no useful *function*". Admittedly, every now and then biologists lecture themselves sternly on their practices. They assure themselves – and us – that their usage is just shorthand. And they try to pretend that they are not doing what they do, by refusing to talk of "teleology" and substituting "teleonomy" or some such thing (Mayr, 1974). But the language does persist (Ruse, 1988).

Using a teleological terminology seems unavoidable in biological discourse, while, at the same time, biologists would question granting those prevailing linguistic usages any positive import. Biologists justify for their seemingly inconsistent practice by pretending that teleological phrasing is but a convenient mode of expression: one could always translate propositions thus phrased into non-teleological statements, and get rid of whatever looks methodologically inadmissible in teleological arguments. Discarding the term "teleology" itself from any methodological assessment of positive scientific practice and replacing it by "teleonomy" bear a clear indication that a distinction is aimed at between an acceptable and an exceptionable use of the teleological mode of speech (Pittendrigh, 1958). It is epistemologically challenging to try and identify what would form the positive aspect of teleological argumentation in biology.

When discussing this issue previously (Duchesneau, 1977, 1978, 1980), I focused on functional analysis as part of an analytical strategy. A functional statement based on the description of a biological process may serve as a starting point in seeking the best explanation of those mechanisms underpinning the dispositional device involved. But such a statement cannot fulfil its methodological role in biology if it is not grounded in a theoretical representation, even a sketchy one, about the way structures and functions are integrated in organisms. In principle, one should ultimately be able to account for the dispositional devices and the more or less elementary structures that combine to achieve the empirically identified functional processes. But, in the meantime, short of reaching that level of understanding, analysis unfolds a provisional account of functional processes by projecting a set of anticipatory sufficient reasons. In due course, these sufficient reasons are expected to get causally determined by researchers eliciting theoretical models that may in turn receive adequate empirical confirmation.

1

M. Marion and R. S. Cohen (eds.), Québec Studies in the Philosophy of Science II, 1–12.
© 1996 *Kluwer Academic Publishers.*

As a matter of fact, the structural properties and activities of organisms may not be efficiently described without appealing to, and relying on, a finalistic semantics. But, without that kind of methodological stratagem, the featuring of phenomena would remain parcelled out, and as a consequence, we would be prevented from attaining a representative synthesis, a systematic representation of the object to analyze. Indeed, at the empirical stage of analysis, a functional relation cannot be considered as explanatory. Its strategical import is to suggest "metaphorically" some mode of integration to be specified for the causal device(s) in need of being accounted for. The search for models which is thus initiated and induced depends upon a theoretical program whose function is to generate those very models. A similar position has been argued for by Cummins (1975).

To sum up, a so-called functional analysis can be warranted when it aims at accounting for the global activity of a system in terms of lower-level devices: it serves to indicate, in a provisional fashion, how those devices may fit together according to an inherent program, and how they can thus match the requirements for higher-level organic activity. Functional analysis sketches a set of sufficient reasons to account for some global phenomena, and these sufficient reasons shape up in the form of programmatic hypotheses. In particular, inventing hypotheses of that kind is at the heart of theory-building in molecular genetics. But, in general, functional relations will probably serve as a means of conceptual determination for all theories that purport to represent the specific order of biological phenomena. It would evidently be the case with central concepts in physiological theories: for instance, concepts such as "internal milieu", "integrative action of the nervous system", "metabolic regulation", etc. It seems as if every theory concerned with the combination and integration between various levels of organic elements – organisms, tissues, cells, corpuscles, genes, etc. – frames up a set of sufficient reasons analyzable in terms of integrative mechanisms and dispositional devices. Hence, my previous query:

Insofar as a theory purports to feature the integral order of given phenomena by spinning an explanatory web, should we not grant functional analysis, besides a heuristical role at the empirical level, a kind of regulatory role in the framing of biological theories (Duchesneau, 1980: 267)?

Recently, we have witnessed significant moves concerning the strategies involved in biological theory-building. The irreplaceability and autonomy of teleological argumentation seem to have been specially set into relief by philosophers of biology. On various accounts, works by Rosenberg (1985), Bechtel (1984, 1986), and Campbell (1985) represent this new trend and deserve a critical assessment from the standpoint I have chosen.

Rosenberg states that teleological arguments in description as well as explanation characterize *prima facie* the whole of biology, including domains in which biochemical models seem to prevail. He takes as an example the analysis accounting for the presence of thymine as one of the four structural bases of

DNA, while the corresponding element for RNA is uracil. Thymine enables a faithful replication of the DNA structure, even if this process involves a higher energy cost. The same requirement of faithful replication does not constrain the transcribing of RNA. Appealing to this type of argument to account for a structural difference underlines the role of a given type of effect as a specific functional ingredient in the sought for explanation (Rosenberg, 1985: 40). If we suppose that biological phenomena are thus identified by their effects independently of structural analysis, and that appealing to functions affords an explanatory argument, the question remains whether these methodological traits can be eliminated without jeopardizing the possibility of providing a theory for those phenomena.

The attempt to concede teleology a positive role in theorizing, entails two apparently antinomic theses. One consists in holding that biological phenomena cannot imply a different type of causality from that of physico-chemical phenomena: there is no question of admitting any "vital principle" acting in a sort of intentional fashion. If there is evidence of goal-directedness in biological phenomena, the teleology involved should be compatible with an ontology that would restrict itself to admitting efficient causes of a mechanistic type behind phenomena. But, on the other hand, one must justify the fact that teleological description and explanation may not be dispensed with. Biological systems seem to require functional arguments at the very level of theoretical representation in order that phenomena may be accounted for.

The strategy which Rosenberg favours draws on the analyses of Cummins (1975) and Nagel (1977). The problem at hand boils down to the question of conceiving functional devices as belonging to self-regulating systems; these upper-level systems result from layers of lower-level systems so contrived and combined as to interact causally and issue goal-directed effects. Ultimately, the lowest-level systems are non-teleological: hence the compatibility of the explanation with the norms of cybernetic mechanism. This strategy makes it possible to underline the specific analytical role devoted to the unfolding of functional effects. Such an analysis serves a heuristical function by focusing on the boxed-in layers of dispositional devices: it sets the target for unveiling those elementary structures and their modes of combination, which warrant self-regulation and vital organic integration.

Apart from that empirical role, a question remains, however, about the reduction of laws of the teleological kind insofar as they involve non-strictly empirical arguments. Indeed, these hypothetical law statements play a heuristical role when one is concerned with identifying the components of lower-level systems and their effects in the framing up higher-level self-regulating systems. But identifying those components does not afford a set of necessary and sufficient conditions for their production that would enable getting rid of the provisional and hypothetic approach based on purported functional effects. Researchers are confronted with the almost infinite diversity of means for attaining or achieving any biological end, especially when dealing with structural levels higher than those dealt with by molecular genetics. In order to get a full retranscribing of hypothetical teleological laws into non-teleo-logical statements about the more elementary integrative devices, retran-scription should state disjunctively the entire set of mechanisms that could account for the same functional effect; and one should not omit the unfolding of subsidiary conditions relating to other lower-level systems interacting with the one under analysis.

We are left to speculate whether the successive stages in such a reduction could be gone through in an accessible process implying a finite number of steps. The postulate which drives modern biology forward seems to be essen-tially that of *de jure* reductionism. But, in terms of research strategy, teleo-logical analysis seems to be practically unavoidable even at elementary levels – intracellular or molecular. For it warrants the efficiency of explanation in terms of plain intelligibility as well as systematic consistency. Teleological analysis matches at least pragmatic criteria for the validity of hypotheses, by contrast with more formal, and properly logical ones. However, on its positive side, that kind of research strategy should be understood to imply the view that functional traits depend upon an architecture of integrated lower-level systems making for the goal-directedness of the more global system. At the theo-retical level, this postulate would ultimately impose the task of character-izing the type(s) of dispositional devices that can "mechanically" produce the functional effect. Rosenberg simply argues that biological explanation sets itself between two poles, that of predicting and that of providing an adequate representation of the empirically revealed order: (1) predicting implies investigating the operative modes of those mechanisms which intervene in the interconnected lower-level systems; (2) explaining the order of phenomena in a consistent and pragmatic way – i.e. symbolizing it in a systematic fashion – requires an approach through functional analyses and arguments. Rosenberg's point is that persisting theoretical arguments of a finalistic type relate to an intent to systematize and explain, which overrules the objective of exact pre-dicting. This prevalence seems to hold for a large part of the biological corpus (Rosenberg, 1985: 67).

Rosenberg's thesis does not quite satisfy those who are concerned with the structure of biological theories. We are left to determine how reductionist and teleological models intervene respectively in the theorizing. This question is generally left in abeyance by philosophers of science, because it relates to actual practice, which seems a matter for historical concern rather than for normative appraisal. But, on the one hand, it should prove useful to inquire into the heuristics of biological theories as they formed historically since the early nineteenth century. On the other hand, one ought also to consider the specific models which contemporary molecular biology developed, and resorted to, in order to escape the methodological limitations of strict reductionism.

Indeed any investigation about teleological analysis will benefit from inquiring about the past research strategies which framed up for a large part the discursive texture of biology as a science. Bechtel (1986) considers that teleological models exemplify the more fundamental aspect of biological theories, since they alone would account for the organic and functional integration featured by vital phenomena. In support of this view, he refers in particular to the historical trend which has been identified as "teleomechanism" (Lenoir, 1982). This methodological and theoretical trend grew out of eighteenth century vitalism and organicism (Duchesneau, 1982, 1985), but it essentially developed in the first part of the nineteenth century as an alternative to *Naturphilosophie*. Contrary to *Naturphilosophie* which fostered teleological arguments of a more profoundly metaphysical sort, teleomechanism would combine experimental methods eliciting specific physico-chemical laws with functional analyses destined to account for the integrative aspect of biological phenomena. Teleomechanism dominated German physiology from J. F. Blumenbach (1752–1840) to J. P. Müller (1801–1858), roughly from 1790 to 1850, before it was overrun by the advent of more reductionist and positivistic methodologies. All theories pertaining to teleomechanism featured a heuristics based on a reflexive concept of functional integration: such a heuristics was called upon to guide the analysis of biological processes in terms of interdependent mechanisms and to account for the causal relations featured by vital phenomena. This reflexive concept signified a *sui generis* cause elicit-ing the emergence of complex structures capable of autonomous functioning; the same cause would regulate processes in a functional manner within organisms. Historically, the teleological arguments embodied in the teleomechanist research programme would present two different forms. In the first version, Blumenbach's *Bildungstrieb*, for instance, or K. E. von Baer's *Gestaltungskraft* represented an emergent capacity to generate certain effects that formed specific features of living organisms; that capacity was conceived as related to an internal organization plan for developing the various elements constitutive of organs and, at a higher level, of organisms. Later on, the second version suspended direct reference to a vital principle eliciting the emergence of morphotypes, i.e., of specific living organizations and their ensuing functional effects. Henceforth, the focus was on the specific determinations caused by the internal disposition of organisms and the integration of organic structures: those constraints would determine the way physico-chemical laws apply to components and structures so that properly identified functional effects come about. This second phase was illustrated by the concepts and theories of J. Liebig, C. Bergman, and R. Leuckart. In the history of teleomechanism, the advent of the cell theory with M. Schleiden and T. Schwann in 1838–1839 (cf. Schwann, 1969) would mark the transition from one phase to the next. But Lenoir's analysis, which I followed in establishing the specifics of teleomechanism, is nevertheless to be faulted for its distorting interpretation of the Schwannian cell theory (cf. Duchesneau, 1985). This theory cannot fit the teleomechanist scheme, at least when one focuses on the essentials of the tradition it induced. Since the cell theory probably formed the initial leading research programme for contemporary biology, its features should not be underestimated; and they are of special interest because they seem to illustrate the methodological limits involved in resorting to teleological arguments.

The history of cell theory up to the present sets in relief the fact that it consistently afforded some schematic representation for the basic integrative devices of organisms (cf. Duchesneau, 1987). Schwann himself tried to link the production of this structural device with physico-chemical processes which he took to be analogous to crystallization. On the one hand, he seemed to reserve the concepts of a teleological type for expressing in a descriptive way the general morphogenetic and metabolic features of vital phenomena. But, on the other, in the Schwannian programme, means for achieving analysis and explanation consisted more precisely in combining mechanistic models. Such mechanistic models were expected to become progressively more sophisticated, in particular those concerning embryogenesis and the derivation of the more complex anatomical structures from plain cells. The purpose of the theory was to warrant the compatibility and congruence of those models with the structural-functional schematic representations, i.e. the descriptive teleology, which serve as a guide in theory-building. We may indeed interpret those representations as regulative functional arguments; but we shall not hold that these arguments played an explanatory role proper. Such a role was assigned to the evolving set or combination of mechanistic models.

I am not therefore ready to agree with Bechtel's notion of functional explanation, which he derives directly from teleomechanism (Bechtel, 1984): it bears too strong a theoretical meaning in contrast with the more limited acceptation teleological concepts received in the history of cell morphology and physiology. Biological science seems to have discarded this strong meaning in the course of its history for the sake of a more relative teleological schematism which played an ancillary, though indispensable, role in a theory-building process based on eliciting progressively more refined reductionist models. Conversely, one of Rosenberg's statements seems specially warranted in this context: the schematic representation by means of which the spotting of functional features operates, cannot get reduced to the mechanistic models framing up the theory. And one should add that the schematic representation itself does not remain statically provisional: it evolves with the invention of explanatory mechanistic models. Correlatively, these models also vary in their structure, due to their relationship with the regulative concepts guiding the theory-building process. From my viewpoint, finalistic arguments seem to relate to structural-functional schematic representations: these seem to be grounded in a kind of regulative idea biology makes its profit with, without getting alienated to it. Finalistic arguments of a descriptive kind always need to be interpreted by means of mechanistic models, since their purpose is to guide the search towards experimentally warranted causal explanations.

Evidently, the historical case ought to be further explored; and a more exacting investigation along those line still ought to be undertaken. But I shall turn now to a differently focused inquiry. My aim is to check whether the same general conception of functional analysis is likely to fit certain features of contemporary theories at the upper and lower boundaries of cell biology. This means taking interest in analyses that pertain to gene structures and genetic mechanisms on the one hand, in analyses that pertain to the global functioning of organisms and their interaction in the external milieu on the other. We are therefore referred back to the theoretical make-up of molecular biology and evolutionary theory, and it is worthwhile attempting to discern there new forms of schematic teleological representation destined to complement and integrate the reductionist and mechanistic models.

J. H. Campbell has produced interesting analyses on that very topic. According to him, biological explanation is essentially characterized by models created to account for the correlation between structures and functions at various levels of empirical investigation (Campbell, 1985). To sum up, biologists are concerned with functional processes as these depend upon complex structures resulting from evolutionary adaptation. In neo-darwinian perspective, these complex structures emerge from purely mechanical processes without inherent goal-directedness. In line with this postulated reductionist substratum, biological theory had kept building theoretical models that correlated the features of structural devices and their functional meaning for the integrative organic activity. Molecular biology seemed able to reabsorb those correlations by unfolding the genetic mechanisms at the basis of vital organization. Campbell's point is that functional analysis still rules over the interpretation of those mechanisms and their relating to environmental evolutive determinations.

Three aspects of contemporary theorizing in molecular biology drive us away from previous neo-darwinian methodological reductionism. Genes show up as possessing an integrative structure. Gene replication as a substratum for the metabolic activity of organisms implies coding processes that depend on multigenic families: these processes make it possible to stock and use information in a more stable, more adaptive, and more functional way to generate phenotypic features. Traditionally, genes were conceived as molecular structures issuing messages which were rigidly replicable from one cell generation to the next. It seems as if cells possess a manifold of enzymes capable of interacting with representatives of the various multigenic families so as to modify the originally coded message according to what might be viewed as an adaptation project. The modifications of DNA molecules by means of enzymes are subordinate to some functional regulation at the level of multigenic families due to control devices termed *governors*. But the basis on which evolutionary mechanisms act, consists precisely in gene variations. Within the functional integration due to enzymatic regulation, those variations that seem significant for evolution result from an integrative structuring of shifting alleles whether these result directly from the devices producing the alteration or have been merely induced through those devices. Another meaningful element for functional representation in molecular genetics is the *operon* structure which characterizes genes endowed with a differential sensitivity to surrounding information. This structure combines a repressor with a gene that can code for the transcribing of a specific enzyme (Jacob and Monod, 1961).

In light of those major analytic features of molecular genetics, Campbell underlines that theoretical research in such a domain consists in building adequate representations of the mechanisms underpinning the progressively more complex systems of structural-functional correlations: the models thus conceived imply that the functional properties emerging from integrative processes may be accounted for. It is not enough to observe linear series of causal determinations starting from the replication of basic molecular structures; one should also explain how the various structural levels are organized so as to produce complex effects which express the emerging functional order. This appears particularly relevant when looking for genetic sufficient reasons to account for complex regulatory systems in eukaryotes. Those systems are indeed conceived through analogy with similar, though basically less complex, systems in prokaryotes; but, beyond this analogy, one may acknowledge that there is increasing goal-directedness within the structural determinants, with the result that they elicit more complex self-regulatory processes.

Up to this point, an epistemological analysis like Campbell's does not diverge considerably from what we learn from the progressive unfolding of the essentials of biological theory, at least starting from the advent of cell theory at mid-nineteenth century. It has the merit of signalling that no causal explanation of biological processes can avoid making room for describing and analyzing structural-functional relationships at various complexity levels. In this context, resorting to teleological language is a means of characterizing such relationships. When analyzing organic devices, the concepts resorted to point to the functional significance of the processes those devices may accomplish. Indeed, all concepts of this kind are not equally admissible: biologists should only retain those relating to models that may transcribe functional operations in analytically adequate fashion. This is the case, for instance, with models that account for molecular devices through biochemical processes.

In this context, one can indeed pretend that the neo-darwinian, reductionist methodology is reinforced in its *de jure* prerogatives: devices generating functional effects are presumed to derive from mechanical processes involving material determinants of the vital milieu. Thus, the mechanism of natural selection would operate on phenotypic traits as produced through contingent variations affecting genomes: adaptive structures would result without implicit goal-directedness. Teleological argumentation would intervene without causal implication in the analysis of emerging effects, and it would afford a description phrased in "as if (*als ob*)" mode: such a description would serve the purpose of translating the "phenomenal" order of integrative structures and processes. But, under these conditions, such a phenomenal order can never mean the cause of functional integration itself.

However, Campbell envisages extending the teleological model of structural-functional correlations to the evolutionary process itself. Under certain circumstances, the gene structures would possess the capacity of exerting an evolutionary function. Thus, *transposon* structures can be observed shifting place within the organic device they help build and control. Hence, they would help produce rearrangements and evolutive variations in the genome. They form elementary structures, but they also imply a functional disposition: they increase the evolutive capacity of organisms by granting them additional capacity for phenotypic adaptation to change, in comparison with concurrents in the same ecosystem. Such structures possess a supplementary function, that of programming, so to speak, their own transformation for the sake of further adaptations. An interesting example is the transposon integrating the gene responsible for streptococcus resistance to erythromycin, a transposon named *Tn917*. In that structure, the repressor plays an adaptive role by coding the replication of the antibiotic-resistant gene; it also plays an evolutive role by changing – for the sake of the phenotype, so to speak – the way the transposition gene operates and codes for enzymes of the transposase type. Thus, we are confronted with gene structures endowed with evolutive functions: these gene structures serve as *evolutionary drivers*, or even more significantly, as *evolutionary directors*. Such regulatory systems account for the existence of crucial alternatives in pathways for adaptive development; they also account for the reinforcement of given sequences of adaptive variations in different possible ways according to the group of mutants.

From thereon, it seems difficult to resist being seduced by psychomorphic metaphors: the terms involved tend to overdetermine theoretical concepts denoting goal-directed processes. Campbell has shown that a number of complex gene structures possess *sensors* which enable evolutionary directors to code for adaptive processes. For instance, DNA metabolism in the *Escherichia coli* bacteria is governed by such a system: this system, so to speak, "induces" mutations in response to alarm signals. This is what Campbell means by *sensory evolution*. But, at that point, one cannot but wonder whether the type of structure-function correlation described in prokaryotes can serve to explain infinitely more complex systems in eukaryotes: for example, immunitary regulation systems, or systems endowed with powers for prospective adaptation, such as the endocrine and nervous systems of the more complex metazoa (Mathen and Levy, 1986; Levy, 1988). At all levels, sets of metaphors can intervene and render it very ambiguous to determine the type of causality involved. A good instance of that tendency is afforded by Campbell himself in one of his theoretical or speculative conclusions:

[. . .] Evolution is more than just adaptation. More important, it is a progressive process that teaches the species how to evolve. A gene pool learns by adapting. It learns the strategies and directions that were successful in the past. Those species and lineages that develop inappropriate or ineffective strategies are eliminated, while the surviving ones develop evolutionary directors that bias them to continue their successful modes of adaptation (Campbell, 1985: 146).

Even if there is an easy abuse of metaphorical language, biologists are spontaneously reluctant to admit that an epistemologically valid model for causal explanation may be grounded on functional analysis. This is the case, even if functional analysis rules significantly over discovery strategies, and possibly theory-building. Indeed, we use such models to interpret structure-function correlations in a heuristical fashion, with the proviso that this practice will facilitate access to mechanistic explanatory models. I believe on that account that discussion should refocus on the distinction between acceptable descriptive or analytic models and those that would not qualify and should be discarded.

At first sight, resorting to a finalistic argumentation in contemporary biology ought not to mean adopting some explanatory system based on theoretical entities that would embody *a tergo* determination by final causes. We should be more interested with the way some theoretical entities, when legitimately agreed to, relate to analytical models which account for the synchronic as well as diachronic organization of organic devices and the ensuing processes. In the way they form, as well as in the effects they generate, these processes are subject to representation by means of functional terms; but such representations must be correlated with analytical models expressing an objective transcription of phenomena. Under these conditions, a certain number of finalistic models might embody a mere system of subjective representations. The difficulty is with determining the criteria of methodological and theoretical validity for "as if" conceptual constructs in which protracted metaphorical phrases intervene. The epistemological control can only come about with a critical inquiry concerning the analytic reference models which serve in ordering, and accounting for, empirical data. The philosophy and history of biological science may probably testify to the fact that not one truly sanctioned conception about the validity of analytic models has ever ruled exclusively over any phase in the recent development of that science. Any present epistemological analysis about those models should take into account a basic fact: a plurality of research and explanation strategies coexist within such research programmes as those that organized the progressive development of analytic models; and teleological symbolization will play a significant part in determining this manifold of strategies.

To sum up, biologists do not seem to be able to dispense with finalistic terminology. Under which conditions is a teleological argumentation justified? Initially, contemporary philosophers of science would insist that teleological arguments may be entirely translated in terms of "mechanistic" models. Later on, functional analysis did appear as a methodological process that serves to support the linking of functional processes with devices consisting in integrative structures. Hence what we identified as an architectonic role in theorizing, that of proposing a schematic representation to account for structural-functional integration.

For instance, Rosenberg (1985) stressed the heuristical import of an analysis that points to an architecture of structural devices that would serve in accounting for the processes and functional behaviors of organisms. But, for him, the full potential translation of functional statements into causal law statements shall remain suspended, and, as we need to systematize presently our representation of phenomena, we shall be led to favour analyses in terms of functional properties, instead of trying to unravel all the involved mechanisms. The combination modalities among teleological and non-teleological models in the same theory would then be left undetermined, and the apparent antinomy between the two methodological approaches would tend to persist. But I think that by setting the appropriate order of correlation or subordination between the two types of models, we could suppress at least part of the antinomy. According to Bechtel (1984, 1986), teleological models have formed the main thread of biology in the course of its history. He cites as an example the impact of *teleomechanism* on the development of morphogenetic theories. My own analysis of the cell theory from Schwann onwards challenges the notion of an unconditional prevalence of teleological arguments in this regard. In this case, if theorizing requires a functional schematic representation, the research programme aims at developing analytic tools, namely a set of mechanistic models which can be adjusted and perfected. A stronger thesis on behalf of finalistic models has been proposed by Campbell (1985) in the context of molecular genetics. According to him, functional properties originate from genetic devices. Hence the requirement that a causal account be provided for structure-function correlations. Campbell stresses that these sufficient reasons must be specific and reflect the evolutive functions that rule over organic structuring. But then he runs the risk of "ontologizing" functional determinations beyond their provisional usage in characterizing phenomena. To avoid that trap which Campbell tends to fall in, we should focus on the admissible analytical models biologists develop to account for the teleologically depicted structure-function correlations. We should try to assess the way functional theoretical concepts are transcribed by means of such models in accordance with specific research strategies. Subjected to this methodological test, teleological arguments will probably manifest themselves as useful, possibly indispensable, analogies for biologists in drawing up, adjusting or relativizing their analytic explanatory models.

Université de Montréal

BIBLIOGRAPHY

Bechtel, W., 1984, 'The Evolution of Our Understanding of the Cell: A Study in the Dynamics of Scientific Progress', *Studies in the History and Philosophy of Science* 15, 309–356.
Bechtel, W., 1986, 'Teleological Functional Analyses and the Hierarchical Organization of Nature', in N. Rescher (ed.), *Current Issues in Teleology*, University Press of America, Lanham, pp. 26–48.
Campbell, J. H., 1985, 'An Organizational Interpretation of Evolution', in D. J. Depew and B. H. Weber (eds.), *Evolution at a Crossroads: The New Biology and the New Philosophy of Science*, MIT Press, Cambridge (Mass.).
Cummins, R., 1975, 'Functional Analysis', *Journal of Philosophy* 72, 741–765.
Duchesneau, F., 1977, 'Analyse fonctionnelle et principe des conditions d'existence biologique', *Revue Internationale de Philosophie* 31, 285–312.
Duchesneau, F., 1978, 'Téléologie et détermination positive de l'ordre biologique', *Dialectica* 32, 135–153.

Duchesneau, F., 1980, 'Analyse fonctionnelle et causalité biologique', *Revue Internationale de Philosophie* **34**, 229–267.

Duchesneau, F., 1982, *La Physiologie des Lumières. Empirisme, modèles et théories*, Nijhoff, The Hague.

Duchesneau, F., 1985, 'Vitalism in Late Eighteenth-Century: The Cases of Barthez, Blumenbach and John Hunter', in W. F. Bynum and R. Porter (eds.), *William Hunter and the Eighteenth-Century Medical World*, Cambridge University Press, Cambridge.

Duchesneau, F., 1987, *Genèse de la théorie cellulaire*, Bellarmin, Montréal; Vrin, Paris.

Jacob, F. and Monod, J., 1961, 'On the Regulation of Gene Activity', *Cold Spring Harbor Symposium on Quantitative Biology* **26**, 193–211.

Lenoir, T., 1982, *The Strategy of Life: Teleology and Mechanics in Nineteenth Century German Biology*, Reidel, Dordrecht.

Levy, E., 1988, 'Networks and Teleology', *Canadian Journal of Philosophy*, Supplementary Volume 14, 159–186.

Matthen, M. and Levy, E., 1986, 'Organic Teleology', in N. Rescher (ed.), *Current Issues in Teleology*, University Press of America, Lanham.

Mayr, E., 1974, 'Teleological and Teleonomic: A New Analysis', in R. S. Cohen and M. W. Wartofsky (eds.), *Boston Studies in the Philosophy of Science* **14**, 91–117.

Nagel, E., 1977, 'Teleology Revisited', *Journal of Philosophy* **74**, 261–301.

O'Grady, R. T., 1986, 'Historical Processes, Evolutionary Explanations, and Problems with Teleology', *Canadian Journal of Zoology* **64**, 1010–1020.

Pittendrigh, C. S., 1958, 'Adaptation, Natural Selection and Behavior', in A. Roe and G. G. Simpson (eds.), *Behavior and Evolution*, Yale University Press, New Haven, pp. 390–416.

Rescher, N., 1986, *Current Issues in Teleology*, University Press of America, Lanham.

Rosenberg, A., 1985, *The Structure of Biological Science*, Cambridge University Press, Cambridge.

Ruse, M., 1988, *Philosophy of Biology Today*, State University of New York Press, Albany.

Schwann, T., 1969, *Microscopical Researches into the Accordance in the Structure and Growth of Animals and Plants*, Kraus Reprint, New York.

Woodfield, A., 1976, *Teleology*, Cambridge University Press, Cambridge.

Wright, L., 1976, *Teleological Explanations*, University of California Press, Berkeley.

NATURAL SELECTION AND SELECTION TYPE THEORIES[1]

1. In a relatively recent article, Lindley Darden and Joseph Cain (1989) propose a general characterisation of selection type theories, which is to say, according to them, of all theories which are of the same type as the theory of natural selection. Their goal, they tell us, is to construct an abstraction, a schematic outline that can be filled in by any instance of a theory of a certain type. In this case, it is to be filled by a selective theory. According to them, types of theories solve types of problems. For example, selection type theories solve adaptation problems.[2] Among the examples of selective theories Darden and Cain mention are the theory of natural selection, the clonal selection theory of antibody production (Burnet, 1957), the theory of neural group selection (Edelman, 1987), and the theory of epigenesis of neural networks by selective stabilisation (Changeux, 1983). Darden and Cain are interested in analogous theories which can be found in different fields or domains, but their point is not to unify these fields under a unique paradigm, understood as an all-inclusive theory. Rather, their ultimate aim is to provide scientists guidance in theory construction by allowing them to specify in advance the minimum requirements of an adequate theory when faced with a certain type of problem (1989, p. 125). Here, natural selection plays the role of a paradigm in the Kuhnian sense of an exemplar, that is, something which allows a scientist to see a new problem as like a problem encountered before. (Kuhn, 1969) Darden and Cain's contribution can be seen as an attempt to analyse this exemplary quality of a theory in relation to a certain set of problems.

In a slightly different sense of what a type of theory is, many scientists feel the same way as Darden and Cain do about recent selective theories. That is: they consider that current appeals to selection in immunology, neurology or development indicate that, in these domains also, the same type of theories are at work as the theory of natural selection. Jerne (1965), Changeux (1983), Edelman (1987) all recognize in the selective theories of which they are the fathers a shift to "population thinking" and an extension of the "Darwinian" approach, method or paradigm, to new domains of inquiry.

Doubts as to the adequacy of this characterization arise from reading some of Massimo Piattelli-Palmarini's attacks upon the instructive approach to learning in cognitive psychology (1986; 1989). Piattelli-Palmarini argues that the rise of selective theories in biology – he mentions to a large extent the same theories as Darden and Cain do – and in linguistics marks the beginning of a new era in both fields. But one which, according to him, ruptures with "Darwinian adaptation", at least as conceived of in the framework of

13

M. Marion and R. S. Cohen (eds.), Québec Studies in the Philosophy of Science II, 13–24.
© 1996 *Kluwer Academic Publishers.*

what Gould and Lewontin named the panglossian paradigm.[3] Contrary to
Darden and Cain, Piattelli-Palmarini associates selective theories not with neo-
darwinism but with Gould and Lewontin's (1979) criticism of adaptationism,
with the theory of punctuated equilibria (Gould and Eldredge, 1972) and
with the concept of exaptation (Gould and Vrba, 1982). He takes all of
these theoretical proposals to be signs that in evolutionary biology itself
major conceptual changes are taking place which are driving it away from
neo-darwinism. He believes that selective theories pursue this more recent
line of theoretical development rather than extend the domain of adaptive
explanation.

These two apparently divergent evaluations of the same phenomenon are
not necessarily contradictory. Darden and Cain's conception of a theory type
does not entail that two theories of the same type when applied to the same
domain will not have opposing consequences. Hence their proposal should
not be understood as excluding competition between theories of the same
type within one domain.[4] Nonetheless, if types of theories are to some extent
construed in the sense of exemplars, and it seems clear that they are, then it
is a reasonable desideratum that the analysis of theories into types be sufficiently precise to distinguish large paradigm shifts and major changes of
explanatory principles. My contention is that the passage from the theory of
natural selection to what are commonly termed selective or selection type
theories constitutes such a major shift in explanatory principles and an important change in the problems to be explained.

In what follows I will argue that some of the examples of selective theories,
common both to Darden and Cain and Piattelli-Palmarini,[5] belong to a different
type of theory than the theory of natural selection, in any relevant sense of
type of theory.[6] My claim is not simply that Darden and Cain have mis-
applied their own criteria and inadvertently classified as belonging to the selection type some theories which do not so belong. Rather, I submit that their
method of classifying theories into types is insufficiently precise to capture
some important changes in what is explained and in how it is explained.[7] It
is clear that, in order to sustain my claim, I do not need to show that all of
the theories which Darden and Cain consider as examples of selection type
theories belong to a family of explanation different than that of natural selection. To suggest that their classification is inadequate, it is enough for me to
show that some of the theories which for them are clear cases of theories of
the same type as the theory of natural selection, but applied in another domain
than the evolution of species, are in fact theories of an altogether different type.
Whether or not some other theory, also used as an example by them, is of
the same type as the theory of natural selection does not affect my argument.
This said, my main target is not Darden and Cain's conception of a type of
theory, but the conceptual changes currently taking place in biology. I think
that the shift from natural selection to selective theories indicates a change
in the questions which are being asked in biology, and in turn, in what seems
in need of explanation.

2. Such claims rest, first of all, on a definition or analysis of the theory of natural selection.[8] This is always a dangerous task since no such characterisation, once it has past the level of an elementary description, seems able to satisfy everyone. Luckily enough, for our purposes, an elementary description is sufficient. Such a description contains at least the following elements:

1. A population characterized by the existence of variations in certain traits or properties of the individuals. This implies the existence of a mechanism of some sort which is responsible for the production of variation in the population.

2. The fact that individuals also vary in fitness and reproduction because, or as a result, of their variation in certain traits or properties.

3. The traits or properties of the individuals which cause variation in fitness and reproduction are heritable.

That such a characterisation is insufficient is clearly brought to light by the fact that, as David Hull (1988, p. 403) noticed, it does not allow one to distinguish between natural selection and "Lamarkian" selection. That is, it cannot discriminate a selective theory from an instructive theory! In order to separate instructive selection from non-instructive selection, one must introduce a further rule which excludes that steps two and three have a directional effect upon the mechanism producing variations.

4. The mechanism producing variations is independent of steps 2 and 3.

Selection drives the system by favoring the fitter variants; it also drives it by inducing the production of on average fitter variants, that is to say by ensuring that the more adapted genotypes will have a greater statistical representation in the next generation, but natural selection does not make it more likely that the next mutation will enhance fitness. As is commonly said, mutation is random or blind.

The result of the process of natural selection is taken to be twofold: adaptation and evolution. That is what natural selection explains, the evolution of species and the fact that individual organisms are relatively well adapted to their environment. Adaptation follows from the fact that natural selection favors the fitter variants in each generation. In consequence, given a sufficient lapse of time, organisms tend to be adapted to their environment. Given that the environment changes, one should not expect organisms to be perfectly adapted. Evolution results from the fact that as populations are invaded by fitter variants and as changes in the environment favor different traits, reproductively linked populations will exhibit changing characteristic traits.

These two consequences of natural selection are, I believe, neatly captured by the doctrine that fitness supervenes on both properties of the organism and properties of the environment.[9] Two identical organisms placed in different environments would not have the same fitness, just as two different organisms placed in an identical environment do not have the same fitness.[10] Which is to say that fitness is a relational property, rather than an intrinsic property attached to an individual object or class of objects. As a consequence, one should expect that biological universals are not nomological. Common properties of organisms are assumed to be shared by descent through historical accidents, rather than for some in principle reason linked to the selection process. Different histories could have produced different properties. Vertebrates are tetrapods in virtue of a historical contingency, not because of some inherent superiority of that body plan.[11] The supervenience of fitness upon both organismic and environmental properties and the absence of nomological biological universals indicates the way in which natural selection explains adaptation. Adaptation, as Sober (1984) reminds us, is an historical concept. To say that some trait is an adaptation is to say something about its origin, that it evolved as a result of natural selection.

3. Consider now a selection type theory like Changeux's theory of selective stabilisation of neural networks. To some extent it satisfies our elementary description of natural selection. At the origin there are populations of neurons characterised by a variety of synaptic contacts. Through interaction with the environment, some of these synaptic contacts are favored, in some unspecified way. As a result of repeated encounters with external stimuli the favored synaptic connections are selectively stabilised. But, of course, there is a lot more to Changeux's theory. This part alone, which tells us why it is called the theory of "selective stabilisation of neural networks", explains little more than the fact that there are more synaptic connections at birth than five years later. This is not a trivial achievement, if it turns out to be true, since it certainly is important to know why Mother Nature seems bent on producing a lot more dendrites than will actually be used and also a lot more neurons than apparently seem necessary.

In order to get a better grasp of Changeux's theory, it is useful to consider the historical problem situation out of which it arose. Changeux himself (1979; 1983) reminds us that, at the beginning of the 1970's, one of the major difficulties for neurology was to account for the intricate pattern of synaptic relations in the brain in the absence of sufficient genetic information to specify these connections. The first step in the solution came from results in automata theory showing that with exactly two (yes/no) signals and very few branching rules it was possible to generate extremely complex networks. Although no two networks produced by the same rules would be identical they would exhibit similar patterns of connectivity and related behaviour. Transposed to a neurological setting this meant that the organism could be relatively sloppy in establishing synaptic connections as long as the resulting network fell within a relatively broad class of systems. High redundancy could favor that outcome. Changeux then proposed that the organism's encounter with its environment would result in a differential activation of synaptic connections causing the stabilisation of certain connections characteristic of a mature network. He further proposed that this reduction in redundancy, following Atlan's principle of complexity from noise (1972; 1979), constituted a gain in the complexity of the system which could be interpreted as a form of learning. (Changeux, 1983) He thus provided an interesting answer to the question of why we originally have more synaptic connections than later on. High redundancy of the original neuronal network increases fitness directly by allowing us to learn and indirectly by helping us to end up with the pattern of connectivity characteristic of our species.

If we ask, now, what is the cause or the origin of the fact that the nervous system is adapted, either in the sense of being able to learn, or in the sense of having the characteristic behaviour of a human brain, or of having the right connectivity in the absence of sufficient information to code for specific connections, the answer, at least within the theory, is not so much selection, as the fact that the system belongs to a class or family of highly complex systems. It is because of certain characteristics of self-organising systems that reduction of redundancy within the systems can be equated to a gain in complexity. It is because of certain characteristics of the class of complex systems to which it belongs that the nervous system exhibits its particular pattern of activity. Finally it is because the dynamics of many different network structures end up in the same attractor set[12] that the nervous system can be sloppy and redundant in original synaptic connections. In short, though selection does play a role in this theory, most of the explanatory power has been passed on to certain characteristics of complex systems and to the body of knowledge associated with these systems.

Of course, one may want to respond that if the nervous system exhibits these characteristics, typical of a class of abstract systems, perhaps is it because there was selection for them over a long period of time? This is most probably true, but note that this is not the role which selection plays *within* our selective theory. That is to say, that is not the reason why Changeux's theory is called selective as opposed to something else. If we had a theory of the organisation of neural networks in which genes coded for every synaptic connection in its finest detail, a perfectly instructive theory as it is, we still could, and probably would, attribute the presence of such genes to natural selection. That would not make this theory selective in any relevant sense.

Selection plays a role in Changeux's theory but it is a highly constrained role. One in which it is relatively difficult to identify what could constitute a selection pressure. A selection pressure can be identified when there is selection for a certain trait or characteristic within a population.[13] Variants bearing the trait for which there is selection are favored until some equilibrium frequency is reached or until all other genotypes (at the same loci) are eliminated from the population. This means that different selection pressures entail different genotype distributions in the population. In other words, selection shapes or moulds the system. The distribution of traits in the population, and the characteristic traits of the organisms which form the population, all of which are a function of the distribution of genotypes, result from the selection pressures sustained by the population. Different selection pressures would have given different traits. Nothing of the sort happens in Changeux's model. Quite to the contrary it is made so as to ensure that in spite of different selection pressures neural networks of the same class end up in the same connectivity pattern. Selection drives the system towards a predetermined attractor set, but the attractor set, the distribution of traits in which the system ends up, is independent of the precise selection pressure to which the population of neurons was subject. Selection plays here a very different role from that which it plays in the theory of natural selection.

4. It may be objected that Changeux's theory constitutes an easy target. As Gerald Edelman, the proponent of one of the most articulate and complete selective type theory, reminds us:

the mere application of any general idea of selection to the nervous system, while suggestive, is inadequate. This statement is *a fortiori* true of selection by Platonic reference (Jerne 1967), eliminative selection (Young 1965, 1973, 1975), *and various mechanisms of microselection (Changeux and Danchin 1976) at the level of synapses that do not define a higher-order unit or offer a detailed consideration of the relationship necessary and sufficient to yield categorisation.* (1987, p. 321) (my emphasis)

Under this reading, Changeux's contribution (along with Jerne or Young's) is just a bad example of what selective theories are or should be.

Edelman's theory of neuronal group selection, also known under the title of *Neural Darwinism* (1987), is certainly one of the most ambitious selective type theories today. It is a remarkable intellectual achievement, by the breath of information it brings under its sway, by the ingenuity of the mechanisms it postulates, by its overall coherence, and by the scope of its implications. Like Darwin's original contribution, the theory of neuronal group selection is essentially a historical hypothesis rather than a nomological proposition. Its aim is to explain how the complex brain of certain animals operates, not to break new ground in automata theory or in parallel processing networks, even though both of these fields are relevant for what it tries to do. The neuronal group selection theory is a hypothesis about the nervous system of real animals in nature. Whether or not it turns out to be true depends on how certain natural objects are made and not simply on some forthcoming results in complexity theory. Selection then, in this context, could receive more than the purely logical role it obtains in Changeux's synaptic stabilisation theory.

"Neural Darwinism" can be divided into three parts or sub-hypotheses, all of which, according to Edelman (1987, pp. 315–320), are logically dependent on each other. The first component or sub-hypothesis is what Edelman calls the "regulator hypothesis". It is an hypothesis in embryology and development according to which very few rules expressed by a limited number of adhesion molecules[14] are responsible for the anatomic organisation of the nervous system, a process which Edelman terms: the formation of the primary repertoire (1987, pp. 73–104). The regulator hypothesis serves many roles or functions in the general economy of Edelman's work; we will concentrate on two. First it ensures that, in the absence of sufficient genetic information to code for all detailed synaptic connections, "the cells involved exhibit interactive and cooperative spatial ordering" that is nonetheless genetically constrained (1987, pp. 73–74). This first role of the regulator hypothesis is to show how the general architecture of a very complex network can be reached through the use of very few simple construction rules.[15] As a consequence of this, one can understand how minor changes in the genetic material determining the rules can result in major reorganisations of the final network. The second role of the regulator hypothesis concerns this specific ordering. The adhesion molecules bind cells into collectives where neurons are highly connected. These collectives are in turn connected to each other but more loosely in the sense that the density of synaptic connections between collectives is much lower than within each collective. "But because the main function of CAMs is to regulate dynamic cellular processes and not to specify cell addresses exactly, variability is also introduced during development." (1987, p. 76) According to Edelman, this makes it likely that one important desiderata in the organisation of neural networks will be satisfied: degeneracy. Degeneracy in this context means, as it usually does in medicine, lack of specificity, and it is considered by Edelman as playing, in the neuronal group selection theory, a role equivalent to that of variation in natural selection. Because of the highly redundant connections existing within each collective, every collective contains numerous groups of neurons more or less able to respond coherently to the same pattern of stimulation. It is these groups of neurons which form the secondary repertoire which will be selected through interaction with the environment. The primary repertoire is an anatomic characteristic of the brain (under)determined by the regulator hypothesis and defined by a certain pattern of connectivity, which has evolved through a process of natural selection. The secondary repertoire is selected in somatic time, through a process of interaction with the environment. In this case what changes are not the (anatomic) connections between the cells or groups of neurons, but the weight attached to each synaptic connection. What evolves, what is selected, are particular patterns of neuronal activity.

The second set of hypotheses concerns the role of reentrant connections and of topological maps in the nervous system. According to Edelman, "(i)t seems necessary that there be in the brain at least one level of representation of sensory input that has the feature of a topographic or spatial map corresponding to at least parts of real-world objects in space." (1987, p. 108) He also suggests that such mapping is the result, at the anatomical, level of many feedback events rather than the expression of a "preestablished detailed pattern." (1987, p. 120) Reentrant connections are another feature which Edelman presumes appears early in the formation of nervous networks and toward which there is a general tendency during evolution. Its importance is for the temporal correlation and for the classification of inputs (1987, pp. 143–149). This second set of hypotheses, contrary to the first one, the regulator hypothesis, does not serve to satisfy any requirement of a model of natural selection, like the presence of variability. But it does indicate that certain demands of parallel processing are fulfilled by the nervous system.

The third set of hypotheses concerns perceptual categorisation and may be considered as *the theory of neuronal group selection*, inasmuch as it constitute an application of the general principles of the theory to the problem of perceptual categorisation. This last set of hypotheses argues that perceptual categorisation is the result of the correlation through reentrant connectivity of selection processes taking place in local cortical sensory maps and local motor maps. Both kinds of maps result of the selection of groups of neurons responding to various stimuli and categorisation is the result of the coordination of such response.

As in the case of the theory of the selective stabilisation of neural networks, it seems that in the theory of neural group selection most of the explanatory weight has shifted from selection itself to the particular characteristic of the system involved. Notwithstanding what Edelman himself seems to imply it is not so much selection which yields a solution to the problem of categorisation as the correlation of various maps through a system of reentrant connectivity.[16] Selection plays a role here, but once again, one which is highly constrained by the nature of the system within which it takes place.

5. Now it may be objected, as is more or less implied in Piattelli-Palmarini's approach, that what is at work here is not really a change in explanatory strategy but a return to Darwin's own sober evaluation of the limited power of natural selection. In other words, selective type theories are perhaps not representative of the explanatory strategy of the adaptationist programme or the panglossian paradigm, but these later approaches are not a correct construal of the theory of natural selection. In conclusion I want to suggest that more than that is at stake by asking what happens when the typical explanatory strategy of selective theories, in opposition to that of natural selection, is applied to biological evolution.

This is more or less the research programme in which Stuart Kauffman has been engaged for the last twenty years.[17] Kauffman has been interested in the fact that complex systems, such as genomic regulatory networks underlying ontogenesis or the epigenesis of the neuronal system exhibit strong structural and dynamic self-organised properties. (1986a, p. 67) Precisely the type of properties which play an important explanatory role in both Changeux's and Edelman's theories. Struck by the fact that the spontaneous order of such systems is highly reminiscent of the order found in organisms, Kauffman postulated that much of that order might be derived from these self-organisational properties rather than selection. He then asked if selection was able to avoid the generic properties of systems able to sustain such self-organisational properties. The results he obtained suggest that many features of organisms might be there not because, but in spite of, natural selection.[18] They also indicate that apparently only one class of such self-organisational systems may be able to evolve adaptively. He takes these results as indications that there may be some nomological biological universals and that from these characteristics of self-organisational systems we may be able to predict a lot about actual organisms without knowing the details of their evolutionary history.

Whether these preliminary results turn out to be as sound and as important as they seem is still an open question. But clearly by shifting the explanatory power for the order we observe in nature from natural selection to the characteristics of complex systems, Kauffman has taken a step which has major consequences. One of these consequences is that, if what he says turns out to be true, fitness will tend to appear more as a dispositional property of certain types of systems than as a property supervening on both organismic and environmental properties. Organisms could still turn out to be more fit in one environment than another, but just as a sugar cube dissolves better in a cup of coffee than in its box on a shelf in a supermarket. In both environments its solubility remains the same. The shift from natural selection to the characteristics of certain complex systems also indicates a change in the questions asked and in what seems in need of explanation. If fitness is a dispositional property of some well-defined type of system, then it follows that adaptation, understood either as the fact that organisms are adjusted to their environment or as the fact that certain traits are the historical result of selection, is no longer a problem in need of explanation. Fitness as a dispositional property entails that adaptation is a direct consequence of a certain form of organisation in whatever environment. If natural selection explains adaptation problems, selective type theories suggest that particular adaptations are no problems at all. The goal of such theories is to construct a system like the immune network, the brain, or genomic networks, which will be adapted in every circumstances. This is usually done by designing a self-organisational network which retains its generic properties even in the presence of severe selection. Selection in this context plays a role reminiscent of that of noise in information theory, rather than of a process which ensures an ever greater adjustment between the organism and the environment. The end result is a system which is always adapted. Selective type theories suggest that adaptation problems are pseudo-problems.

This, I believe, is an alternative programme to the theory of natural selection, not just a reformulation of it in a new domain, though selection still plays a role here.

Université du Québec à Montréal

NOTES

[1] A preliminary version of this text was presented at the 9th International Congress of Logic, Methodology and Philosophy of Science in Uppsala, Sweden, in August 1991. Two other versions were then presented, one at the University College of Cape Breton, Sydney, Nova Scotia in March 1992 and a second one in Paris, at the CREA, in June 1992. I received useful comments on all these occasions. I especially want to thank M. Anspach, M. Baker, A Boyer, L. Darden, R. P. Le Scouarnec, D. Mary, R. Nadeau, A. Orléan, S. Stuart, L. Scubla, H. Zirn.

[2] Of course, more than one type of theory can correspond to the same type of problems. For example, both instructive and selective theories can be applied to adaptation problems. Nonetheless, the construction of a "type" associated a group of theories (or of potential theories) with a family of problems, so that, when confronted with a problem a scientist should know towards which type (or types) of theory he better direct his research effort.

[3] Cf. Gould and Lewontin (1979). It is not clear if, for Piattelli-Palmarini, selective theories simply bring out the truth of natural selection against panglossian deformations, or if Darwin himself is guilty of the instructivist fallacy.

[4] Clearly a desirable characteristic of any conception of a type of theory.

[5] Some examples used by Piattelli-Palmarini, like the Chomskian theory of language acquisition, are presumably not selection type theories following Darden and Cain's account. In consequence I will limit myself to examples which are common to both analyses of selective or selection type theories.

[6] Both the theory of natural selection and the selective theories under scrutiny are "extremal" theories in the sense given to that term by Alexander Rosenberg (1985), hence they belong to the same type of theory, but as Newtonian mechanics also belongs to that type this is not a relevant sense of type in this context.

[7] I will not attempt to analyse why Darden and Cain's classification of theories in type is inadequate. Since my main goal is not to criticise them but to document what I take to be a major conceptual change in biology, I will rest satisfied with the fact that my demonstration suggests that their mode of classification is insufficiently precise.

[8] It is known that some authors like David Hull (1988) have challenged whether a conceptual characterisation of natural selection is in any way important to a historical or epistemological understanding of what constitutes Darwinism. For a good counter-example to that claim, see Gayon (1992, especially pp. 1–60) where he offers an enlightening historico-conceptual analysis of Darwin's selection hypothesis.

[9] Cf. Rosenberg (1978) and Sober (1984).

[10] To the contrary, it is typical to believe, in philosophy of mind, that mental states supervene on brain states alone, or rather on the states of a functionally defined class of computational systems. A doctrine which is currently challenged by Putnam's thought experiment (1975) and by variants of that experiment found in Burge (1986) and Dennett (1987).

[11] See Gould (1989) for a particularly eloquent statement of that position.

[12] For an interesting epistemological and quantitative analysis of this phenomenon see Atlan (1989).

[13] For the concept of "selection for" see Sober (1984).

[14] CAM or SAM, Cellular Adhesion Molecules or Surface Adhesion Molecules.

[15] Remember that this was one of the role played by Changeux's theory of the stabilisation of neural networks.

[16] At this point it seems far from clear that selection has any role to play in the fact that the selection of various neuronal groups gives rise to maps. It is the connectivity pattern of the neuronal network which ensures that, whatever the particular stimulation, this will be the case.

[17] For two succincts presentation of this overall programme see Kauffman (1985; 1986a). For more technical results see Kauffman (1969; 1986b) and Kauffman and Levin (1987); Kauffman and Weinberger (1989).

[18] See Kauffman (1985; 1986a); Kauffman and Levin (1987) and Weisbuch (1991) for these results.

REFERENCES

Atlan, H., 1972, *L'organisation biologique et la théorie de l'information*, Hermann, Paris, 300 pp.

Atlan, H., 1979, *Entre le cristal et la fumée*, Seuil, Paris, 288 pp.

Atlan, H., 1989, 'Automata Network Theories in Immunology: Their Utility and Their Underdetermination', *Bulletin of Mathematical Biology* 51(2), 247–253.

Burge, T., 1986, 'Individualism and Psychology', *The Philosophical Review* XCV(1), 3–46.

Burnet, F. M., 1957, 'A Modification of Jerne's Theory of Antibody Production Using the Concept of Clonal Selection', *The Australian Journal of Science* 20, 67–69.

Changeux, J. P., 1979, 'Déterminisme génétique et épigenèse des réseaux de neurones: existe-t-il un compromis biologique possible entre Chomsky et Piaget?', in M. Piattelli-Palmarini (ed.), *Théories du langage Théories de l'apprentissage: Le débat entre Jean Piaget et Noam Chomsky*, Seuil, Paris, pp. 276–291; English translation, 1980, *Language and Learning: The Debate between Jean Piaget and Noam Chomsky*, Harvard University Press, Cambridge MA.

Changeux, J. P., 1983, *L'homme neuronal*, Artheme Fayard, Paris, pp. 379; English translation, 1985, *Neuronal Man: The Biology of Mind*, Random House, New York, Pantheon.

Changeux, J. P. and Danchin, A., 1976, 'Selective Stabilization of Developing Synapses as a Mechanism for the Specification of Neuronal Networks', *Nature* 264, 705–711.

Darden, J. and Cain, J. A., 1989, 'Selection Type Theories', *Philosophy of Science* 56, 106–129.

Dennett, D. C., 1987, *The Intentional Stance*, MIT Press, Cambridge MA, 388 pp.

Edelman, G. M., 1987, *Natural Darwinism: The Theory of Neuronal Group Selection*, Basic Books, New York, 371 pp.

Gayon, J., 1992, *Darwin et l'après-Darwin: Une histoire de l'hypothèse de sélection naturelle*, Kimé, Paris, 453 pp.

Gould, S. J. 1989, *Wonderful Life: The Burgess Scale and the Nature of History*, Norton Company, New York, 347 pp.

Gould S. J. and Eldredge, N., 1972, 'Punctuated Equilibria: An Alternative to Phyletic Gradualism', in T. J. M. Schopf (ed.), *Models in Paleobiology*.

Gould S. J. and Lewontin, R., 1979, 'The Spandrels of San Marco and the Panglossian Paradigm: A Critique of the Adaptionist Programme', *Proceedings of the Royal Society* B205, 581–598.

Gould S. J. and Vrba, E., 1982, 'Exaption – a Missing Term in the Science of forms', *Paleobiology* 8, 4–15.

Hull, D. 1988, *Science as a Process: An Evolutionary Account of the Social and Conceptual Development of Science*, The University of Chicago Press, Chicago, 586 pp.

Jerne, N., 1966, 'The Natural-Selection Theory of Antibody Formation: Ten Years Later', in *Phage and the Origins of Molecular Biology*, Cold Spring Harbor Laboratories of Quantitative Biology, Cold Spring Harbor NY.

Jerne, N., 1967, 'Antibodies and Learning: Selection versus Instruction', in G. C. Quarton, T. Melnechuk, and F. O. Schmitt (eds.), *The Neurosciences: A Study Program*, Rockefeller University Press, New York, pp. 200–205.

Kauffman, S. A., 1969, 'Metabolic Stability and Epigenesis in Randomly Constructed Genetic Nets', *Journal of Theoretical Biology* 22, 437–467.

Kauffman, S. A., 1985, 'Self-Organisation, Selective Adaptation, and Its Limits: A New Pattern of Inference in Evolution and Development', in D. J. Depew and B. H. Weber (eds.), *Evolution at Crossroads: The New Biology and the New Philosophy of Science*, MIT Press, Cambridge MA, pp. 169–207.

Kauffman, S. A., 1986a, 'A Framework to Think About Evolving Genetic Regulatory Systems', in W. Bechtel (ed.), *Intergrating Scientific Disciplines*, Martinus Nijhoff Publishers, Dordrecht.

Kauffman, S. A., 1986b, 'Autocatalytic Sets of Proteins', *Journal of Theoretical Biology* 119, 1–24.

Kauffman, S. A. and Levin, S., 1987, 'Towards a General Theory of Adaptive Walks on Rugged Landscapes', *Journal of Theoretical Biology* 128, 1–35.

Kauffman, S. A. and Weinberger, E. D., 1991, 'The NK Model of Rugged Fitness Landscapes', in A. S. Peterson and S. A. Kauffman (eds.), *Molecular Evolution on Rugged Landscapes: Proteins, RNA and the Immune System*, Addison-Wesley Publishing Company, Redwood City CA.

Kuhn, T. S., 1970, *The Structure of Scientific Revolutions*, The University of Chicago Press, Chicago, 210 pp.

Piattelli-Palmarini, M., 1986, 'The Rise of Selective Theories: A Case Study and Some Lessons from Immunology', in W. Demopoulos and A. Marras (eds.), *Language Learning and Concept Acquisition*, Ablex Publishing Corporation, Norwood NJ, pp. 117–130.

Piattelli-Palmarini, M., 1989, 'Evolution, Selection and Cognition: From "Learning" to Parameter Setting in Biology and the Study of Language', *Cognition* **31**, 1–44.

Putnam, H., 1975, 'The Meaning of "Meaning" ', in *Mind Language and Reality: Philosophical Papers Volume 2*, Cambridge University Press, Cambridge, pp. 215–271.

Rosenberg, A., 1978, 'The Supervenience of Biological Concepts', *Philosophy of Science* **45**, 368–386.

Rosenberg, A., 1985, 'Adaptationist Imperatives and the Panglossian Paradigms', in J. H. Fetzer (ed.), *Sociobiology and Epistemology*, Kluwer Academic Press, pp. 161–180.

Sober, E., 1984, *The Nature of Selection: Evolutionary Theory in Philosophical Focus*, MIT Press, Cambridge MA, 383 pp.

Weisbuch, G., 1991, 'Systèmes complexes et comportement générique', in F. Fogelman Soulié (ed.), *Les Théories de la Complexité: autour de l'oeuvre d'Henri Atlan*, Seuil, Paris, pp. 171–181.

Young, J. Z., 1965, 'The Organization of a Memory System', *Proceedings of the Royal Society of London (Biological Sciences)* **163**, 285–320.

Young, J. Z., 1973, 'Memory as a Selective Process', *Australian Academy of Science Report: Symposium on Biological Memory*, Australian Academy of Science, Canberra, pp. 25–45.

Young J. Z., 1975, 'Sources of Discovery in Neuroscience', in F. G. Worden, J. P. Swazey, and G. Adelman (eds.), *The Neurosciences: Paths of Discovery*, MIT Press, Cambridge MA, pp. 15–46.

DANIEL LAURIER

FUNCTION, NORMALITY AND TEMPORALITY

Davantage d'avantages
avantagent davantage
Boby Lapointe

I examine different ways of analysing the concept of function in terms of natural selection, showing that it is possible to do so while admitting a non-statistical notion of normality and taking into account the possibility that a thing may acquire new functions, or loose functions previously acquired. There are several possible solutions, but I believe that the best one makes the concept of function a concept "directed towards the present", and not towards the past as most selectionists seem to think. I conclude by raising a point that would seem to be a difficulty in principle for all forms of naturalisation of normative concepts.

1. INTRODUCTORY REMARKS

Proposals aimed at using the concept of natural function in the service of naturalising semantic/intentional phenomena have multiplied in the past decade or so.[1] I would like to determine whether this proliferation can be accounted for by the "adaptive superiority" of this type of approach, and more specifically, to clarify the relation between the "teleosemantic" program and that of radical interpretation.

The main motivation of the teleosemantic program, or at least the one that, in my opinion, provides its best justification, comes from the conviction that the normative character of teleological notions might make it possible to account for the fact that an agent can have false beliefs, even though the content of a belief is its truth-condition. Just as, intuitively, a thing's function is that which the thing is *supposed* to do, even if it does not do it, or indeed is incapable of doing it, so the content of a belief can be identified as (or defined as a function of) that which the belief is supposed to do, even if it cannot do it. In this manner, one hopes to exploit the analogy between the fact that an agent can have a belief even in the case where its truth-condition is not realised and the fact that a thing can have a function without fulfilling it. Now it so happens that the program of radical interpretation is founded precisely on the irreducibly normative character of intentional notions, so it is not far-fetched to suppose that there may be interesting relationships between the two types of program. But the program of radical interpretation is also based on the principle that agents ought to be interpreted in such a way as to make them appear as rational as possible, which precludes, at least

25

M. Marion and R. S. Cohen (eds.), Québec Studies in the Philosophy of Science II, 25–52.
© 1996 *Kluwer Academic Publishers.*

in Davidson's mind, that they could be generally irrational (or generally in error). Now it is not clear that this principle of rationality accords with the principles of teleosemantics. It is possible, furthermore, that the teleosemanticians and the "radical interpreters" would not be able to share the same optimism regarding the naturalisation of intentionality.

Since the aim is that of evaluating the relevance of teleofunctional notions in the perspective of an explanation of intentional phenomena, I will ignore as much as possible the question of what it is for an artefact to have a function, to concentrate on the notion of *natural* function, meaning by that any function that does not derive directly from the intentions of a designer or a user. For the same reason, the notion of "goal-directed behaviour", which I consider an intentional notion, will not be discussed.

There are many ways to approach the notion of function. First of all, there is a general distinction between the approaches that aim at producing an analysis of the ordinary language, or common-sense, concept of function, and those that attempt to develop a concept of function that enjoys certain theoretical virtues, that is to say (no pun intended), a concept capable of fulfilling certain theoretical functions. A theoretical notion can depart from the corresponding common-sense notion in various ways, and maintain with the latter a mere family resemblance. This has the methodological consequence that an analysis of this second kind cannot always be criticised by pointing out that certain of its consequences are contrary to our intuitions regarding the application of the concept of function. Such an analysis must above all be criticised on the basis of the fact that it does or does not see to it that the concept of function can play the role that it is presumed it must play. It must be clear that my interest here for the concept of function is of this second type, which complicates things a little, since two theoretical analyses of the concept of function are not in competition unless they aim at making the concept fulfil the same theoretical role, whereas there are, at least in principle, a multiplicity of roles one may want to make it fulfil. As indicated earlier, my goal is to identify or to clarify a notion of function that could (at least *prima facie*) play a role in the characterisation of the notion of content, and/or that allows one to evaluate the respective merits of teleosemantics and radical interpretation. From this point of view, the essential trait of the concept of function is its normative character; but even limiting attention to the analyses that satisfy this condition, there are several proposals to examine, not counting those that concern, not the characterisation of the concept of function, but the manner of putting it to work in the explanation of the notion of content. It is necessary, in particular, to make a distinction between the approaches that try to show that the concept of function is an intrinsically explanatory concept, and the ones that consider it to have only a descriptive, heuristic or evaluative role. Most of these seem to be variants or modifications of the traditional conception according to which a thing has a function insofar as it contributes to the attainment of a goal of the system of which it is a part (Boorse, 1976, Prior, 1985 and Woodfield, 1976), and will be set aside here.

2. WRIGHT'S ANALYSIS AND STATISTICAL NORMALITY

The conception that seems presently to enjoy the favour of a majority of teleosemanticians, although each one has his/her own version of it, derives mainly from the etiological analysis of Wright (1973, 1976). Two aspects of Wright's analysis are generally retained, namely, that a function-attribution is intrinsically explanatory and that it makes reference to a certain etiology of the thing to which the function is attributed. Wright's proposal is, in substance, that Y is a function of X if and only if (i) X causes or tends to cause Y and (ii) the existence of X is caused (in part) by the fact that (i); or in other words, Y is a function of X if and only if X exists in part *because* X causes or tends to cause Y.

This analysis has been the target of several criticisms, in particular those of Boorse (1976), Prior (1985), Bigelow and Pargetter (1987) and Bedau (1991, 1992b), which notably induced Millikan (1989a) and Neander (1991a, 1991b) to come to the defence of the etiological conception, while modifying it somewhat. The confusion arising from the exchanges seems to be due in part to the fact that protagonists apparently do not agree on the conditions that must be satisfied by an adequate analysis of the notion of function, and in part to the fact that they do not all distinguish systematically between functions of types and functions of tokens. Although Wright's analysis was intended to apply to all kinds of functions, I will only mention here the criticisms that attempt to show that the etiological conception does not suitably account for the notion of *natural* function, as opposed to those that attempt to show that it does not fit the notion of *non-natural* function.

I will begin with a question raised by Prior (1985: 318–320) concerning the capacity of the etiological conception to account *simultaneously* for the distinction between a thing that has a function and one that has not, and for the distinction between a thing that functions normally and one that does not function normally. It will be noted, however, that this objection aims less at finding fault with the etiological conception than at showing that, contrary to what its defenders claim for it, it does not account for the normative character of the notion of function any better than a goal-contribution or systemic account does.

It is clear that according to Wright's analysis, an organ cannot have a function unless it *presently* has the tendency to produce the effect that it has the function to produce (condition (i)). This accords with the judgement according to which the appendix, for example, has no function in humans, because although the appendix was selected (presumably) because it digested cellulose, contemporary appendices no longer have this effect. Condition (i) of Wright's analysis thus seems essential for distinguishing between organs that have a function and those that do not, or no longer do. However, this same condition apparently prevents it, according to Prior, from accounting for the distinction between (functional) organs which function normally and those which do not function normally, unless it is reinterpreted in a certain way.

Indeed, if my pancreas becomes incapable of producing a sufficient quantity of insulin, it no longer has the tendency to produce a sufficient quantity of insulin, and therefore can no longer, according to Wright's analysis, have as a function the production of insulin, even if the fact that I have a pancreas is explained in part by the fact that pancreases of my ancestors produced insulin. One can remove this difficulty by supposing that Wright's analysis implicitly refers not to what the thing to which a function is attributed tends to cause, but to that which the majority of things *of the same type* cause or tend to cause. But this is not quite enough, because in a situation where a sufficient proportion of the human population suddenly became blind, the eyes would no longer have the tendency to produce, in the brain for example, the effects they normally produce, and even thus modified, Wright's analysis would imply (in virtue of condition (i)) that the eyes would, quite simply, no longer have the function of producing these effects (nor, apparently, any other function). It seems that we cannot avoid this consequence without widening Wright's characterisation again in such a way as to make reference not only to the effects which the entities of type X produce presently, but also to the effects they had in the past. In the situation we have just imagined, the analysis thus modified would allow us to maintain that the eyes presently have the function of stimulating the visual cortex in such and such a way, even though the majority of them are incapable of it, by reason of the fact that the vast majority of eyes present and past have or had the tendency to produce that effect and that the eyes exist presently because they generally have or had that effect.

Here we must take note of two things, namely that by reinterpreting Wright's analysis in this way (1) we admit that first of all it is types, and not tokens, that are the bearers of natural functions (and that a token has no natural function except in so far as it is of a certain type), and (2) we invoke an apparently purely statistical notion for distinguishing between functional items which function normally and those which do not function normally. Prior (1985: 320) concludes from this that the etiologists like Neander and Millikan are wrong to claim that the etiological conception is able to found the notion of function on a non-statistical notion of normality. She maintains on the contrary that the etiological approach has no other choice than to have recourse to such a conception of normality if it wants to be able to distinguish between things that do not fulfil their functions and things that do, without thereby compromising its capacity to distinguish between things that have a function and things that do not or do no longer. From that point of view, this approach would not, therefore, have any advantage over the systemic approach that Neander and Millikan criticise on precisely this point. Since Prior believes she can show that etiological analyses cannot account for the possibility that (1) a thing's function changes with time, or that (2) a thing that has no ancestors (that is to say, no causal history) has natural functions, she concludes that the systemic approach is superior. Her rejection of the etiological approach therefore rests essentially on these two supplementary arguments (which we find also in

Boorse (1976) and Bigelow et Pargetter (1987)). Before examining them, it is necessary to ask whether the etiological approach cannot rest on anything other than a statistical definition of normality, and whether that is problematic.

3. NATURAL SELECTION AND STATISTICAL NORMALITY

Millikan (1984, 1989a) and Neander (1991a, b) defend two versions of the etiological conception which, although different, both generate a non-statistical notion of normality and make no reference to the present capacities or dispositions of functional items. According to Neander's (1991b) version, Y is a function of X if and only if X was selected because its ancestors had Y as an effect, or in simpler terms, the functions of a thing are those of its effects for which it (that is to say that *type* of thing) was selected.[2] A function is *natural* when it results from a non-intentional process of selection and non-natural otherwise. It will be noted that in order to obtain this characterisation from that of Wright, it suffices to omit from it the first condition and to make explicit reference, in the second, to a process of *selection.*

By abandoning Wright's first condition, Millikan and Neander are apparently in a position to admit that a type of thing can have a certain function even if only a minority of the things of that type fulfil it or are capable of fulfilling it. Even if the entire population of the globe suddenly became blind (or if the earth were suddenly plunged into permanent darkness), the eyes would still have the function of stimulating the visual cortex in the usual way, since according to this purely selectionist conception, the function of a thing has absolutely nothing to do with its present dispositions, but depends exclusively on its causal origin. Of course, for a thing to have the function of producing Y, it is necessary that a certain number of its ancestors did actually produce Y; but it may be a question of a minority; it could even, in principle, be the case that a minority of all the items of that type, past and present, produce or are capable of producing Y.

One might be tempted to dispute this last possibility, by pointing out that a type of item cannot be selected unless it has effects which confer an adaptive advantage to those who possess it and that the effects it eventually acquires the function to produce must be those which confer such an advantage, so that one must expect the number of items of that type which actually fulfil their function to end up exceeding the number of items that do not fulfil it.

Even if this reasoning were correct, it would show at most that an analysis based on a (temporally indexed) statistical notion of normality can lead to the same function attributions as a selectionist analysis, without touching on the question of what criteria legitimise a function-attribution (that is to say, the question of what is the nature of the norms on which function-attributions are founded). But this reasoning is not correct and only derives its appearance of plausibility from the fact that it rests on implicit hypotheses that are not always true.

If it is true that a type of item cannot be selected unless it has certain effects which confer an adaptive advantage to those that possess it, and that the effects that it can acquire the function to produce can only be those for which it was selected, it does not in anyway follow that the number of items of that type that fulfil that function must finally exceed that of those that do not, but at most that the number of *individuals* that possess an item of that type (which functions or not) must ultimately exceed the number of those that do not possess it. Even that is true only if (i) a sufficient number of items of that type continue to confer an adaptive advantage for a sufficiently long period (that is to say, if selection in favour of that type of item lasts for a sufficient period), (ii) the fact of having an item of that type is passed from generation to generation, and (iii) the pressure of natural selection is not thwarted by other evolutionary forces like migration, genetic drift or mutation. A type of item can cease to produce the effects for which it was selected, for example, because of a change in the environment, and continue to be passed on by inertia (or because it henceforth confers another type of advantage), so that over the whole population, all generations taken together, only a minority of items of that type produce or produced these effects.

Perhaps one can at least claim that if the selection in favour of a type of item is maintained for a long enough time, and for the same reason, there will then be a (non-empty) series of generations such that the number of individuals belonging to one of those generations[3] which possess an item of that type fulfilling its function exceeds the number of individuals belonging to one of those generations which possess an item of that type not fulfilling its function. But, besides the fact that it is not obvious that this would be very useful to those who wish to defend a statistical conception of normality, it is probably not entirely sound, either.

Millikan readily cites the example of spermatozoa to illustrate the possibility that a type of item can have a function although the majority of items of that type do not fulfil it. Indeed it is evident that the vast majority of spermatozoa historically produced have not fertilised any egg, and even that in each generation there are more spermatozoa that do not fertilise an egg than there are that do so, although that plainly seems to be one of their functions. I do not know to what extent one must be convinced by this particular example, since it is not clear that the selectionist analysis allows one to say that each spermatozoon has, individually, the function of fertilising an egg. Millikan and Neander seem generally to take it as given that when a type of item is selected, it is from then on *fixed* in the population in question, in the sense that all the members of the population end up possessing it, which is not always the case, as notably when the adaptive advantage conferred by the item depends on its frequency in the population. But this is exactly the case with spermatozoa, since only males produce them. Even ignoring this difficulty, it is not clear that the functions of spermatozoa should not be considered to be derived functions in Millikan's sense: functions, that is, which are inherited by dint of being produced by organs (namely the testicles) that have

the function of fertilising eggs, even of fertilising eggs by producing spermatozoa. In this case, we can surely say that males that had testicles which
produced spermatozoa had, in that, an adaptive advantage (and, perhaps, that
the ones who produced more spermatozoa had more advantages), which allows
us to conclude that the testicles were selected because they produced spermatozoa, and because they thereby fertilised eggs, and therefore that they have
the function of producing spermatozoa and fertilising eggs. Perhaps this
is enough to conclude that spermatozoa *collectively* have the function of
fertilising eggs, but we cannot thence conclude that each spermatozoon individually has that function without invoking something else than natural
selection, as for example the principle according to which if a device has
the function of producing Y and if the device normally produces Y by producing Z, then Z has the (derived) function of producing Y.

If this example strikes me as objectionable, it is nevertheless in the service
of a just cause, since it helps to focus attention on the fact that, in principle
(and to the extent that only natural selection is concerned), there is no limit
to the number of tokens of a given type of item that an individual can possess.
This means that it would be very risky to draw general conclusions concerning
the proportion of items of a certain type which fulfil their function from
hypotheses concerning the proportion of individuals that possess items of
that type. But what about the case where each individual possesses at most one
item of the type under consideration? Could not one maintain that there should
then be a period during which the majority of items of that type have the effects
it is their function to produce?

Note first of all that there is a sense in which two items cannot be of the
same type if they do not have the same effects in the same environments. Since
by hypothesis all the members of a population subjected to the same process
of selection evolve in the same environment, it would seem that a type of
item can only be selected for producing Y if all the items of that type actually
produce Y. But this conclusion rests on a mis-appreciation of what counts as
an environment. An environment is a set of relatively stable but incompletely
determined characteristics, in the sense that individuals that evolve in the same
environment can nevertheless find themselves in different *situations*. It is
difficult to say exactly at what moment different situations begin to belong
to different environments, but there is no doubt that such a distinction is
necessary.

A given individual may never find itself in a situation in which the item
under consideration will produce Y, or in a situation in which the fact that
this item produces Y confers an advantage. Imagine, for example, a mechanism for the detection of predators. It is possible that the detectors belonging
to certain individuals never have the occasion to function, because those individuals have the good luck never to find themselves in the vicinity of a
predator, or that they only have occasion to function in situations where there
is not, in fact, a predator. To simplify, suppose it happens that this detector
does not produce Y when there is a predator, but that it produces Y only if

there is a predator nearby, so that the second possibility is excluded. The population is divided in two groups: the members of the first sometimes find themselves in proximity to a predator, but not those of the second, but almost all of them have a mechanism that produces or would produce Y only if there is or was a predator nearby. Since it happens sometimes that there is a predator without it being detected, it seems that the members of the second group have a greater chance of surviving and reproducing than those of the first, and that the item in question should not, therefore, be selected for producing Y. But this does not take into account the fact that even the members of the second group, which evolve, by hypothesis, in the same environment as the others, run the risk of finding themselves in the proximity of a predator (in other words, belonging to one or the other group is random, or else they could be considered as two distinct populations, and there would be two distinct selection processes). In these conditions, all the members of the population that have an item of the type in question have an adaptive advantage over those that do not have it, even if the members of the first group are in the minority. It is therefore possible that this type of item could be selected for producing Y even if a minority of items of that type actually produce Y.[4] It goes without saying, however, that the more the proportion of individuals that find themselves sometimes in the proximity of a predator diminishes, the less it is advantageous to possess the item in question; but the advantage can be sufficient for selection to favour the individuals that possess that item even in the case where the probability of meeting a predator is less than 1/2. This means that the hypothesis evoked earlier must be rejected, even in the favourable case where no individual possesses more than one item of the type in question. In other words, it is plainly possible for a type of item to be or have been selected for producing a certain effect without it being the case, at any time, that most items of that type have this effect.

It should also be noted that there is an ambiguity in the notion of statistical normality brought up in this discussion. We have supposed until now, that when an item has a function, it fulfils it or does not fulfil it, *simpliciter*. Now this does not take account of the fact that a given functional item can sometimes produce the effect it has the function to produce, and not produce it at other times. A purely statistical approach should therefore lead to the conclusion not simply that most of the items that have a function fulfil it, but that most of the items that have a function fulfil it *most of the time*. But there is apparently no reason to believe that the effect for which a type of item is selected must be among those that the items of that type produce most of the time, or even among those that the items of that type produce most often. The capacity to produce a certain effect (such as frightening an intruder or capturing prey) once in a while can be enough to confer an adaptive advantage to those that possess it.

It surely seems then that the selectionist conception is distinct from all the others (including Wright's) in breaking with the tradition according to which things that have a function generally fulfil that function, in admitting, that is, that deviant cases can be more numerous than normal cases, while still preserving the normative character of the notion of function.

Prior (1985: 319) does not dispute this conclusion, but nevertheless maintains that this analysis is inferior to Wright's, and that every version of the etiological approach that would try to account for the distinction between things that have a function and things that do not is bound to depend on a statistical notion of normality. By abandoning any reference to the present dispositions of functional items, Neander[5] has indeed no other choice but to assert that the appendix presently has the function of digesting cellulose, since it is (presumably) because appendices digested cellulose that they were selected; whereas Prior's intuition (which apparently accords with the physiologists' point of view) is that this organ no longer has that function.[6] In other words, the selectionist conception apparently does not admit the possibility that an item can lose one of its functions. Indeed, if a type of item was selected because its ancestors had a certain sort of effect, that remains true quite a long time after the items of that type have ceased to produce that sort of effect.

4. TEMPORALLY INDEXED FUNCTIONS

Prior's judgement helps to focus attention on the fact that a functional item does not generally have a given function except during a certain period, and that a function-attribution must consequently be temporally indexed. According to the purely selectionist conception mentioned in the last section, a type of item can acquire a function, but cannot cease to have a function once it has acquired it. This accords poorly not only with common sense, but also with the practice of physiologists.[7] But there are other ways to do justice to the intuition according to which a function must be the result of a process of selection, without thereby admitting this consequence. The most obvious would consist simply in reinstating Wright's first condition, which would however bring us back to the idea that a type of item cannot have a function unless most of the items of that type fulfil it, and would thereby support Prior's contention. The latter claims that no version of the selectionist conception can simultaneously avoid both of these snags. This is what we cannot begin to verify before having determined what are the possible selectionist analyses (or since it is doubtless impossible to be exhaustive, what are, among them, the most plausible).

Note, first of all, that a selectionist approach is not bent on limiting the process of selection that confers a function to a type of item to the past, or to retain the whole past, even the most distant. Bigelow and Pargetter (1987) even proposed a selectionist analysis that completely abstracts away from the histories of functional items in order to rest solely on the disposition or propensity to be selected. This proposal was severely criticised by the etiologists (Godfrey-Smith, 1991a, Millikan, 1993b, Neander, 1991a, and Mitchell, 1993), mainly on the ground that it does not account for the explanatory value of the concept of function. But the problem of explanatory value will be tackled later on.

Bigelow and Pargetter maintain that the concept of function is fundamentally oriented towards the future ("forward looking"), and not towards the past, as the etiologists would have it. According to them, a functional item does not get its function by dint of having been selected for producing a certain effect but by dint of conferring on that which possesses it a propensity that *augments* the chances of survival (a "survival enhancing propensity") *in its natural habitat*, or by dint of having a propensity to be selected in virtue of the fact that it produces or is disposed to produce a certain effect (which it thereby has the function to produce). Bigelow and Pargetter seem to take these two formulations as equivalent, although the fact that an item contributes positively to the chances of survival (or better, to the fitness) of certain individuals does not mean that it will be selected, or even that it would be selected if those individuals were in their natural habitat, at least according to a certain interpretation.

Indeed, the notion of positive contribution to fitness seems rather problematic; it is a comparative and counterfactual notion which seems as if it could be interpreted in at least two ways. Intuitively, an item contributes positively to the fitness of the individual that possesses it if this individual has a greater chance of surviving and reproducing than if it did not possess it. But if the individual did not possess that item, there are an indefinite number of items that it could possess in its place. According to this interpretation, an item does not contribute positively to the fitness of an individual unless it is superior to all the items that the individual could have in its place, which seems rather counter-intuitive, but clearly accords with the idea that to contribute positively to fitness implies a propensity to be selected. It goes without saying that we cannot compare the effects of different items on fitness except in relation to a given environment, whence the need for Bigelow and Pargetter to invoke the notion of natural habitat.

But according to the most natural interpretation, it should be possible for two items to contribute positively to fitness albeit in different degrees. This poses no difficulty as long as it is a question of two items that the same individual can possess simultaneously, but in that case the one is not selected in preference to the other. It is tempting to think that an item contributes positively to the fitness of the individual that possesses it when this individual has a greater chance of surviving and reproducing that if it were simply deprived of that item, that is, when it has an adaptive advantage, not over all those that do not possess the item, but simply over those that differ from it only in so far as they do not possess the item. But the idea that one individual could be identical with another save for the fact that it lacks a certain trait (that is, the idea that all the individuals which differ minimally from a given individual do not significantly differ from each other, for example from the point of view of their chances of survival and reproduction) is fairly dubious. On the other hand, if we keep the idea that in order to be selected it is enough to contribute positively to fitness, this interpretation leads to the conclusion that an item can be selected even if it is in fact less advantageous than another item also present in a given population, which accords poorly with the theory of natural selection. Indeed it is entirely possible that the individuals that possess a certain item A would have a greater chance of surviving and reproducing than those that only differ from them insofar as they do not possess item A, but a lesser chance of surviving and reproducing than those that possess item B in place of item A. In that case, we would have to admit that even if there were individuals in the population that possessed item B, item A would be selected (just like item B)! This interpretation could not be held, then, unless one rejected Bigelow and Pargetter's second formulation, according to which an item has a function when it has a propensity to be selected. I am thus inclined to think that the first interpretation corresponds most closely to the intentions of Bigelow and Pargetter, or seems at least to be the most charitable.

However, as Godfrey-Smith (1991a: 36) and Millikan (1993b: 38–40) remark, an item is not selected simply in virtue of the fact that it contributes positively to the fitness of those that possess it, that is (according to the interpretation we have retained) in virtue of the fact that those that possess it have better chances of surviving and reproducing than all those that would not possess it, but rather in virtue of the fact that those that possess it have an adaptive advantage over those that do not *in fact* possess it. In other words, there is no selection without competition, and a given item is in competition with the items that certain members of the population actually have in its place, and not with all the items that they could have in its place, whatever they may be.

To suppose that to say that an item has a propensity to be selected comes to the same thing as saying that it would be selected if the population in question were in its natural habitat *and* if it never came in competition with an item over which it had no advantage, would clearly have the effect of trivialising Bigelow and Pargetter's proposal. In these conditions, we must understand that to say that an item has a propensity to be selected is not simply to say that it would be selected if the population in question were in its natural habitat, but also to suppose that it would never then come into competition with an item over which it would not have the advantage. But short of being able to show *a priori* that this item is "the best possible", nothing can exclude the possibility that evolution might one day chance on an item more advantageous still, or that the natural habitat modify itself in such a way that another item becomes more advantageous (not to mention the possible effects of other evolutionary forces). It follows that, correctly interpreted, Bigelow and Pargetter's approach depends as much on contingent historical hypotheses as the etiological approach, although in this case it is a question of hypotheses concerning the future.

It is true that as long as we do not identify the natural habitat with the given historical environment, these historical hypotheses remain counterfactual; but there is really no other plausible way to understand the notion of natural habitat. It would certainly be circular to suppose that the natural environment of a population is that in which its members would have the best chance of surviving and reproducing, that is, that in which the traits they possess would be selected (and would therefore be more advantageous than all their possible competitors). On the other hand, Bigelow and Pargetter are implicitly committed to the claim that the function of an item cannot change unless the natural habitat of the given population can change. But if the natural habitat can change, what would be the sense of saying that a certain item would be selected if those that possess it were in their natural habitat? What counts as the natural habitat of a population at a given moment is certainly historically contingent.

I conclude from this that the most coherent interpretation of Bigelow and Pargetter's suggestion, although perhaps not the most faithful, is the one according to which an item presently has the function to produce Y if it presently has a propensity to be selected for producing Y, in the sense that it *will be* selected for producing Y if the given (historical) environment does not change significantly. As I emphasised, this implies that one hypothesises that no more advantageous competitor will in fact present itself (without a change in the environment), and that natural selection will not be thwarted by other forces. This analysis differs from the etiological analysis not only by the fact that it makes reference to the future where the latter makes reference to the past, but also by the fact that it explicitly postulates the stability of the *present* environment.

By simply replacing, in the etiological analysis, the reference to the past with a reference to the future, we would be led to the idea according to which an item has the function to produce Y if it is true that it will eventually be selected for producing Y, that is, if the given population finds itself one day in an environment in which that item will be selected for producing Y. According to this analysis, an item would presently have the function of producing all the effects for which it will eventually be selected, that is to say, all the effects in virtue of which it will confer an adaptive advantage at one time or another, and this before being selected for producing them and even before having ever produced them. The problem with this analysis is the inverse of that of the etiological analysis: an item can lose a function, since it can cease to be selected for producing a certain effect, but it cannot acquire a new function. It already has the function of producing all the effects for which it will be selected.

Just as the approach of Bigelow and Pargetter suggests a selectionist analysis that stretches the present environment into the future, so the etiological approach could be modified in such a way as to stretch the present environment into the past, by saying, for example, that an item has the function of producing Y if and only if it was recently selected for producing Y in an environment similar to the present one. We could also completely ignore the present environment, and only make reference to the recent past or near future, or again limit ourselves to the present. We therefore have at least seven possible selectionist analyses:

<blockquote>
an item (presently) has the function of producing Y if and only if
</blockquote>

(1)   it has already been selected for producing Y

(2)   it will eventually be selected for producing Y

(3)   it was recently selected for producing Y

(4)   it will soon be selected for producing Y

(5)   it was recently selected for producing Y and continues to be

(6)   it is presently selected for producing Y and will continue to be for a certain time

(7)   it is presently selected for producing Y.

It has already been remarked that proposals (1) and (2) cannot account for the fact that an item can both lose one function and gain another. It should be noted, however, that they must be understood in such a way as to implicitly make reference to the past or future environment, without presuming that it resembles the present environment. Strictly speaking, they are equivalent to:

(1')   an item (presently) has the function of producing Y if and only if there is an environment in which it has already been selected for producing Y

and

(2')   an item (presently) has the function of producing Y if and only if there is an environment in which it will someday be selected for producing Y.

Proposals (3) and (4) must be understood in the same way, and therefore leave open, contrary to (5) and (6), the possibility that the environment of the recent past or near future might be different from the present one. Finally, proposals (5), (6) and (7) reveal that in spite of the fact that many authors use the expression "was selected" to mean that natural selection led to the fixation of a certain item within the population, natural selection is a *process* that extends over a certain period (and that does not necessarily lead to the fixation of the selected item, that is, of the item favoured by that process). Proposals (1)–(4) all potentially suffer from this ambiguity between the result and the process. We will call the interpretations which refer to the result of selection (3r) and (4r), and designate by (3p) and (4p) those which refer to the process of selection, that is to say, to the fact that the item in question is favoured by selection during the period under consideration (or in an equivalent way, to the fact that the item in question confers an adaptive advantage to those that possess it, in virtue of the fact that it produces Y).

Now, consider (3r). According to this proposal, an item presently has the function of producing Y if and only if it was recently fixed in the population in virtue of the fact that it produced Y, that is, if and only if it was recently favoured by natural selection because it produced Y *and* if the process of selection led to its fixation. It is clear that this analysis allows an item to loose a function it acquired earlier: an item no longer has the function of producing Y when the process of selection which led to its fixation is remote. It does not allow, however, an item already fixed in the population to acquire a new function (unless it should cease to be fixed in order to become fixed again). Now there is no reason to believe that this is not possible. Such a situation corresponds, for example, to the conjecture of Gould and Vrba (1982: 7) regarding the function of feathers in birds, according to which birds would have had feathers long before being able to fly, although nowadays feathers have the function of helping them to fly. But in this case it is not the fact that they help birds to fly that explains the fixation of feathers in birds, and since this fixation is not recent, feathers cannot have any function, according to the analysis (3r).

Analogous remarks apply to (4r), which can therefore be rejected as well. This leaves five proposals to consider ((3p)–(4p), and (5)–(7)), that is, five proposals which all seem to allow an item to lose and acquire functions, without depending on a statistical notion of normality.

5. PAST, PRESENT, OR FUTURE?

First of all, consider an analysis which is very close to the one defended by Godfrey-Smith (1991a),[8] namely (3p). According to this proposal, an item has the function of producing Y if and only if it was recently favoured by natural selection in virtue of the fact that it produced Y. This proposal only distinguishes itself from (5) in the case where the environment has recently transformed itself in such a way that the item in question is no longer favoured *because* it produces Y (either because it no longer produces it, or because producing it no longer constitutes an advantage). So one ought to prefer (3p) to (5) only if one admits that an item can keep the function of producing Y even when it is no longer selected for producing Y. If one admits that, one must eventually ask oneself for how long it can keep the function. One cannot say that it keeps it indefinitely without rendering it, for all practical purposes, equivalent to (1p). Many people seem to have an intuition that an item which was until recently favoured by selection for producing Y, but which has now ceased to be so selected, keeps the function of producing Y, at least for a certain period of time. This intuition can rest on the fact that selection always trails the evolution of the environment, and does not seem unreasonable. But suppose that the moment a certain item ceases to be selected for producing Y, it begins to be selected for producing Z:

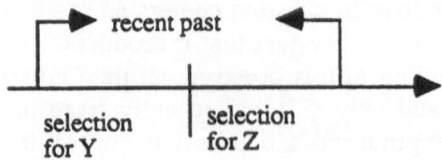

Since there is generally no problem with supposing that a thing has several functions,[9] we could perhaps say that the item continues to have the function of producing Y, and begins to have the function of producing Z. But it might be the case that the production of Z is incompatible with that of Y, which raises the problem mentioned in note 5, namely that the item in question could be "supposed to do" two mutually incompatible things, at least for a certain period of time. Perhaps this objection is not decisive, to the extent that there is no reason to think that Mother Nature should be coherent.

However it seems clear that there is no incoherence in this case, since the two functions in question are in a certain sense necessitated by two different environments. If I am supposed to do two incompatible things because a certain person asked them of me, that person is incoherent; but if two different people asked me to do these things, neither of them is necessarily incoherent. In the same way, an item cannot confer an adaptive advantage in virtue of the fact that it produces Y and at the same time in virtue of the fact that it produces Z, if it is impossible for it to produce both Y and Z, so that if an item is selected for producing Y and for producing Z, it can only be relatively to two different environments. Now proposal (3p) clearly conceals this fact, which seems to me to be a reason to prefer (5). To modify (3p) in such a way as to ensure that the item be selected for producing Y from the beginning to the end of the period in question could help overcome the difficulty, but it would have the effect that in the situation described earlier, the item in question would have no function, which is hardly more acceptable.

By parity of reasoning, (6) seems preferable to (4p). But there might also be a further reason in favour of this preference. (4p) allows an item to have the function of producing Y not only before beginning to be selected for producing it, but even before ever having produced it. Of course, (3p) symmetrically allows an item to have the function of producing Y not only after having ceased to be selected for producing it, but even after having ceased to produce it; but that is not particularly shocking, nor does it lead to any difficulty in itself. This seems linked to the fact that it is intuitively more acceptable to conclude that a certain item is *presently* supposed to do Y from the fact that it *was* selected for doing so, than it is to conclude this from the fact that it *will be*. This asymmetry is surely explained by the fact that it is natural, in many cases, to presume that an item which was recently selected for doing Y continues to be presently, and therefore by the fact that (3p) is easily confused with (5).

It seems judicious therefore to concentrate on proposals (5), (6) and (7) from now on. These three proposals differ only in their temporal orientations; (5) is oriented towards the past, (6) towards the future, and (7) towards the present. None implies a statistical notion of normality. However they do not all accord with the same conception of the explanatory role of function-attributions. According to Neander, Millikan and Godfrey-Smith, only an analysis oriented towards the past allows us to do justice to the intuition according to which attributing a function to something is the same as causally explaining its existence. Indeed it seems clear enough that if there is a sense in which a function-attribution based on (6) or (7) explains the existence of a thing, it is probably not a question of a causal explanation. But we can ask if, and in what sense, a function-attribution based on (5) causally explains the existence of the functional item in question.

According to that analysis, a type of item presently has the function of producing Y if and only if it was recently selected for producing Y and presently continues to be. Now a type of item is selected *for* producing Y only if at least certain items of that type produce Y and if the fact that they produce Y confers an adaptive advantage to those that possess them, that is, only if it is selected because it produces Y. Since an item cannot produce any effect whatsoever without existing, the fact that it produces such and such an effect clearly cannot be the (or even a) cause of its existence, if we mean by that the fact that it comes to exist. It is natural to interpret a function-attribution corresponding to proposal (1) as an explanation of the fact that a certain type of item is presently widespread (or even fixed) in a certain population. But an item can have a function in the sense of (5) even if it is not yet very widespread, and even if it never becomes widespread.

The one thing that the fact that a certain type of item is selected seemingly allows us to explain, is that those that possess an item of that type reproduce more than the others; but this explanation is independent of the specific effect *for* which that type of item is selected. To say that a type of item is selected because it produces Y, is to say that the fact that items of that type produce Y confers an adaptive advantage to those that possess them. The relation between the fact that this type of item produces Y and the fact that those that possess an item of that type have an advantage is not a causal relation, but, as it were, a *constitutive* one: the fact that these individuals have an advantage (or that they have such and such a degree of fitness) is not caused by, but consists in the fact that they possess an item that produces Y. Thus, the fact that the item in question produces Y is (partially) constitutive of the fact that the individuals that possess it have an adaptive advantage over those that do not possess it, which explains, in its turn, why these individuals reproduce more. It is not certain that the explanatory link between the fact of having an advantage and the fact of reproducing more is a causal link, but even if that were not the case, it could be possible to maintain that the fact that the item produces Y is (partially) the cause of the fact that those that possess it reproduce more than the others. Be that as it may, let us suppose for the moment that to say that an item is selected for producing Y causally explains (when it is true) why the individuals that possess it survive and reproduce more than the others.

What is remarkable about this conclusion is that it is oriented neither towards the past nor towards the future, without for all that infringing on the principle according to which a cause must precede its effect. As long as we insist on saying that a function-attribution must causally explain the *existence* of the functional item in question, it is natural to exclude analyses oriented towards the present or towards the future; since if one can say that a certain item exists because it is selected or because it will be selected, that cannot be in the causal sense. But the fact that an item is selected at time t only explains why that item exists at time $t + x$ in virtue of the fact that it explains why the individuals that possess it reproduce more than the others, and it is plainly possible to interpret this explanation as a causal explanation.

This conclusion allows one to claim that (7) is preferable to (5) and (6). Note, first of all, that (7) and (6) have exactly the same explanatory force, as (6) only differs from (7) by the fact that a function-attribution conforming to (6) includes a prediction according to which individuals that possess the item in question will continue to reproduce more than others. But what is most awkward here is that according to analysis (6) there might come a time when the item under consideration ceases to have the function of producing Y, even though it is still selected for producing it (namely when it is no longer the case that it will continue to be selected for producing Y). In the same way, according to analysis (5), there is a time when the item in question does not yet have the function of producing Y, although it is already selected for producing it (namely when it begins to be selected for producing Y). Let t be the moment when the item in question begins to be selected for producing Y. Supposing that this item continues to be selected for producing Y until the time $t + x$, we would then find ourselves in the following situation. According to analysis (5), the item has the function of producing Y at time $t + x$, but not at time t, and according to analysis (6), it has this function at time t, but not at time $t + x$; whereas according to analysis (7), it has this function at time $t + x$, *and* at time t. Thus analyses (5) and (6) establish an asymmetry between time t and time $t + x$, even though from the point of view of explanation, there is no significant difference. The fact that the item in question is selected for producing Y at time t explains why the individuals that possess it reproduce more than the others at time t $(+\Delta t)$, in exactly the same way as the fact that it is selected for producing Y at time $t + x$ explains why the individuals that possess it reproduce more than the others at time $t + x$ $(+\Delta (t + x))$.[10] But in these conditions, should not the concept of function apply or not apply in the same way in both cases? This seems to provide some reason to conclude that (7) is the analysis that best accounts for the explanatory role of the notion of function: a thing has the function of doing Y from the moment it begins to be selected for doing it, and as much and as long as it is selected for doing it.

6. SOME UNRESOLVED QUESTIONS

I think I have shown that there are at least three ways to analyse the notion of function in purely selectionist terms, while avoiding a statistical notion of normality and accounting for the possibility that an item can gain and lose functions. Of these three analyses, two introduce an explanatory asymmetry that seems unwarranted, namely the one which is oriented towards the future, and the one which is oriented towards the past. An analysis completely oriented towards the present, such as (7), therefore seems preferable. This shows that (7) satisfies the constraints that have been accepted until now, but not that it constitutes an acceptable analysis. Now it is a matter of examining some of the characteristics of this analysis, and to determine whether they militate in its favour.

First, let us briefly examine another objection raised by Prior (1985: 314) against the etiological approach. This objection, which is probably the most popular one, is anticipated by Wimsatt (1972: 15) and one can find formulations of it in Boorse (1976: 373) and Bigelow and Pargetter (1987: 188). It is based on the intuition according to which a type of item which would not have been selected, and which would therefore have no causal history, might nevertheless have functions. Thus it is tempting, at first sight, to claim with these authors that the heart would have the function of circulating the blood even if the world had been created only five minutes ago, in exactly its present state. Neander and Millikan admit that their respective analyses conflict with this intuition, but conclude that in this case it is the intuition that must be corrected by the theory, and not the reverse. Moreover, Neander claims that only a purely etiological analysis allows the non-statistical nature of biological norms to be accounted for; but I have shown that this is not the case.

All else being equal, it is preferable to be in agreement with an intuition then to be in conflict with it. Now it seems clear that analysis (7) allows us to do at least partial justice to the intuition put forward by this "creationist" objection. Indeed one can assume that the heart begins to be selected for circulating blood as soon as the world begins to exist, if it is true that it is created in exactly the state in which it presently finds itself (and that the heart is presently selected for this, at least in part). Of course, there can be no selection in the case where the world is created only to be destroyed the next minute, because an item cannot be selected for producing a certain effect unless it at least has the time to produce it, so that there cannot, according to this analysis, be any function without a certain duration. Analysis (7) agrees in this sense with the idea that the concept of function is partly historical, and not purely dispositional. If those who advance the "creationist" objection (as seems to be the case with Bigelow and Pargetter) mean to claim that the function of the heart would be to circulate the blood simply in virtue of the fact that it would be disposed to circulate it (and that would lead to it being selected, if the world continued to exist), then it is clear that (7) does not accord with *this* intuition. It would be wrong, therefore, to claim that (7) marks a return to a dispositional conception, although it comes closer to it.

According to (7), a type of item presently has the function of producing Y if and only if it is presently selected for producing Y. Earlier I noted that, contrary to some of the other selectionist analyses that I mentioned, the expression "to be selected" could only refer in this formulation to the *process* of natural selection, and not to an eventual result. I took for granted, to all intents and purposes, that to say that an item is selected comes to the same thing as saying that it is favoured by natural selection, or that it confers on those that possess it an adaptive advantage over those that do not possess it. But these reformulations may not be equivalent. Should we understand that an item is selected only if those that possess it actually multiply more than others, that is, only if the population actually evolves under the pressure of natural selection, or could we admit that an item is selected as soon as it confers an adaptive advantage to those that possess it, even if other evolutionary forces see to it that the proportion of these that possess it does not increase? In other words, can we say that an item has a function as soon as the pressure of natural selection favours it, even if this pressure is cancelled out or reversed by those of other forces? It will be noted that this question is relevant to the issue of the explanatory role of function-attributions. Thus if we admit the second possibility, to say that an item is selected does not always constitute an explanation, even a partial one, of the fact that those that possess it reproduce more than others, since that might not be the case.

It is well known that the course of evolution is not determined solely by the pressure of natural selection, but also depends in part on certain "random" factors such as mutation and genetic drift, which can see to it that an item is eliminated from a population even though it is actually more advantageous than its competitors. Suppose that at time t, item A is in competition with item B, over which it has the advantage, but that the individuals that possess item B reproduce more than those that possess item A. According to the way in which (7) is interpreted, it leads in this case to different verdicts concerning item A. If it must be understood that natural selection must actually find expression in a relative increase in the rates of reproduction of the individuals that possess the most advantageous item (interpretation (i)) over that of all the others, then it must be said that item A has no function; whereas if it suffices for the item in question actually to be more advantageous than its competitors (interpretation (ii)), then one can say that the item A has a function. According to interpretation (ii), an item that has the function of producing Y therefore ceases to have that function as soon as there appears in the population an item that is superior to it, even if the latter is quickly eliminated as a result of other evolutionary factors. These two interpretations could be (approximately) formulated in the following way:

(i) an item (presently) has the function of producing Y if and only if it actually wins against its competitors in virtue of the fact that producing Y gives it an adaptive advantage over them.

(ii) an item (presently) has the function of producing Y if and only if the fact that it produces Y gives it an adaptive advantage over its competitors.

Neither of these interpretations of (7) allows a function to be attributed to item B, despite the fact that there are circumstances in which one might be tempted to do so; since the fact that natural selection is not the cause of the fact that B outclasses A does not prevent it from being the cause of the fact that B outclasses its other competitors, and is therefore not incompatible with

B being more advantageous than its other competitors. And if this was the case, one might be inclined to concede to it the function of producing those of its effects in virtue of which it gets the better of its competitors (other than A), or in virtue of which it has the advantage over them (especially if this situation persists long enough, and if its other competitors are in sufficient number). In other words, this raises the question: over how many competitors must an item have the advantage, or how many competitors must it outclass, in order to have a function? Until now I have supposed that it is a matter of all the (actual, not possible) competitors, but perhaps one ought to weaken this requirement, or even agree to say that an item only has a function relative to some of its competitors (namely either those it actually gets the better of because of natural selection, or those over which it has an adaptive advantage, whether or not this expresses itself in a relative increase in the rate of reproduction). This reveals at least four possible ramifications for each of the interpretations distinguished above:

(ia) an item (presently) has the function of producing Y if and only if it actually gets the better of *all* of its competitors in virtue of the fact that producing Y gives it an adaptive advantage over them.

(ib) an item (presently) has the function of producing Y if and only if it actually gets the better of *most* of its competitors in virtue of the fact that producing Y gives it an adaptive advantage over them.

(ic) an item (presently) has the function of producing Y if and only if it actually gets the better of *some* of its competitors in virtue of the fact that producing Y gives it an adaptive advantage over them.

(id) an item X (presently) has the function of producing Y *relative to* a certain item X', if and only if it actually gets the better of X' in virtue of the fact that producing Y gives X an adaptive advantage over X'.

(iia) an item (presently) has the function of producing Y if and only if the fact that it produces Y gives it an adaptive advantage over *all* its competitors.

(iib) an item (presently) has the function of producing Y if and only if the fact that it produces Y gives it an adaptive advantage over *most* of its competitors.

(iic) an item (presently) has the function of producing Y if and only if the fact that it produces Y gives it an adaptive advantage over *some* of its competitors.

(iid) An item X (presently) has the function of producing Y *relative to* a certain item X', if and only if the fact that it produces Y gives it an adaptive advantage over X'.

It is important to note that these diverse possibilities are entirely independent of the fact that (7) is an analysis "oriented towards the present" and that they do not therefore call into question the reasons put forward earlier for preferring (7) to (5), for example. Selectionists generally take it for granted that the process of selection to which reference is made does in fact express itself by tangible effects on the rate of reproduction (which corresponds to formulations (ia)–(id)), probably because the other interpretation would, in certain cases, deprive function attributions of explanatory value.[11]

Godfrey-Smith (1991a: 73) briefly considers, albeit in relation to an analysis oriented towards the past that is fairly close to (5), the possibility of saying that an item only has a function *relative to* those of its competitors that it actually gets the better of *because* of natural selection. He remarks that this solution is undoubtedly the one that best respects the logic of selectionist explanations, but rejects it, raising the points that it is not in keeping with the usage of the concept of function generally made by biologists and that it would make it difficult to manage. Added to these two reasons for avoiding as far as possible such a relativisation, is the fact that it would imply a corresponding relativisation of the normative notion of "being supposed to do such and such a thing". But this notion does not seem to have this kind of relativity.

After having considered options (ia) and (ic), Godfrey-Smith (1991a: 76–77) finally retains option (ib), namely the one according to which a functional item must actually get the better of *most* of its competitors in virtue of the fact that certain of its effects give it an adaptive advantage over them. Options (ia) and (ic) are respectively judged to be too exacting and not exacting enough. The first implies that an item lacks a function as soon as it gets the better of one of its competitors for a reason other than the pressure of natural selection, and the second that an item has a function as soon as it has got the better of at least one competitor by reason of natural selection (even if it gets the better of all the others because of other factors). But (ib) could also raise a difficulty, since it makes it possible (as do (ic) and (id) for the matter) that two items in competition simultaneously have a function. Indeed, it is theoretically possible that A and B get the better of all other competitors by reason of natural selection, and that B gets the better of A because of another factor, even though it is less advantageous than A. In these conditions, (ib) would grant a function to both A and B; but perhaps this is not an unacceptable consequence, particularly since if this trend continues, there will probably come a time when A will disappear, and when B will get the better of all its competitors because of natural selection.

Remarks similar to those of the last two paragraphs apply to (iia)–(iid).

I agree with Godfrey-Smith that options (ic) and (iic) are probably too weak, and that (id) or (iid) should only be accepted as a last recourse. It may well be, however, that the concept of function is simply not well enough determined for the choice of one of the four remaining options not to be at least partially arbitrary.

7. SCEPTICAL CONCLUSIONS

If the difficulty of choosing between (ia) and (ib), or between (iia) and (iib) can be attributed to the relatively vague or non-rigorous nature of the concept of function, there could be a way to approach the opposition between (ia) and (ib) on the one hand, and (iia) and (iib) on the other, which allows one to justify a preference. One will say that it is enough to recall that only (ia)–(ib) guarantee the explanatory character of function-attributions, and that this is an argument in their favour. But I am not entirely convinced that this is a necessary requirement, especially from the perspective of a teleofunctional analysis of mental states, if it is true that such an analysis would aim at uniting the concepts of function and meaning: because the idea that to specify the content of a mental state must amount to providing a (causal) explanation of the fact that it exists or persists, or of the fact that those that possess it reproduce more than others, seems rather confused to me. What matters, from the teleosemantic perspective, is first and foremost the normative content of the notion of function. But there could be, from this point of view, a significant difference between (i) and (ii).

Indeed, the fact that X has the function of producing Y must allow one to conclude that X is supposed to produce Y. If we keep to analysis (ii), this means that one must admit that:

(Pii)   if the fact that an item produces Y confers it an adaptive advantage over (all or most of) its competitors, then it is supposed to produce Y,

whereas if we keep to (i), one must admit that:

(Pi)   if an item actually gets the better of (all or most of) its competitors in virtue of the fact that it produces Y, then it is supposed to produce Y.

Note that the antecedent of (Pi) implies that of (Pii), so that if (Pii) is not acceptable even though (Pi) is, the validity of (Pi) must rest, at least partially, on that part of its content that exceeds that of (Pii), that is to say, on the existence of a causal relation between the fact of being more advantageous and that of reproducing more. This means that (i) will be preferable to (ii) if (but not only if) there are reasons to believe that the existence of such a causal relation is essential to the conclusion that a certain item is supposed to do such and such a thing. In the opposite case, it will be necessary to admit that this causal relation contributes only to the explanatory role of function-attributions. I strongly doubt that one can find such reasons. But I am also sceptical about the possibility of maintaining that (Pii) is sufficient.

Indeed it seems to me that all efforts to naturalise a normative notion of function raise a difficulty of principle. Admitting the distinction between descriptive and normative statements, we can pose the problem by saying that the link between the analysans and the fact of being "supposed" to do something, that is to say (Pi) and (Pii), can only the descriptive or normative. If they are descriptive, they should apparently be either purely logical, or analytic, or nomological, or extensional (material), or "metalinguistic" (in a sense which I will specify in a moment). Supposing that the analysans conforms to naturalism, the first two options are clearly excluded, because they would exemplify the naturalistic fallacy. The next two options are possible only on condition of being able to determine what a thing is supposed to do, independently of its natural properties; but in that case there is not really any naturalisation of the normative dimension of the concept of function. A principle such as (Pi) or (Pii), is only plausible, from a naturalist perspective, if it is understood in a "metalinguistic" sense, that is, as a way of recording the fact that it is largely accepted within a certain community. Such an interpretation can have a certain interest, depending on which community we are dealing with, but leaves intact the question of whether the principle in question is legitimate.

There remains the possibility that the principles in question are normative, that is, that they have the status of recommendations concerning the manner in which it is suitable to use the concept of function. But then there is a serious risk that the explanation of the concept of function might be circular, which will only be avoided if we make use of an independent naturalisation of the notion of norm that itself avoids this kind of circularity. But the prospects of finding such a naturalisation are discouraging to say the least, which inclines me to think that a coherent naturalism can probably only be a form of eliminativism.[12]

Université de Montréal

NOTES

[1] One can cite, among the best-known works, those of Dennett (1987a), Dretske (1988), Godfrey-Smith (1991a, b), McGinn (1987), Millikan (1984, 1993), and Papineau (1987).

[2] It should be noted that Neander only proposes the first formulation for natural functions. According to Neander (1991a: 461–462), non-natural functions also result from a process of selection, but not necessarily from a process of selection based on past effects of things that have these functions. She is speaking, in that case, of "intentional selection".

[3] To be rigorous, it would surely be necessary to limit oneself, here, to individuals that reach a certain age.

[4] One will notice, however, that this discussion leaves intact the question of whether a type of item can be selected for producing Y when only a minority of items of that type are *capable* of producing Y. The principle stated earlier according to which two items cannot be of the same type if they do not have the same effects in the same environments, seems to exclude this possibility automatically. But this principle is probably formulated incorrectly, since a thing does not belong to a single type, and it is certainly possible that, of two organs of the same type, only one should be capable of fulfilling its function. We can surely conclude from this that there is a physical difference between them and thus that there is a physical type of which they do not both belong, but maybe not that they are not of the same biological type. In the context of the present discussion, it would be circular to consider the relevant notion of biological type as a functional notion. To say that an item is selected is to say (among other things) that the items that share certain *physical/causal* characteristics (and not those that share a certain function) confer an adaptive advantage on those that possess them. If the notion of "heart", for example, is a functional notion, then (there is a sense in which) it is circular to say that the *heart* is (or was) selected for circulating the blood.

For it to be possible that a given heart has the function of circulating the blood even if it should happen to be incapable of doing it, it is necessary that it belong to the type "heart" in virtue of characteristics other than that of circulating the blood. Is it possible that the items of that type (that is, those that have those characteristics) could be selected for circulating the blood even though only a minority of them are capable of doing so? It depends, of course, on the way in which the type is defined. But suppose that the type of item X (that is, the items having the characteristic X) are selected because some of its members produce Y, and that the vast majority of items of that type are capable of producing Y. Unless all the entities are of type X, there will always be a type X' containing all the members of X, plus a certain number of others that are themselves incapable of producing Y and that can be more numerous than the first. As long as the individuals that possess an item of type X' have, on average, an adaptive advantage over the others, there is nothing, apparently, to stop us from admitting that the type X' is selected, as well as the type X. And as far as I can see, it is not impossible that this is the case even when the majority of items of type X' are incapable of producing the effects for which they are selected. I am inclined, therefore, to respond affirmatively to the question asked above.

[5]   This goes for Millikan as well, although Prior does not seem to be aware of her works.

[6]   As a matter of fact, this example may be objectionable. According to Karen Neander, there are reasons to believe that our present appendix never had the capacity to digest cellulose, and comes from a genotype which has been in competition with that of some other organ, which did digest cellulose. But the example is not essential. It remains true that one and the same trait cannot, on this analysis, loose a function it has acquired.

[7]   On the other hand, this leads us to admit the possibility that a type of item could simultaneously have several incompatible functions, which runs counter to the principle of deontic logic according to which what is obligatory must be possible. The "logic" of functional statements would therefore seem to deviate from that of normative statements in general. Of course, one cannot say that an item that has the function of doing Y thereby has the "obligation" to do Y, but it is legitimate to say that it is "meant" or "supposed" to do Y, and it seems *at first glance* than an item could not be supposed to do the impossible.

[8]   And to which Millikan (1993b: 40) seems finally to come around when she identifies the function of an item with that which, in 1984, she called its *most proximal* proper function.

[9]   One might think that the objection I am about to formulate is not well founded, since (3p) apparently does not permit one to say, in the imagined situation, that the item in question acquires the function of producing Z as soon as it begins to be selected for producing Y. But it is easy to eliminate this worry by slightly modifying the case. It suffices to suppose that the recent period in question includes the moment when that item ceases to be selected for producing Y and begins to be selected for producing Z.

[10]   Naturally, one must not forget that the "moments of time" mentioned in such contexts are rather plump, and would not be measured in nanoseconds. But this does not affect the reasoning.

[11]   Formulations (iia)–(iid) apparently introduce a counterfactual element which could be troublesome. Indeed, supposing that "to be selected" is used in the "material" way privileged by the selectionists, (iia) could apparently be stated in the following way:

(iia')   an item (presently) has the function of producing Y if and only if it would be selected for producing Y over all its competitors, if natural selection were the only evolutionary force in effect.

But (iib)–(iid) can hardly be read in an analogous way:

(iib')   an item (presently) has the function of producing Y if and only if it would be selected for producing Y over most of its competitors if natural selection were the only evolutionary force in effect.

(iic')   an item (presently) has the function of producing Y if and only if it would be selected for producing Y over some of its competitors if natural selection were the only evolutionary force in effect.

(iid')   an item X (presently) has the function of producing Y relative to a certain item
            X', if and only if it would be selected for producing Y over X' if natural selec-
            tion were the only evolutionary force in effect.

This is because if natural selection *is* the only force in effect, then an item is selected over another
if (and only if) it is more advantageous. In this hypothesis then, (iib')–(iid') reduce to (iia').

[12] The material in this paper has been used in talks given at the CRÉA, in Paris, at the Université
Paul-Sabatier, in Toulouse, and at the Université du Québec à Trois-Rivières, in November
and December 1993. I would like to thank Éric Audureau, Pascal Engel, Nicolas Kaufman, Pierre
Jacob, Martin Montminy, Karen Neander and Claude Panaccio for having had some (and mixed)
reaction. I owe special thanks to Lawrence Deck for having translated the present article from
the French.

REFERENCES

Bedau, Mark, 1991, 'Can Biological Teleology be Naturalised?', *J. of Phil.* **88**, 647–655.
Bedau, Mark, 1992a, 'Goal-Directed Systems and the Good', *The Monist* **75**, 34–51.
Bedau, Mark, 1992b, 'Where is the Good in Teleology?', *Phil. and Phenom. Research* **52**, 781–806.
Bigelow, John and Pargetter, R., 1987, 'Functions', *J. of Phil.* **84**, 181–196.
Bogdan, Radu (ed.), 1986, *Belief*, Oxford University Press, Oxford.
Boorse, Christopher, 1976, 'Wright on Functions', Sober ed. (1984), 369–385.
Bunge, Mario (ed.), 1973, *The Methodological Unity of Science*, Reidel, Dordrecht.
Canfield, John V., 1990, 'The Concept of Function in Biology', *Philosophical Topics* **18**, 29–54.
Charles, David and Lennon, Kathleen (eds.), 1992, *Reduction, Explanation and Realism*, Clarendon Press, Oxford.
Charlton, William, 1991, 'Teleology and Mental States', *Proc. of the Arist. Society Suppl.* **65**, 17–32.
Colodny, Robert G. (ed.), 1977, *Logic, Laws, and Life*, University of Pittsburgh Press, Pittsburgh.
Cummins, Robert, 1975, 'Functional Analysis', Sober ed. (1984), 386–406.
Cummins, Robert, 1989, *Meaning and Mental Representation*, MIT Press, Cambridge (Mass.).
Davidson, Donald, 1980a, *Essays on Actions and Events*, Oxford University Press, Oxford.
Davidson, Donald, 1980b, 'Towards a Unified Theory of Meaning and Action', *Grazer Philosophische Studien* **2**, 1–12.
Davidson, Donald, 1983, 'A Coherence Theory of Truth and Knowledge', Lepore ed. (1986), 307–319.
Davidson, Donald, 1984, *Inquiries into Truth and Interpretation*, Oxford University Press, Oxford.
Davidson, Donald, 1990a, 'The Structure and Content of Truth', *J. of Phil.* **87**, 279–328.
Davidson, Donald, 1990c, 'Representation and Interpretation', Said, K. A. M. *et al.* eds. (1990), 13–26.
Dennett, Daniel C., 1983, 'Intentional Systems in Cognitive Ethology: The Panglossian Paradigm Defended', Dennett (1987a), 269–286.
Dennett, Daniel C., 1987a, *The Intentional Stance*, MIT Press, Cambridge (Mass.).
Dennett, Daniel C., 1987b, 'Evolution, Error and Intentionality', Dennett (1987a), 287–322.
Dennett, Daniel C., 1989, 'Cognitive Ethology: Hunting for Bargains or a Wild Goose Chase?', Montefiore and Noble eds. (1989), 101–116.
Dennett, Daniel C., 1990, 'The Interpretation of Texts, People and Other Artifacts', *Phil. and Phenom. Research* **50**, Supplement, 177–194.
Dretske, Fred, 1981, *Knowledge and the Flow of Information*, MIT Press, Cambridge (Mass.).
Dretske, Fred, 1983, 'Précis of Knowledge and the Flow of Information', *Behavioral and Brain Sciences* **6**, 55–90.
Dretske, Fred, 1986, 'Misrepresentation', Bogdan ed. (1986), 17–36.
Dretske, Fred, 1988, *Explaining Behavior*, MIT Press, Cambridge (Mass.).

Dretske, Fred, 1990, 'Précis of Explaining Behavior', *Phil. and Phenom. Research* **50**, 783–793.
Dretske, Fred, 1993, 'Mental Events as Structural Causes of Behaviour', Heil and Mele eds. (1993), 121–136.
Dupré, John (ed.), 1987, *The Latest on the Best*, MIT Press, Cambridge (Mass.).
Enc, Berent, 1979, 'Function Attribution and Functional Explanations', *Phil. of Science* **46**, 343–365.
Enc, Berent, 1982, 'Intentional States of Mechanical Devices', *Mind* **91**, 161–182.
Feldman, Richard, 1988, 'Rationality, Reliability, and Natural Selection', *Phil. of Science* **55**, 218–227.
Fetzer, James H. (ed.), 1985, *Sociobiology and Epistemology*, Reidel, Dordrecht.
Fetzer, James H., 1990, 'Evolution, Rationality and Testability', *Synthese* **82**, 423–439.
Fodor, Jerry A., 1984, 'Semantics, Wisconsin Style', *Synthese* **59**, 231–250.
Fodor, Jerry A., 1986, 'Why Paramecia Don't Have Mental Representations', French, Uehling and Wettstein eds. (1986), 3–24.
Fodor, Jerry A., 1987, *Psychosemantics*, MIT Press, Cambridge (Mass.).
Fodor, Jerry A., 1990a, *A Theory of Content and Other Essays*, MIT Press, Cambridge (Mass.).
Fodor, Jerry A., 1990b, 'Psychosemantics, or: Where Do Truth Conditions Come From?', Lycan W. C. ed. (1990), 312–337.
Fobes, Graeme, 1989, 'Biosemantics and the Normative Properties of Thought', Tomberlin J. ed. (1989), 533–548.
French, Peter A., Uehling, T. E., and Wettstein, H. K. (eds.), 1984, *Midwest Studies in Philosophy 9: Causation and Causal Theories*, University of Minnesota Press, Minneapolis.
French, Peter A., Uehling, T. E. and Wettstein, H. K. (eds.), 1986, *Midwest Studies in Philosophy 10*, University of Minnesota Press, Minneapolis.
Gauker, Christopher, 1986, 'The Principle of Charity', *Synthese* **69**, 1–25.
Godfrey-Smith, Peter, 1989, 'Misinformation', *Can. J. of Phil.* **19**, 533–550.
Godfrey-Smith, Peter, 1991a, *Teleonomy and the Philosophy of Mind*, thèse de doctorat, University of California, San Diego.
Godfrey-Smith, Peter, 1991b, 'Signal, Decision, Action', *J. of Phil.* **88**, 709–722.
Godfrey-Smith, Peter, 1992, 'Indication and Adaptation', *Synthese* **92**, 283–312.
Gould, S. J. and Vrba, E. S., 1982, 'Exaptation: A Missing Term in the Science of Form', *Paleobiology* **8**, 4–15.
Grandy, Richard, 1987, 'Information-based Epistemology, Ecological Epistemology and Epistemology Naturalized', *Synthese* **70**, 191–204.
Grandy, Richard E., 1973, 'Reference, Meaning and Belief', *J. of Phil.* **70**, 439–452.
Hahlweg, Kai and Hooker, C. A. (eds.), 1989, *Issues in Evolutionary Epistemology*, SUNY Press, Albany.
Heil, John and Alfred Mele (eds.), 1993, *Mental Causation*, Clarendon Press, Oxford.
Knowles, Dudley (ed.), 1990, *Explanation and its Limits*, Cambridge University Press, Cambridge.
Loewer, Barry and Rey, G. (eds.), 1991, *Meaning in Mind: Fodor and His Critics*, Blackwell, Oxford.
Lycan, William G., 1987, *Consciousness*, MIT Press, Cambridge (Mass.).
Lycan, William G. (ed.), 1990, *Mind and Cognition*, Blackwell, Oxford.
MacDonald, Graham, 1989, 'Biology and Representation', *Mind and Language* **4**, 186–200.
Machamer, Peter, 1977, 'Teleology and Selective Processes', Colodny, R. G. ed. (1977), 129–142.
Matthen, Mohan, 1988, 'Biological Functions and Perceptual Content', *J. of Phil.* **85**, 5–27.
Matthen, Mohan, 1991, 'Naturalism and Teleology', *J. of Phil.* **88**, 656–657.
Matthen, Mohan and Levy, E., 1984, 'Teleology, Error and the Human Immune System', *J. of Phil.* **81**, 351–372.
McGinn, Colin, 1989, *Mental Content*, Blackwell, Oxford.
McLaughlin, Brian (ed.), 1991, *Dretske and his Critics*, Blackwell, Oxford.
Millikan, Ruth G., 1984, *Language, Thought and Other Biological Categories*, MIT Press, Cambridge (Mass).

Millikan, Ruth G., 1986, 'Thoughts Without Laws; Cognitive Science with Content', Millikan (1993), 51–82.
Millikan, Ruth G., 1989a, 'In Defense of Proper Functions', Millikan (1993c), 13–30.
Millikan, Ruth G., 1989b, 'Biosemantics', Millikan (1993c), 83–102.
Millikan, Ruth G., 1990, 'Compare and Contrast Dretske, Fodor and Millikan on Teleosemantics', Millikan (1993c), 123–134.
Millikan, Ruth G., 1991, 'Speaking Up for Darwin', Loewer and Rey eds. (1991), 151–164.
Millikan, Ruth G., 1993a, 'Explanation in Biopsychology', Millikan (1993c), 171–192.
Millikan, Ruth G., 1993b, 'Propensities, Explanations, and the Brain', Millikan (1993c), 31–50.
Millikan, Ruth G., 1993c, *White Queen Psychology and Other Essays for Alice*, MIT Press, Cambridge (Mass.).
Mitchell, Sandra D., 1993, 'Dispositions or Etiologies? A Comment on Bigelow and Pargetter', *J. of Phil.* **90**, 249–259.
Montefiore, Alan and Denis Noble (eds.), 1989, *Goals, No-Goals and Own Goals*, Unwin Hyman, London.
Nagel, Ernest, 1977, 'Teleology Revisited', *J. of Phil.* **74**, 261–301.
Neander, Karen, 1991a, 'The Teleological Notion of "Function" ', *Aust. J. of Phil.* **69**, 454–458.
Neander, Karen, 1991b, 'Functions as Selected Effects: the Conceptual Analyst's Defense', *Phil. of Science* **58**, 168–184.
Papineau, David, 1984, 'Representation and Explanation', *Phil of Science* **51**, 550–572.
Papineau, David, 1987, *Reality and Representation*, Blackwell, Oxford.
Papineau, David, 1990, 'Truth and Teleology', Knowles ed. (1990), 21–44.
Papineau, David, 1991, 'Teleology and Mental States', *Proc. of the Arist. Society Suppl. Vol.* **65**, 33–54.
Pietroski, Paul M., 1991, 'Intentionality and Evolutionary Error', inédit.
Prior, Elizabeth W., 1985, 'What Is Wrong with Etiological Accounts of Biological Function?', *Pacific Phil. Quarterly* **66**, 310–328.
Putnam, Hilary, 1992, *Renewing Philosophy*, Harvard University Press, Cambridge (Mass.).
Rescher, Nicholas (ed.), 1986, *Current Issues in Teleology*, University Press of America, Lanham.
Rosenberg, Alexander, 1985a, *The Structure of Biological Science*, Cambridge University Press, Cambridge.
Rosenberg, Alexander, 1986, 'Intentional Psychology and Evolutionary Biology', *Behaviorism* **14**, 15–27 and 125–138.
Ruse, Michael, 1973, *The Philosophy of Biology*, Hutchinson, London.
Ruse, Michael, 1981, *Is Science Sexist?*, Reidel, Dordrecht.
Said, K. A. M. and al. (eds.), 1990, *Modelling the Mind*, Oxford University Press, Oxford.
Schoemaker, Paul J. H., 1991, 'The Quest for Optimality: A Positive Heuristic of Science?', *Behavioral and Brain Sciences* **14**, 205–214.
Sober, Elliott, 1981, 'The Evolution of Rationality', *Synthese* **48**, 95–120.
Sober, Elliott, 1984a, *The Nature of Selection*, MIT Press, Cambridge (Mass.).
Sober, Elliott (ed.), 1984b, *Conceptual Issues in Evolutionary Biology*, MIT Press, Cambridge (Mass.).
Sober, Elliott, 1985a, 'Methodological Behaviorism, Evolution, and Game Theory', Fetzer ed. (1985), 181–200.
Sober, Elliott, 1985b, 'Panglossian Functionalism and the Philosophy of Mind', *Synthese* **64**, 165–193.
Stich, Stephen, 1981, 'Dennett on Intentional Systems', Lycan, W. G. ed. (1990), 167–184.
Stich, Stephen, 1984, 'Relativism, Rationality and the Limits of Intentional Description', *Pacific Phil. Quarterly* **65**, 211–235.
Stich, Stephen, 1985, 'Could Men Be an Irrational Animal?', *Synthese* **64**, 115–135.
Stich, Stephen, 1990, *The Fragmentation of Reason*, MIT Press, Cambridge (Mass.).
Stich, Stephen, 1991, 'The Fragmentation of Reason: Précis of Two Chapters', *Phil. and Phenom. Research* **51**, 179–183.
Thagard, Paul and Nisbett, R. E., 1983, 'Rationality and Charity', *Phil. of Science* **50**, 250–267.

Tomberlin, James E. (ed.), 1989, *Philosophical Perspectives 3*, Ridgeview, Atascadero.

Villanueva, Enrique (ed.), 1990, *Information, Semantics and Epistemology*, Blackwell, Oxford.

Williams, Mary B., 1970, 'Deducing the Consequences of Evolution: A Mathematical Model', *J. of Theor. Biol.* **29**, 343–385.

Williams, Mary B., 1973, 'The Logical Status of the Theory of Natural Selection and Other Evolutionary Controversies', Sober, E. ed. (1984b), 83–98.

Wimsatt, William C., 1972, 'Teleology and the Logical Structure of Function Statements', *Studies in History and Phil. of Science* **3**, 1–80.

Woodfield, Andrew, 1976, *Teleology*, Cambridge University Press, Cambridge.

Woodfield, Andrew, 1987, 'Two Categories of Content', *Mind and Language* **1**, 319–354.

Woodfield, Andrew, 1990, 'The Emergence of Natural Representations', *Phil. Topics* **18**, 187–213.

Wright, Larry, 1972, 'Explanation and Teleology', *Phil. of Science* **39**, 204–218.

Wright, Larry, 1973, 'Functions', Sober ed. (1984), 347–368.

Wright, Larry, 1976, *Teleological Explanations*, University of California Press, Berkeley.

Wright, Larry, 1978, 'The Ins and Outs of Teleology: A Critical Examination of Woodfield', *Inquiry* **21**, 223–245.

NATURAL SELECTION AND
INDEXICAL REPRESENTATION

The claim that true belief is a useful commodity should strike no one as perverse. If I am to eat that red berry on that shrub it had better be the case that its consumption will not terminate my life. There may be exceptions to the above claim, cases where 'ignorance is bliss' or 'a little knowledge is a dangerous thing,' but a generally false belief set just cannot, by and large, be an advantage for humans. In fact, I am almost embarrassed to have to defend the claim that true belief is a useful commodity: so entirely preposterous does its negation seem to me. But defend it I will, in what follows.

1. STICH'S ARGUMENT AGAINST RELIABLE INFERENTIAL SYSTEMS

Let me first introduce a distinction between false positives and false negatives. To infer *P* when *P* is not the case is to utter a false positive. To infer *not P* when *P* is the case is to utter a false negative. Consider the example I began with. If I have a false negative belief about poisonous red berries, I will not live to write about the experience if I eat such berries. That is, if I believe that "that red berry is not poisonous" when it is poisonous then I am a dead duck. If, on the other hand, I endorse the false positive that "that red berry is poisonous" when it isn't then I pay no penalty. Stephen Stich uses this sort of example to argue that a more reliable inferential system may trail a less reliable one in external fitness. But what is external fitness?

Elliot Sober distinguishes the internal from the external fitness of a genetic program by noting that internal fitness has only to do with the efficiency of the program considered by itself. Comparing a good genetic program with a good computer program, he notes that: "A good program will not only generate the right output for the right input, it will do so economically. It will not use up too much of the computer's memory, nor will it be a drain on the energy source on which the computer runs." (Sober, 1981, p. 105) External fitness, in contrast, is such that one genetic program is fitter than another if, *ceteris paribus*, its input/output pairings are more conducive to survival and successful reproduction (Stich, 1990, p. 60).

Stich's claim is that adopting a very cautious, risk-aversive inferential strategy that results in mostly false positive beliefs will tend to produce a higher level of external fitness than adopting a risky inferential strategy that results in mostly false negative beliefs. Hence, the risks attached to adopting a more reliable inferential strategy result in a less externally fit organism. His point is that one false negative combined with many true negatives may result in a generally reliable inference system. But this system will not be as exter-

M. Marion and R. S. Cohen (eds.), *Québec Studies in the Philosophy of Science II*, 53–62.
© 1996 *Kluwer Academic Publishers.*

nally fitness enhancing as a system where one has many false positive beliefs and so a generally unreliable inference system where the false negative in the reliable system is something like: "that red berry is not poisonous." It follows that it is exceedingly unlikely that there has been selection for reliable inferential systems. Stich concludes that someone who wishes to provide an evolutionary foundation for reliabilism cannot successfully run this line of argumentation.

Let us grant that there may be cases where believing false positives will be more fitness enhancing than believing false negatives. Believing that red berries are poisonous when they are not is less dangerous than believing that red berries are not poisonous when they are, where consumption of such berries follows. Of course, if all one will eat are red berries then one loses either way: one starves or dies from poisoning. The issue is: does Stich's argument show that evolution would not favor reliable inferential systems as they will be, generally, less externally fit? The answer, I think, is NO.

The fact that believing lots of false positives can be safer than believing lots of false negatives does not begin to prove that reliable inferential systems will not be favored by natural selection. At best, it shows that particular false beliefs can be very bad for your health. What it does not show is that, generally, false beliefs will be more externally fitness-enhancing than true beliefs. Stich tries to finesse us when he adds that

... a very cautious, risk-aversive inferential strategy – one that leaps to the conclusion that danger is present on very slight evidence – will typically lead to false beliefs more often, and true ones less often, than a less hair-trigger one that waits for more evidence before rendering a judgement. Nonetheless, the unreliable, error-prone risk-aversive strategy may well be favored by natural selection. (Stich, 1990, p. 62.)

Here, as elsewhere in philosophy, the careful selection of examples can make all the difference. For consider the case where I quickly form true beliefs about buses in my natural habitat and my friend slowly comes to false conclusions in this regard. I quickly step out of the way of oncoming buses with the aid of a true belief, my friend eventually concludes that the bus he sees is a bus facsimile carefully constructed to make a deep philosophical point by a brilliant, if slightly deranged, philosopher. He holds his ground only to buy the farm. In such a case it would seem true belief or accurate representation has some purchase on survival and external fitness.[1] Since evolution cannot select for true belief directly, reliable processes are selected for. We need not require that I always get the right answer in such situations for my point to hold, only that I generally get the right answer. After all, reliabilists have never held that reliable processes are infallible.[2] Such a strong requirement would be implausible. Rather, the idea is that cognitive processes that tend to produce the truth will be selected for. One need not fear, in other words, that evolutionary Panglossianism has been carried to absurdity.

2. SELECTION FOR AND SELECTION OF

One needs to refine the sense in which the term 'natural selection' is being used here. Following Steven J. Gould and Richard Lewontin, Elliot Sober distinguishes between selection of an object and selecting for a property.[3] As Sober notes:

To say that there is selection for a given property means that having that property causes success in survival and reproduction. But to say that a given sort of object was selected is merely to say that the result of the selection process was to increase the representation of that kind of object. (Sober, 1984, p. 100.)

For instance, the selection of the human chin was a byproduct of selection for certain jaw structure properties. The human chin is what Sober calls a 'free rider' because it was not selected for, but of.[4] As he notes: " 'Selection of' pertains to the effects of a selection process, whereas 'selection for' describes its causes." (Sober, 1984, p. 100) Fred Dretske has recently employed this distinction to suggest that there has been selection for reliable processes but that, indirectly, this has led to the selection of a very special kind of object: true indexical representations. In humans the form that these representations take, according to Dretske, is that of a belief. Let me illustrate what is meant by a true indexical belief. Suppose that Hannibal Lector is approaching me with his barbecue apron on and I exclaim: "There's Hannibal the Cannibal now!" Such a proposition provides us with an example of a true indexical belief. It is a belief about the here and now and it ought to move me to action. Held in the appropriate circumstances, I think such beliefs can be crucial to our survival and survival is a necessary condition for reproductive success. I agree with Dretske that there has been selection for reliable processes. But I deny that there has been selection of true indexical representations. To see why Dretske cannot be right about true beliefs, we need to return to the example that Sober borrows from Gould and Lewontin to illustrate the selection for/selection of distinction.

The human chin was not selected for. Rather, certain jaw structure properties were selected for because they made possible proper mastication for our ancestors. An inevitable architectural consequence of the selection for these jaw structure properties was the chin. Objects that are selected of are termed "free riders" because they confer no evolutionary benefit in terms of survival and reproduction. The chin falls in this latter category. In such cases, the increasing frequency of such a free rider trait is not tied to some evolutionary advantage that possession of that trait confers. Instead, at the outset these traits enter as a nonadvantageous by-product of some other trait that was advantageous. Now lets consider Dretske's claim that reliable processes were selected for, while true beliefs were selected of. If the analogy is to be exact, then the chin is to jaw structure properties as true beliefs are to reliable processes. That is, both the chin and true beliefs must be free riders that confer no evolutionary advantage for humans. But Dretske's motivation for suggesting that reliable processes were selected for was that this would provide an indirect way of getting mostly true beliefs. Dretske thinks true beliefs do confer an evolutionary advantage for humans. As he notes: "Getting-things right is not just a useful skill. It is a biological imperative." (Dretske, p. 89) Since indexical representations are not heritable because they are not traits that could be passed on, the only way to select truth is to select for reliability. Reliable processes, such as good reasoning, perceptual capacities and other sensory modalities, being traits were selected for because they produce something that confers a tremendous evolutionary advantage: accurate indexical representations. True beliefs are the product of reliable processes but they cannot have the status of a free rider, such as the chin. Possessing a chin is not a biological imperative, possessing true beliefs is. Possessing a chin is not an evolutionary advantage, possessing true beliefs is. Hence, half of Dretske's analogy breaks down.

Here is what Dretske should have said: the need for true beliefs explains why reliable processes were selected for just as the need for mastication explains why certain jaw structure properties were selected for. True beliefs were not selected of. In fact, true beliefs cannot be selected for or of because true beliefs are not traits and only traits can be selected for or of. For example, the chin was selected of but it might, at some time, be selected for. Suppose that sexual selection results in the favoring of certain chins in so far as the opposite sex is attracted to males possessing such traits. One might find that there is an increasing frequency of "Kirk Douglas style" chins in subsequent generations. The point is that the kind of thing that can be selected of is exactly the sort of thing that might be selected for: traits. True beliefs are not traits, they are semantic properties. It follows that true beliefs are not the sort of thing that could be selected for or of. Dretske is mistaken.

But what, one might ask, was selected of in relation to the reliable processes that were selected for? The first thing to note here is that there need be no free riders vis a vis reliable processes. It simply is not the case that for every trait that is selected for there will be some object that is selected of as a free rider. It is plausible to suppose that the ability to enjoy a beautiful sunset or the ability to do modal logic are by-products of, or free riders with respect to, respectively, perception and good reasoning. Elliot Sober has, in fact, suggested that the ability to do calculus and modal logic may be free riders with respect to the more primitive reasoning abilities that they have their foundation in. As Sober insists: "Granted, there was no selection for the ability to do calculus, but that ability could nevertheless emerge via natural selection. Suppose that the ability to do science and mathematics were correlated with certain simpler skills of reasoning and communication that were beneficial in our ancestral past. Then the capacity for more abstract reasoning would have evolved as a free rider." (Sober, 1984, p. 24) It is clear that humans employ their native abilities in ways in which such abilities were not originally employed. For instance, perception may have been selected for because this trait made possible the pursuit of food, the avoidance of one's predators, and so forth. The fact that we now employ perception to view a splendid sunset confers no evolutionary advantage on us, though it may result in an aesthetic benefit.

3. HUMAN IRRATIONALITY AND THE FREE RIDER EFFECT

This approach promises to clarify much that has seemed recalcitrant, if not impossible, for cognitive scientists to explain. Consider the cognitive science literature on human rationality. Kahneman and Tversky, Nisbett and Ross, and many others have over the last few decades produced an enormous number of inventive experiments that collectively seem to prove that humans standardly violate a variety of inferential norms. That is, canons of inductive reasoning and deductive reasoning that are sancrosanct among logicians and statisticians are standardly violated by subjects in various experimental set-ups. I have, in the past, suggested that where modern science, complex social phenomena and logic meet that one should not be surprised to find that humans will make a variety of errors even if errors would not have been made in the specific environments wherein these abilities were selected for.[5] But such an explanation, despite its intuitive appeal, must be fleshed out by appealing to evolutionary mechanisms if it is to be taken seriously by psychologists, evolutionary biologists, and naturalistically-inclined philosophers. If Sober is right to think that the ability for abstract reasoning required for science and mathematics was a free rider on the more primitive reasoning capacities of our ancestors then a solution to the human irrationality problem is available to us. We now have a plausible explanation for both the fact that the results of pure research in science often do not matter much for our evolutionary success and for the failure of many experimental subjects to satisfy the standard canons of good reasoning embedded in deductive and inductive logic.

Here is the solution. The fact that reasoning was and is a reliable process in those domains in which it was selected for is consistent with the fact that alternative uses of these same abilities in science and mathematics may result in systematic errors in psychological experiments and false theories in science. The fact that we have difficulty teaching our students modal logic can be explained in the same way. The sort of abstract reasoning involved in science, mathematics and logic is the result of a free rider effect. But free rider effects confer no evolutionary advantage. It follows that where our original capacities usually resulted in true beliefs, such free rider capacities need not result in true beliefs very often. Certainly, there is no biological imperative for true representations in these domains in contrast to the domains in which our more primitive abilities originally functioned. In our original habitat, true indexical representations were crucial for the survival and reproduction of our genotype. If I am wrong about that there being no bus on the road that I am about to step on to, then I die. To the extent that reasoning was involved in the acquisition of such accurate indexical representations, a primitive reasoning capacity that was reliable must have been selected for. But if I go astray in the employment of abstract reasoning abilities in pure science, modern mathematics or formal logic, I pay no such immediate price. That is not to say that there will be no price to pay, but the price is not likely to be so high as to endanger my life or cripple my capacity for reproductive success.

There is an issue that needs resolution concerning the nature of the fundamental reasoning capacity that was selected for and how it differs from the secondary, free rider, usages to which such capacities later gave rise. As a first approximation, my suggestion is that our underlying psycho-logic involves deductive rules of inference, like *modus ponens* and *modus tollens*, and inductive rules of inference such as the straight rule. A deductive rule like constructive dilemma or an inductive rule like the conjunction rule for independent events from classical probability theory are not likely on the list of rules that were the output of our original reasoning capacity. But there is no hard evidence that we have and there may never be hard evidence concerning the exact nature of such a primitive reasoning ability. In general, those inference rules that are crucial for the acquisition of accurate indexical representations are likely to constitute the output of such a primitive reasoning capacity. Such a capacity may also have included some primitive analogical reasoning ability.[6]

4. CONNECTING SELECTION AND KNOWLEDGE

But there are other benefits that accrue from the employment of the selection for/selection of distinction in the way that I have suggested. If one takes the standard view that knowledge is justified, true belief then a typical move is to suggest that the reason we want justified beliefs is that such beliefs have a good chance of being true. In fact, the defining feature of epistemically justified beliefs, as opposed to morally or prudentially justified beliefs, is that such beliefs are likely to be true. If that were not the case, it seems unlikely that anyone would care about having justified beliefs. Goldman's externalist, truth-linked historical reliabilism provides a way of securing such a linkage between justification and truth. Unfortunately, I think the lottery paradox, the generality problem and the sceptical concerns of Barry Stroud cause severe problems for Goldman's account. Internalists, on the other hand, have fallen prey to the problem that giving reasons, or being able to give reasons, does not seem necessary for knowledge if one thinks that young children and dogs know things. Moreover, if doxastic voluntarism is a false doctrine as many have argued then the whole internalist framework is in jeopardy.[7] It is unlikely that we choose our beliefs based on good reasons or any occurrent justificational thought at all. Worse yet, internalist justification can never ensure that a connection to truth has been obtained and so it cannot stop the epistemic regress problem that it was designed to put an end to.

Dretske and, more recently, Sartwell have argued that justification of any kind is not a part of our definition of knowledge. For Sartwell, justification is something acquired in the process that eventually leads to knowledge but it is not, strictly speaking, a part of the definition of knowledge. Knowledge just is true belief.[8] Dretske, on the other hand, thinks that reliability theories of justification are 'implausible.' (Dretske, 1989, p. 100.) Divorcing justification from knowledge is, I think, a mistake. It is a mistake because it makes mysterious why it is that we should be so concerned to provide reasons and evidence for our beliefs. One answer that issues from Austin and others is that justification has to do with public assertability conditions. But this view is problematic. Is it just to pacify our peers that we strive to defend our assertions? Such a social explanation can only be part of the story. The rest, if I am right, has to do with the need to know. It is instructive, I think, to see how Goldman and Dretske fail in their attempts to provide a complete epistemology. The weaknesses inherent in their accounts, as strong as these accounts are, can be traced back to where they begin their analysis.

Goldman begins with the concept of 'justification' and ends by explicating the concept of 'knowledge.' The former has some plausibility, the latter cannot succeed due to the lottery paradox, the generality problem and so on. Dretske, conversely, begins with the concept of 'knowledge' and, later, explains away the concept of 'justification' as unconnected to knowledge. The former account appears to have a great deal of merit; the latter, much less merit because it is unclear as to why anyone would care about a notion of epistemic justification that is unconnected to truth and knowledge. Justification gets explained away as only having to do with assertability conditions. But, as I mentioned above, the pervasive role of justificatory considerations in all human endeavours is reasonable precisely because we think there is a connection between justification and truth. The scientist who spends volumes of effort gathering evidence for her theory does so with the hope that the way to truth lies in that direction. *Pace* Laudan, Barnes, and Bloor, truth may well be the central epistemic goal of science. Neither Goldman nor Dretske provides an adequate account of justification and an adequate account of knowledge. This is because neither account succeeds in showing the evolutionary connection between justification and truth that leads to knowledge. Dretske denies that justification is related to knowledge in two senses. He thinks reliable process accounts of justified belief are implausible. He also thinks that internalist accounts of justified belief are only relevant when assertability conditions are at stake. Dretske and I emphasize the indirect role of evolution in producing true belief. But we part company insofar as he fails to allow justified belief simpliciter, construed as the output of a reliable process, its proper place and unique role. Simply put: justification is a necessary condition on knowledge. Dretske's account of knowledge fails to do justice to this fact as he denies that reliabilist justification plays any role in reliabilist knowledge. Goldman accepts the evolutionary connection between justification and truth but fails to explain the exact nature of the evolutionary connection that make this possible and so gives us no reason to believe that the processes that he deems reliable, are reliable.[9] The result is that his account fails to explain how knowledge is possible.

It is here that the selection for/selection of distinction can be of some help. It provides a framework within which we can begin to see how justification and truth are naturalistically related. Consider perception. On my view, perception was selected for. Our perceptual abilities constitute a property we possess that causes success in survival and reproduction. Natural selection is not the only cause of evolution in a species. Mutation, migration and random genetic drift also are causes of evolution. But natural selection is the dominant causal factor in evolution. Standardly, an increase in gene frequency in a population is identified as what constitutes evolution. (Sober, 1984, p. 29) My claim is that at some remote point in our ancestor's evolution, those with perceptual abilities were selected for. What was it that was so beneficial about perception? No doubt the acquisition of food, awareness of predators and such were involved here. Jerison has argued that mammals developed their large brains in order to handle specific functional demands on them as small, marginalized creatures competing in a world of dinosaurs. (Jerison, 1973) The first mammals were nocturnal and needed larger brains to transpose olfactory and auditory inputs into spatial patterns that animals active in the day could handle using just perception.

With the development of language and so belief much more recently, the possibility of accurately representing the external world became a reality. I take it that belief is just one form of representation. The plant that bends toward the sun has its own way of representing the conditions that it responds to. Other forms of representation can be witnessed wherever organisms exist. Accurate or near accurate match-ups of propositions and external world situations makes possible the word/world connections necessary to utter true propositions. True propositions were not selected for but they were an effect of the selection process that begins with selecting for perception, construed as a reliable process, and sometimes ends with true propositions. True propositions, therefore, can be seen as an effect of a selection process. This view is also consonant with current views in evolutionary biology concerning the build-up of our native abilities in the following sense. Our linguistic capacity emerges long after our perceptual abilities in evolutionary time (see Dennett, 1992).

The account presented above can also shed light on some thorny, old issues. For instance: Is knowledge extrinsically valuable and, if so, why? Plato's response in *Meno* was to say that true belief comes and goes while knowledge, made of sturdier metal, endures and so must be properly tethered. There is something right about this response but if knowledge does endure, one wants to know how that is possible. Here's my answer. Suppose true belief is extrinsically valuable for our survival and, indirectly, the reproduction of our genes. That is, suppose our first-order inductions tend to come out right most of the time. We believe that X is a threat to us when X is a threat to us. The problem is that we cannot select for true belief directly because selection works on one's genotype. True beliefs, that is, are not heritable. On the other hand, there can be selection for a trait or property. Perception, I think, was selected for. It is our dominant sensory modality for a reason. Perceptually-based beliefs tend to be reliable indicators of external states of the environment. As Dretske says about someone who wants true beliefs about future newspaper headlines upon receiving a gift of a pair of glasses:

There is no way to give the person what is wanted, true beliefs about future events, without giving more than what is wanted. Truth, at least the kind of truth now in question, is like that. It is something you can't buy. The only thing for sale is a means, a reliable process, for producing true belief, a means which, when deployed, thereby produces, not merely true belief, but knowledge.

(Dretske, 'The Need to Know,' p. 94)

Knowledge is extrinsically valuable because it includes true beliefs and true beliefs are extrinsically valuable in the struggle for survival. Our survival makes possible the perpetuation of our genes – which is, after all, the goal of evolution. If beliefs caused by reliable processes just are the beliefs that tend to be true then we have a clear picture of the value of knowledge. Knowledge is what we get when we aim for true beliefs by employing reliable processes and succeed in our quest.

Concordia University

NOTES

[1] See my 'Epistemic Norms and Evolutionary Success', *Synthese* **85**: 231–244, 1990 and Dretske's 'The Need to Know', in Clay and Lehrer's *Knowledge and Scepticism*, for more on this point.
[2] Stich's discussion of optimality ignores Goldman's claim that speed and power vie with reliability as epistemic parameters. No reliabilist would claim that only reliability matters, contra Stich. See, for instance, Goldman's *Epistemology and Cognition*.
[3] Gould and Lewontin draw this distinction in their paper: *The Spandrels of San Marco and the Panglossian Paradigm: A Critique of the Adaptationist Program.*
[4] Free riders can also be produced due to gene linkage where a neutral gene and an advantageous gene may be close together on one chromosome. Selection for one may increase the frequency of both genes. See Sober, p. 101.
[5] For more on this point, see my *Epistemic Norms and Evolutionary Success.*
[6] The fact that the higher primates display proto-reasoning adds weight to the suggestion that a primitive reasoning capacity was selected for among homo sapien sapiens. Gillan has documented what appears to be analogical reasoning among chimpanzees and Premack has noted that chimpanzees seem to appreciate the concept of conservation. Michael Ruse has catalogued a great deal of evidence along these lines in his *Taking Darwin Seriously*. We also find similar senses of logic and mathematics across cultures, this suggests biological roots. See, for instance, Ruse's discussion of Staal and Bockenski. On the negative side, John Anderson provides evidence of deductive inference failures in his *Cognitive Psychology and Its Implications*, 1985, Chapter 10.
[7] For more on this point, see my 'Doxastic Voluntarism and Forced Belief', *Philosophical Studies* **50**: 39–51, 1986.
[8] See Crispin Sartwell's 'Why Knowledge is Merely True Belief', *Journal of Philosophy* **LXXX1X**, number 4 (April 1992), 167–180.

[9] See John Pollock's 'Epistemic Norms' *Synthese* **71**, 61–95, for criticism of Goldman suggesting that there is no reason to think that reliability is a property of our cognitive processes. Goldman does connect the notion of true belief with genetic fitness in his *Epistemology and Cognition*, p. 98. My claim, of course, is that he needs to say much more about the nature of these connections if this story is to be made plausible.

REFERENCES

Dennett, Daniel, 1992, *Consciousness Defended*, Little and Brown, New York.

Dretske, Fred, 1989, 'The Need to Know', in Clay and Lehrer (eds.), *Knowledge and Scepticism*, Westview Press, Boulder, Colorado, pp. 89–100.

Jerison, Harry, 1973, *The Evolution of the Brain and Intelligence*, Academic Press, New York.

Kahneman, D., Slovic, P. and Tversky, A., 1982, *Judgement Under Uncertainty: Heuristics and Biases*, Cambridge University Press, Cambridge.

Nisbett, Richard and Lee Ross, 1980, *Human Inference: Strategies and Shortcomings of Social Judgement*, Prentice-Hall, New Jersey.

Plato, 1961, 'Meno', in Hamilton and Cairns (eds.), *Plato: The Collected Dialogues*, Princeton University Press, Princeton, New Jersey.

Ruse, Michael, 1986, *Taking Darwin Seriously*, Basil Blackwell, London.

Sartwell, Crispin, 1992, 'Why Knowledge is Merely True Belief', *The Journal of Philosophy* **LXXX1X**, number 4, 167–180.

Sober, Elliot, 1981, 'The Evolution of Rationality', *Synthese* **46**, 95–120.

Sober, Elliot, 1984, *The Nature of Selection*, The MIT Press, Cambridge, Massachusetts.

Stich, Stephen, 1990, *The Fragmentation of Reason*, The MIT Press, Cambridge, Massachusetts.

EVAN THOMPSON

ARTIFICIAL INTELLIGENCE, ARTIFICIAL LIFE, AND THE SYMBOL-MATTER PROBLEM*

I. INTRODUCTION

What is the relation between matter and form? This question is of course as old as philosophy itself. But it also arises at the foundations of two recent scientific endeavours – the computational approach to the mind-brain in cognitive science and artificial intelligence (AI), and the synthetic approach to living systems in theoretical biology and artificial life (AL). In these fields the question arises primarily in connection with the status of *symbols*, that is, items that are physically realized, formally identified, and semantically interpretable.

In "Classical" or "Symbol-Processing" cognitive science, the issue about symbols has two sides, one having to do with semantics and the other having to do with syntax. The semantic side is the problem of how symbols as syntactically (formally) individuated tokens get their meaning. The syntactic side (which has been much less discussed) is the problem of how something physical can also be syntactic. In AI the semantic side of the issue has come to be known as the "symbol grounding problem."[1] I think this term is better suited to cover both the semantic and syntactic sides of the issue. Thus the symbol grounding problem could be seen to comprise both a "semantic grounding problem" – what fixes the semantic interpretation of the symbol system? – and a "syntactic grounding problem" – what fixes the syntactic interpretation of the physical system?[2]

In theoretical biology and artificial life, the issues about symbols are much broader. The main issue has been termed the "symbol-matter problem."[3] How can a material structure also be a symbolic form? And what types of inter-action between material structures and symbolic forms are necessary for evolution? Another issue is whether something must have a symbolic genotype and a material phenotype to be a living system. Is this symbolic genotype-material phenotype coupling a necessary feature of life? Or is it only contingent, accidental to life as we know it on Earth? AL seeks to answer questions like these by placing life as we know it within the wider theoretical context of life as it could be.[4] Its research strategy is to discover the organizational principles that define life by simulating the behaviour of actual and possible living systems on a computer. But there is also a more controversial view within AL, known as "Strong AL," that raises another type of issue about symbols. Whereas "Strong AI" holds that the mind is a computer program (that instantiating the right program suffices for having mental states),[5] "Strong AL" holds that the properties necessary and sufficient for life are purely formal, and so it is possible not only to simulate but also to realize

M. Marion and R. S. Cohen (eds.), Québec Studies in the Philosophy of Science II, 63–80.
© 1996 *Kluwer Academic Publishers.*

living systems within a computational medium.[6] How cogent is this notion of a system that qualifies as living yet has only a formal identity? And to what extent can a formal system clarify the symbol-matter relations that are characteristic of life on Earth?[7]

A systematic and thorough treatment of the issues just raised is clearly beyond the scope of this essay. My aim is more modest. I intend to focus primarily on the concepts of *formality* and *syntactic interpretability* in relation to the symbol grounding and symbol-matter problems.

II. SEARLE'S ARGUMENT THAT "SYNTAX IS NOT INTRINSIC TO PHYSICS"

The issues that interest me are raised in a convenient form by John Searle's most recent argument against the view that mental processes involve symbolic computations in an internal language of thought and that the brain is accordingly a digital computer or physical symbol system.[8] Whereas Searle's earlier "Chinese Room Argument" addressed the relation between the syntactic and semantic features of mental states (see note 5), his new argument addresses the relation between the physical and the syntactic features of the brain. In a nutshell the new argument is that syntactic features are essentially observer-relative and so it is not possible for a physical system, such as the brain, to be intrinsically syntactic; consequently, the view that the brain is a "syntactic engine," is incoherent.

The core of Searle's argument has four steps, from which he then draws two further conclusions:

1.   Computation is defined syntactically in terms of symbol manipulations.

2.   Syntactic features are defined formally (roughly, in terms of certain shapes).

3.   The formal and hence syntactic features of physical systems are always specified relative to some mapping or assignment function provided by an observer outside the system. (" 'Syntax' is an observer-relation notion.")

4.   Therefore, physical systems are syntactic and hence computational only relative to the mapping or assignment function, not intrinsically. ("Syntax is not intrinsic to physics.")

On the basis of (1)–(4) Searle then argues:

5.   One cannot *discover* that a physical system (such as the brain) is intrinsically computational, though one can *assign* to it a computational interpretation.

6.   Syntactic features cannot function causally in the production of behaviour. ("Syntax has no causal powers.")

It is important to see how this argument both differs from and tries to strike deeper than the Chinese Room Argument. The target of the Chinese Room Argument is the position, dubbed "Strong AI" by Searle, that the mind is a computer program, and the argument proceeds by trying to show that mental content cannot be gotten simply from the syntactic operations of computer programs. In contrast, the target of the new argument is the position, dubbed "Cognitivism" by Searle, that the *brain* is a digital computer or physical symbol system,[9] and the argument proceeds by trying to show that "syntax" is an essentially observer-relative notion and therefore it is incoherent to suppose that a physical system could be intrinsically syntactic. In Searle's formulation, the Chinese Room Argument shows that "semantics is not intrinsic to syntax," but the new argument shows that "syntax is not intrinsic to physics." Thus the argument tries to strike deeper because it does not challenge merely the idea that mental processes are computational; it challenges the very coherence of the idea of a physical symbol system.[10] If Searle were right about the relation between physics and syntax, the implications for cognitive science and artificial life would be considerable: The hypothesis that the brain is a syntactic engine would be incoherent; consequently, the view that mental processes are computational would not even be able to get off the ground (unless one is a dualist). And Strong AL too would be untenable, for it would be incoherent to suppose that there could be an observer-independent computational medium in which to realize the principles definitive of life.[11]

The crux of the issue as Searle sees it is whether there can be syntactic features that are intrinsic to a physical system rather than based on some outside assignment or interpretation. Thus for Searle the issue is an ontological one: Are syntactic properties intrinsic or observer-relative? And because he insists that there is a principled distinction between ontology and epistemology – questions about what something is must be treated separately from questions about how we know or determine what something is – Searle rejects attempts to answer the ontological issue about syntax on the basis of considerations about what fixes the syntactic *interpretation* of a physical system. His treatment of syntax thus runs parallel to his treatment of semantics and intentionality. Searle insists that there is a distinction between the "intrinsic intentionality" of mental states and the "derived intentionality" of, say, spoken utterances and written marks: As a matter of ontology, the intentionality of a mental state is internal to it and does not depend on how the state is used or interpreted.[12] Thus ontological issues about the intentional content of mental states must be treated separately from epistemological issues about what fixes the semantic interpretation of mental states. Where the parallel breaks down, however, is that for Searle there is no such thing as "intrinsic syntax" – syntactic properties do depend entirely on how the states of a system are used or interpreted.[13]

Searle provides no argument for his problematic division between ontology and epistemology, and of course there is a long tradition going back to Kant that would reject it. But setting such larger philosophical issues aside, one can, even as a straightforward scientific realist, hold that questions about what something is cannot be treated separately from questions about how we know or determine what something is. In the present context, this would mean that the issue about what syntactic properties are cannot be treated separately from the issue of how a system gets syntactically interpreted. On this view, the parallel between syntax and semantics would be complete: Just as when one discovers that a system is semantically interpretable there need be no further question about whether it is "really" semantic,[14] so too when one discovers that a system is syntactically interpretable there need be no further question about whether it is "really" syntactic.

Of course, "really" is not necessarily equivalent to "intrinsically." Something can be really F without being intrinsically F; it might be F only relationally. There is therefore an interesting question that remains about syntactic interpretation: Does what fixes the syntactic interpretability of a system imply that syntax is *assigned* to the system? Or does it imply that syntax is *discovered* in the system?

These questions cannot be answered *a priori*; instead, one must look at the methodological and empirical justifications given on behalf of the syntactic level in computational endeavours such as cognitive science and artificial life. The remainder of this essay will be devoted to this task and it will reveal that there is a straightforward refutation of Searle's claim that "syntax is not intrinsic to physics." The line of argument that I am going to pursue, however, will have some interesting consequences. It will provide little support to the view that Searle is actually arguing against – the Cognitivist or Classical Symbol-Processing research program; and it will raise some questions about the cogency of the Strong AL thesis. But it will also turn out to provide a new line of support for certain general features of the Connectionist research program.

III. SYNTAX AND THE BRAIN

To begin, we need to examine the standard Cognitivist arguments for considering the brain to be, at a certain level of description, a physical symbol system. Two main arguments have been given: The first appeals to *intentionality*, specifically to the intentional or semantic properties of mental states that are implicated in the generation of behaviour and that must somehow be realized in the brain; the second appeals to *complexity*, specifically to the organizational complexity of the brain and nervous system.

The first argument has been extensively presented by Jerry Fodor and Zenon Pylyshyn.[15] According to Fodor and Pylyshyn, the "Classical" or Symbol-Processing model of mental processes rests on two ideas: The first idea is that one can construct languages in which semantic features correspond systematically (within certain well-known limits) to syntactic features; the second idea is that one can devise machines that have the function of operating on symbols, but whose operations are sensitive only to the syntactic structure (physical form) of the symbols. Fodor and Pylyshyn describe how the two ideas fit together:

> If, in principle, syntactic relations can be made to parallel semantic relations, and if, in principle, you can have a mechanism whose operations on formulas are sensitive to their syntax, then it may be possible to construct a *syntactically* driven machine whose state transitions satisfy *semantical* criteria of coherence. Such a machine would be just what's required for a mechanical model of the semantical coherence of thought; correspondingly, the idea that the brain *is* such a machine is the foundational hypothesis of Classical cognitive science (p. 30).

This hypothesis implies that the symbol structures in the Classical model of mental processes correspond to real physical structures in the brain.[16] What Searle's argument challenges is the assumption that there could be any non-observer-relative fact of the matter about this correspondence. Hence what needs to be considered is how the syntactic features of computational states are typically grounded within the Cognitivist or Symbol-Processing framework.

In *Computation and Cognition*, Pylyshyn discusses how computational explanations must be relativized to the mappings provided by two interpretation functions, the *semantic function* (SF) and the *instantiation function* (IF).[17] The semantic function maps from articulated functional states onto some domain of intended interpretation (e.g., positive integers). (It should be noted that this function is required for something to be a computation because a computation is a rule-governed process defined over *semantically* interpretable items.[18] Hence the first premise of Searle's argument, that computation is defined only *syntactically*, is not strictly speaking correct within the context of the Symbol-Processing approach.) The instantiation function maps from physical states to computational states; more precisely, it specifies the equivalence classes of physical states that count as syntactically distinct computational states. This is the interpretation function relevant to syntactic interpretability, for its purpose is to indicate how syntactic states are physically realized.

Unfortunately Pylyshyn's treatment of the instantiation function does not provide much clarification about syntactic interpretability. Here is how he describes IF:

> By mapping from physical to computational states, such a function provides a way of interpreting a sequence of nomologically governed physical state changes as a *computation*, and therefore of viewing a physical object as a *computer*.[19]

This description does not make clear just what is required of an instantiation function for it to be the case that the syntactically distinct states figuring in the interpretation are not simply assigned to the system, but discovered in it.

The problem becomes more pressing when one remembers that computational states are said to be independent of any particular material medium and hence multiply realizable. As Pylyshyn goes on to observe, the physical realizations of a given computational function are essentially open-ended:

Computational sequences "can be realized in devices operating in any imaginable media – mechanical (as in Charles Babbage's Analytical Engine), hydraulic, acoustical, optical, or organic – even a group of pigeons trained to peck as a Turing machine!"[20] Such multiple realizability has typically been considered as a point in favour of applying the computational framework to the mind, for it provides a model of how there need be only weak, token-token identities between mental states and physical states, and consequently how psychology can be autonomous in relation to neuroscience. Searle counters, however, that the multiple realizability of computational states is simply a sign that the states are not intrinsic to the system, but depend on an interpretation from outside. He thinks that a distinction should be drawn between devices whose functions are multiply realizable (e.g., thermostats) but are nevertheless defined in terms of the production of the same physical effects (e.g., regulating temperature), and devices whose multiple realizability is due to the relevant properties being purely formal or syntactical (e.g., Turing machines). For Searle, it is this difference in types of multiple realizability that explains why, for example, nobody would suppose it possible to make a thermostat out of pigeons even though it would be possible to train pigeons to peck as some simple Turing machine.

This distinction between what might be called *functional* multiple realizability, on the one hand, and *formal* multiple realizability, on the other, indicates that there is another concept that is relevant here. This is the concept of *digitality*. One reason that is typically given for how physical symbol systems can be multiply realizable is that they are *digital* systems. There is disagreement among philosophers and cognitive scientists about how to define a digital system, but for our purposes a digital system is one whose states belong to a finite number of types that are perfectly definite – that is, for any given type, a state is either of that type or it is not, and variation among the states that belong to a given type is insignificant. In physical symbol systems, the insignificant variations occur at the physical level (the level of the so-called "physical machine") and the perfectly definite types correspond to the syntactic states of the system (the states at, say, either the level of the "logical machine" or that of the "abstract machine"). Thus syntactic properties can correspond to arbitrarily many physical properties and syntactic state-transitions can involve arbitrarily many physical causal laws. Hence physical symbol systems are as a class independent of any particular material medium.

One rather large shortcoming of Searle's argument is that he does not address this relation between digitality and formal multiple realizability. But certain remarks that he makes do suggest how he would view the relation.[21] In these remarks, Searle appears to admit that digitality is not observer-relative, but he claims that although the states of a digital system might more naturally support a syntactic characterization by an outside observer, they are nevertheless syntactic *only* in relation to such a characterization.

To address this claim we need to turn to the second line of reasoning given by the Cognitivist for supposing that the brain is a physical symbol system. This argument is the one based on appealing to the organizational complexity of the brain.

Ray Jackendoff gives a concise presentation of the argument from organizational complexity in his book, *Consciousness and the Computational Mind*.[22] Jackendoff argues that computational levels of description are required for systems whose components interact combinatorially. In a computer, the state of each component (e.g., binary switch, flip-flop – hence at the level of the logical machine) is independent of the states of the other components; consequently, the activity of larger components in the machine is not a sum or average of the activity of the component parts, but depends rather on the state of each component and their particular combinatorial properties. It is the complexity of the combinatorial properties that is ultimately responsible for the medium-independence of computational systems: "any device with the same combinatorial properties will have the same set of possible states and will go through the same sequence of states. Since the combinatorial properties are formal rather than physical, we will have arrived at a computational description."[23]

Jackendoff claims that a comparable case obtains in the organizational complexity of the brain, though not in other biological organs, such as the stomach. The activity of neuronal groups is not the sum or average of the component neuronal activities, but depends rather on how the neurons interact combinatorially. Such combinatorial properties are both formal and structurally grounded in the brain; hence it is in virtue of its organizational complexity that the brain is a syntactic engine.

The original inspiration for this approach to the brain is of course the seminal 1943 paper by Warren McCulloch and Walter Pitts entitled 'A Logical Calculus of the Ideas Immanent in Nervous Activity.'[24] By taking the neuron as the functional unit of the nervous system and by treating neuronal activity as binary, digital, and synchronous (neurons as threshold devices that could be either active or inactive, and that all change state at the same discrete time-steps), McCulloch and Pitts were able to show formally that the organizational complexity of various neural nets is sufficient for the computation of the logical operations of the propositional calculus. This result has been foundational for the entire field of cognitive science because it shows how the operation of the nervous system might at some level be appropriately described using mathematical logic and hence how the brain could be seen at that level to be a syntactic (and semantic) machine.

Appealing to organizational complexity provides the beginnings of a more satisfactory approach to the issues about syntactic interpretability. Suppose, for example, that McCulloch and Pitts's idealized neural nets reflected the real functional architecture of the brain. It would then be the case that certain equivalence classes of neuronal states would be computational states in virtue of the *roles* that neurons play in nervous system activity. Of course it would still be true that in constructing a computational *model* of the brain *we* would have to map the all-or-none activity of neurons onto, say, 0's and 1's in binary notation, but the model would nonetheless be non-arbitrarily grounded in an intrinsic feature of the brain, namely, the role that neuronal activity plays in the operation of the nervous system and the generation of behaviour.[25]

My invocation of McCulloch and Pitts is *not* intended as a claim about how computational processes are actually grounded in the brain. On the contrary, it is well known that real neurons are not simple binary switches, and although neurons are the fundamental *anatomical* units of the nervous system, the basic *functional* units are probably relatively invariant patterns of activity in neuronal assemblies.[26] Rather, by invoking McCulloch and Pitts I mean to illustrate the point that, contrary to Searle, there is nothing incoherent in supposing that computational processes are so grounded in the organizational complexity of the nervous system. It is for this reason that I said the argument from organizational complexity provides the beginnings of an approach to syntactic interpretability: It answers the charge raised by Searle, while leaving open the substantive (and ultimately more interesting) empirical issues about the syntactic interpretability of complex systems such as the brain.

The line of argument I have been pursuing does have theoretical implications for these empirical issues. We have seen that to ground claims about syntactic interpretability it is not enough to appeal simply to an abstract mapping from the syntactic to the physical, such as Pylyshyn's instantiation function; we must also appeal to the organizational complexity of the realizing system. For this reason I think that Fodor and Pylyshyn are wrong when they claim that it is mistaken for philosophers such as Dennett[27] and connectionists such as Smolensky[28] to appeal to complexity as a way of distinguishing cognitive from noncognitive systems.[29] On the contrary, it is the organizational complexity of certain systems that warrants the hypothesis that there are additional syntactic (and semantic) levels of description for their behaviour.

One consequence of this point is that the purely top-down approach that characterizes the Symbol-Processing treatment of syntactic interpretability is unsatisfactory. What I mean by "purely top-down" is an approach in which hypotheses about the *brain* are made entirely on the basis of claims about *mental representations*. For example, Fodor and Pylyshyn write:

. . . the symbol structures in a Classical model are assumed to correspond to real physical structures in the brain and the *combinatorial structure* of a representation is supposed to have a counterpart in structural relations among physical properties of the brain. For example, the relation "part of," which holds between a relatively simple symbol and a more complex one, is assumed to correspond to some physical relation among brain states . . .[30]

Such a top-down approach cries out for lower-level constraints. Fodor and Pylyshyn simply take for granted the idea that there is some appropriate instantiation function from syntactic structures to physical structures in the brain, even though, to use their example, no one has the foggiest idea how the syntactic relation "part of" is supposed to correspond to some physical relation among brain states. If the claim that the brain is a "syntactic engine" is to be taken seriously, then we must try to find out what type of syntactic engine it is. To do this we need not only top-down hypotheses based on models of mental representations, we also need detailed models of how syntactic features can, in the first instance, emerge as a consequence of a system's organizational complexity, and in the second instance, act to constrain the behaviour of such a system. Nowhere is the necessity of such models provided for within the Symbol-Processing approach.

In contrast, the new research area known as emergent computation holds considerable promise.[31] Emergent computation is the study of complex systems having three general features: (i) they are composed of a collection of agents each of which follows explicit instructions; (ii) the agents interact according to the instructions and thereby generate implicit, emergent global patterns; and (iii) there is an interpretation function that maps the global patterns onto computations.[32] Within such emergent computational systems the low-level agents are themselves devices that have a purely formal specification, but since they are typically rather simple – for example, the on-off cells of a cellular automaton – one can easily imagine biological analogues.

Nevertheless, merely showing how certain kinds of global behaviour can be given a computational interpretation is not in itself sufficient to show that their syntactic interpretability is equivalent to the syntax being discovered in the system, rather than being merely assigned to it from outside. The task specified by the problem about syntactic interpretability is to show how (*contra* Searle) syntax can be intrinsic to physics for certain kinds of systems. Consequently, what we need are not merely implicit (unprogrammed) global patterns that naturally support a computational interpretation from outside (as in for example Conway's "Game of Life");[33] rather, what we need are models of how a system can have an internal syntax that is not merely assigned by an outside observer and thereby qualify as a naturally syntactic system. Such a model would have to specify a particular sort of relation between syntax and physics – *one that emerges as a consequence of the system's own operation and yet serves to limit the degrees of freedom in the system's behaviour.*

IV. THE LIVING CELL

One well understood example of a system that has a syntactic level of operation in this sense is the living cell. In fact, the cell provides a minimal yet paradigmatic example of a system having a naturally syntactic level of operation, and so I will dwell on it at some length.

In a cell, the syntactic level corresponds of course to the so-called "genetic code" whereby genes (lengths of DNA) specify the kinds of proteins a cell can make. More precisely, the "genetic code" refers to the rules that prescribe specific amino acids given specific triplets of nucleotide bases in DNA. Protein synthesis is thus said to involve specifications that are written in

DNA and then decoded in a complex process involving transcription (by mRNA) and transportation (by tRNA).

The specification relation between DNA and proteins has a number of features that justify using the syntactic term "code" to describe it. First, the code is *quasi-universal*: With the exception of mitochondria, nucleotide triplets always specify the same amino acids regardless of the organism; for example, the triplet AAG (adenine-adenine-guanine) specifies lysine in all organisms from bacteria to humans. Second, the code is *arbitrary* in the sense that, as Maynard Smith puts it, "It is hard to see why a code in which GGC means glycine and AAG means lysine is either better or worse than one in which the meanings are reversed."[34] Third, the code is *compositional* in the sense that there are numerous possibilities for nucleotide triplet combinations in a linear array. Fourth, the code is *digital* because what a nucleotide triplet signifies depends on which token it is out of a finite number of types and variation among tokens within a given type is insignificant.[35]

These four features – quasi-universality, arbitrariness, compositionality, and digitality – warrant describing the genetic code as a complex syntactic system. Yet unlike, say, Morse code, the genetic code is a naturally syntactic system in the sense that it is internal to the very operation of the living cell. The genetic code does not depend for its specification on an interpretation function given by an outside observer, not is it a set of instructions applied by a control that is external to the cell. Rather, it is a set of specifications that is and must be embedded in the internal operation of the cell.

To give this point the emphasis that it deserves in this context, it is worth pausing to consider H. H. Pattee's proposed explanation of the relation between the physical and the syntactic in biological systems.[36] Pattee distinguishes between, on the one hand, the laws of nature, which are universal and closed (holonomic), and on the other hand, structures that conform to the laws of nature yet constrain the motion of matter in ways additional to the laws (non-holonomic auxiliary conditions or constraints). In Pattee's view, the presence of such additional structures in a system can ground the syntactic interpretability of a system: We map the structures onto syntactic properties (e.g., formal rules), which can be described without referring to the physical structures that realize them.

So far we have little more than what is also contained in Pylyshyn's notion of an instantiation function that maps from the physical to the syntactic. But Pattee goes further. Following von Neumann's work on self-reproducing automata,[37] he focuses on a certain class of complex system – those that can self-replicate in virtue of containing (among other things) their own description (self-describing systems) – and explicitly raises the key question: "how can we tell if a self-describing system has its own internal language in any objective sense? How do we know we are not interpreting certain structures as descriptions, only because we recognize them as consistent with rules of one of our own languages?" And he gives as an answer: "we must further restrict our model of a complex system to remove the case of the external observer reading a message that is not really in the system itself. This restriction is achieved by requiring *that a complex system must read and write its own messages*."[38]

Again the living cell serves as a paradigm, and thus we return to the genetic code as a naturally syntactic system embedded in the internal operation of the cell. The writing of its own messages corresponds in the cell to the synthesis of the DNA molecules; the reading of its own messages corresponds to the elaborate process of protein formation according to the universal rules of the genetic code and the specific descriptions in the structural DNA. This code is both arbitrary, as previously mentioned, and rate-independent (what a given nucleotide triplet designates is independent of how fast it is written or decoded). But as Pattee goes on to observe, for the code to be read there must ultimately be a transduction from the rate-independent, "linguistic mode," as he calls it, to the rate-dependent "dynamical mode." The transduction happens when the rate-independent linear array of amino acids folds to become a three-dimensional enzyme (a catalyst whose function is to initiate and control the rates of specific reactions). Within the operation of the cell there is thus a transformation from the enzyme as something designated in the genetic code to the enzyme as an operational component. Moreover, as Pattee remarks, this transformation (the protein folding) is not itself described in the linguistic mode; it happens rather according to "the *laws* of nature under the nonholonomic constraints of the *rule-constructed* polypeptide chain described by the cell's structural DNA."[39]

We can now appreciate how the genetic code as a naturally syntactic system is and must be embedded in the internal operation of the cell. In general, nucleotide triplets are capable of predictably specifying an amino acid if and only if there are properly embedded in the cell's metabolism, i.e., in the multitude of enzymatic regulations in a complex chemical network.[40] This network has a "chicken and egg" character at several levels. First, proteins can arise only from a decoding process, but this process itself cannot happen without proteins. Second, these specification and constitution processes must be properly situated within the intracellular environment, but this environment is itself a result of those very processes. Finally, the entire cell is an autopoietic (self-producing and operationally closed) system defined by a network of processes that simultaneously produce and realize the cell concretely.[41]

The embedding of the genetic code in the cellular dynamics has important consequences for our understanding of naturally syntactic systems. When we refer to the "coding relation" between DNA and proteins we are really choosing to focus on one particular sequence of events in the ongoing recursive turnover of the cell. We do so by abstracting away from a number of the necessary intervening causal steps and by implicitly invoking a *ceteris paribus* clause to cover the metabolic processes.[42] Thus when we talk about DNA *coding for* proteins we are not referring to a special type of causal relation; rather we are abbreviating a lengthy but remarkably stable causal sequence of biochemical events. It is precisely the stability and predictability of the entire sequence that grounds treating nucleotide triplets as in effect symbols that stand for amino acids.

The genetic code therefore provides a straightforward counterexample to Searle's assertion that "syntax is not intrinsic to physics." Yet the treatment of syntactic interpretability implied by this example hardly supports the physical symbol system model of the brain against which Searle is arguing. This model simply takes symbols at face value and treats them as if they were (in principle) independent of the neural systems within which they emerge and reside. In the case of the living cell, however – which is our best understood example of a system that has a naturally syntactic level of operation – we know that there is no such independence of the symbol vehicles from the dynamical context in which they are embedded and from which they arise. This point has important implications for the thesis of Strong AL and for the debate between Connectionism and the Classical Symbol-Processing theory.

V. STRONG AL

According to Christopher Langton, "the principal assumption made in [Strong] Artificial Life is that the 'logical form' of an organism can be separated from its material basis of construction, and that 'aliveness' will be found to be a property of the former, not of the latter."[43] The context of this remark is a discussion of machines, and of how, in the development of cybernetics and the mathematical theory of computation, "The 'logical form' of a machine was separated from its material basis of construction, and it was found that 'machineness' was a property of the former, not of the latter." The application of this idea to the corresponding idea in the case of organisms requires only the additional premise that living systems in general are a type or class of machine.[44]

In philosophy this general idea is familiar not so much from discussions of life but from discussions of mind, and is known as *functionalism*.[45] The idea is that the logical form of a mind can be separated from its material basis, and that mentality is a property of the former, not of the latter. As a theory about the nature of the mind, the idea is sometimes known as "metaphysical functionalism," and in its most pure form is the thesis that what makes a state a *mental* state is not anything physical *per se*, but rather simply its *functional role*, that is, the role the state characteristically plays in relation to other states. The functional role of a state is something abstract and formal in the sense that it can be specified as a set of relations without referring to the materiality of the states that happen to embody those relations; and any material that can support the appropriate network of relations will suffice to realize the functional role. Thus the multiple realizability of mind follows from metaphysical functionalism. When metaphysical functionalism is combined with what is sometimes called "computation-representation functionalism" –

the thesis that, in psychological explanation, mental states should be analyzed into component algorithmic processes defined over symbols after the fashion of a computer program – then one arrives at the view that minds are realizable in a purely computational (symbolic) medium (Strong AI).

Adverting to functionalism enables us to specify the sense in which Strong AL claims that materiality and logical form can be separated in an organism. Strong AL is the computational version of metaphysical functionalism as applied to the biological domain of life instead of the psychological domain of mind.[46] Strong AL holds that what makes a state a "vital" state (one involved in, e.g., metabolism, reproduction, etc.) is simply its functional role. Hence the logical form of an organism – the set of relations holding among all its component states and processes – can be specified without referring to the organism's material constitution; and the material constitution can be anything as long as it can support the appropriate set of relations. Thus multiple realizability follows in the biological domain. Strong AL also holds, however, that the logical form of an organism can be captured entirely in a symbolic description, and so we arrive at the view that life is realizable in a purely computational medium.

In evaluating Strong AL, then, the question that needs to be answered is whether it is indeed possible to abstract the logical form of a living system from its material constitution in the form of a symbolic description. The discussion in the previous section suggests this may be impossible. Recall Pattee's distinction between the rate-independent linguistic mode and the rate-dependent dynamical mode in cellular activity. Two central points connected with this distinction are important here. First, as Pattee emphasizes, the transduction from the first mode to the second (i.e., the protein folding) is not itself linguistically described, but rather is accomplished by the dynamic interaction of the cellular components according to the laws of nature. Moreover, as Claus Emmeche has recently noted,[47] the timing of such processes is crucial, and so the logical form of the cell as a dynamical system is not atemporal (as is typical of abstract symbolic forms), but rather time-dependent (temporally constituted). Second, as Pattee also emphasizes, if the transduction were linguistically described, the speed and precision with which it is accomplished would be considerably reduced. The conclusion at which he accordingly arrives is that "we would not expect a complete formal description or simulation of a complex system to adapt or function as rapidly or reliably as the partially self-describing, tacit dynamic system it stimulates."[48] For these reasons, matter and form may not be separable, even in principle, in biological systems such as the living cell.

The general problem with Strong AL, then, is how it conceives the relation between matter and form in the biological realm. Langton writes that "Life is a property of *form*, not *matter*, a result of the organization of matter rather than something that inheres in the matter itself."[49] What is right with this statement is that life is an emergent phenomenon dependent on processes having a certain form or organization. But what is wrong with it is that, in the biological realm at least, form is something that, as Aristotle emphasized long ago, *does* inhere in the matter itself. Susan Oyama puts this well when she writes: "Form emerges in successive interactions. Far from being imposed on matter by some agent, it is a function of the reactivity of matter at many hierarchical levels, and of the responsiveness of those interactions to each other."[50] Indeed, it is precisely this conceptualization of the relation between matter and form that is needed to provide a foundation for the familiar idea that, as Langton puts it, "Neither nucleotides nor amino acids nor any other carbon-chain molecule is alive – yet put them together in the right way, and the dynamic behavior that emerges out of their interactions is what we call life."[51]

VI. CONNECTIONISM AND THE THEORY OF AUTONOMOUS SYSTEMS

The interdependence of matter and form in biological systems also has implications for the debate between Connectionism and the Classical Symbol-Processing theory. As noted above, the physical symbol system model of the brain simply takes symbols at face value and treats them as if they were (in principle) independent of the neural systems within which they are embedded. In the case of biological systems such as the cell, however, there is no such independence of the symbolic level from its more encompassing dynamical context. Except for the biologically unconstrained, top-down hypotheses based on models of mental representation, there is no reason not to expect the same point to hold for the types of formal regularities in the brain and nervous system. Of course, in systems that are so organizationally complex it is much harder to specify the processes responsible for these regularities. Nevertheless, this difference in systemic complexity does not justify the indifference that the Symbol-Processing approach typically shows to the dynamical context.

This criticism of the Symbol-Processing approach will come as no surprise to those engaged in the Connectionist research program. Connectionist models do not take symbols at face value and then make entirely top-down hypotheses about the realization of symbol structures in the brain. Instead, symbols are typically treated in the Connectionist approach as approximate macrolevel abbreviations of operations whose governing principles reside at a "subsymbolic" level.[52] Indeed, this general type of relation between the symbolic and subsymbolic is precisely the one that holds for the genetic code in living cells: To describe nucleotide triplets as "coding" for amino acids is to abbreviate a lengthy causal sequence of complex intracellular processes whose governing principles reside at a subsymbolic biochemical level. Thus my use of the cell as a minimal paradigm of how syntax can be intrinsic to physics turns out to provide considerations in support of Connectionism.

The term "Connectionism" is of course usually applied to models of subsymbolic principles in neural networks. The goal here is both to understand real, biological neural networks and to solve problems in the theory of machine learning. But J. Doyne Farmer has argued in a recent article that the term

"Connectionism" should really be given a much broader signification.[53] He defines a Connectionist model as one in which the interactions between the variables at any given time are restricted to a finite list of connections, and the values of the connection strengths and/or the architecture of the connections themselves can change with time. Farmer then shows that this class of (meta)dynamical systems includes not only neural networks, but also classifier systems in AI, immune networks, and autocatalytic chemical reaction networks.

Within Connectionism thus broadly construed, the orientation most relevant to the concerns of this essay is the theory of autonomous systems.[54] Whereas *heteronomous* systems are defined by inputs and outputs and an external control, *autonomous* systems are defined by internal mechanisms of self-organization. For example, within neural network research, systems whose learning is "supervised" qualify as heteronomous because changes to the connections in the network are directed by an external training signal (as in the learning algorithm known as "back-propagation").[55] In contrast, systems whose learning is "unsupervised" qualify as autonomous because changes to the connections in the network typically depend on cooperative and competitive relations among the nodes without the direction of any external supervisor.[56] Autonomous systems are clearly the more biologically realistic, and so it is to the theory of autonomous neural networks that we should look for models of how symbolic processes emerge in the brain and nervous system.

In fact, the adaptive resonant neural network theory of Stephen Grossberg, Gail Carpenter, and their colleagues provides a paradigm of such models.[57] Their networks embed competitive learning in a self-regulating control structure that contains both attentional and orientational subsystems. The cycles of interaction between these two subsystems enable the network to self-organize stable internal configurations in response to arbitrary sequences of arbitrarily many input patterns. Grossberg and Carpenter call a given set of such stable internal configurations a "recognition code;" the symbols that constitute the code are "compressed, often digital representations, yet they are formed and stabilized through a process of resonant binding that is distributed across the system."[58] Thus adaptive resonance theory provides a paradigm of how stable formal configurations relevant to perception and action can emerge as a result of the distributed, subsymbolic process of resonant binding in the network, and then act to constrain the behaviour of the network.

The theory of autonomous systems thus seems to be in a good position to claim that it is on the way to discovering the "laws of qualitative structure" underlying symbolic activity in complex systems. This idea of a law of qualitative structure was originally invoked in cognitive science by Alan Newell and Herbert Simon on behalf of their physical symbol system hypothesis.[59] But, as we have seen, the purely top-down approach this hypothesis takes toward symbolic processes is unsatisfactory. In contrast, Connectionism and in particular the theory of autonomous systems explicitly consider the dynamical context of symbolic activity, and so hold out the promise of a formal theory of the necessary and sufficient conditions for symbolic activity in natural and artificial dynamical systems.

Concordia University (Present address: Boston University)

NOTES

* Earlier versions of this essay were presented to a symposium on Connectionism at the 1992 meeting of the Canadian Philosophical Association and to the Department of Philosophy at the University of Toronto. The thoughts expressed here have also benefitted greatly from conversations with Kenneth Cheung, Ronald de Sousa, Paul Thompson, Sonia Sedivy, and Francisco Varela.

[1] See Stephen Harnad, 'The Symbol Grounding Problem', in Stephanie Forrest, ed., *Emergent Computation* (Cambridge, Massachusetts: The MIT Press/A Bradford Book, 1991), pp. 335–346.

[2] This way of looking at the issue was originally suggested to me by Kenneth Cheung.

[3] See H. H. Pattee, 'Simulations, Realizations, and Theories of Life', in Christopher G. Langton, ed., *Artificial Life. Santa Fe Studies in the Sciences of Complexity Volume VI* (Redwood City, CA: Addison-Wesley, 1989), pp. 63–75.

[4] See Christopher G. Langton, 'Artificial Life', in *Artificial Life, op. cit.*, pp. 1–47.

[5] The term 'Strong AI' was coined by John Searle who opposes the view. See his 'Minds, Brains, and Programs', *Behavioral and Brain Sciences* 3 (1980): 417–458.

[6] See Langton, 'Artificial Life', *op. cit.* and Harold C. Morris, 'On the Feasibility of Computational Artificial Life: A Reply to Critics', in Jean Arcady Meyer and Stewart Wilson, eds., *From Animals to Animals* (Cambridge, Massachusetts: The MIT Press/A Bradford Book, 1991), pp. 40–49. See also Elliot Sober, 'Learning from Functionalism – Prospects for Strong Artificial Life', in Christopher G. Langton, Charles Taylor, J. Doyne Farmer, and Steen Rasumussen, eds., *Artificial Life II. Santa Fe Institute Studies in the Sciences of Complexity, Proceedings Volume X* (Redwood City, CA: Addison-Wesley, 1992), pp. 749–765.

[7] See Pattee, 'Simulations, Realizations, and Theories of Life', *op. cit.*, p. 69.

[8] John Searle, 'Is the Brain a Digital Computer?', *Proceedings and Addresses of the American Philosophical Association* 64 (1990): 21–37, and *The Rediscovery of the Mind* (Cambridge, Massachusetts: The MIT Press/A Bradford Book, 1992), Chapter 9.

[9] Searle uses the term "digital computer" but I think "physical symbol system" is a more precise designation for the kind of computational system he has in mind. See Alan Newell, 'Physical Symbol Systems', *Cognitive Science* 4 (1980): 135–183.

[10] Searle does not explicitly say that his new argument strikes deeper than the Chinese Room Argument but he suggests as much: 'The Chinese Room Argument showed that semantics is not intrinsic to syntax. I am now making the separate and different point that syntax is not intrinsic to physics. For the purposes of the original argument I was simply assuming that the syntactical characterization of the computer was unproblematic. But that is a mistake', *The Rediscovery of the Mind, op. cit.*, p. 210.

[11] Searle does not discuss AL, and given his notorious invocation of so-called "brute biological facts," it is not unreasonable to conjecture that he is unaware of this more abstract and systemic approach to biological systems.

[12] See John Searle, *Intentionality. An Essay in the Philosophy of Mind* (Cambridge: Cambridge University Press, 1983), p. 20.

[13] Thus Searle's position is in a sense the inverse of Dennett's. Dennett holds that no intentionality is intrinsic; all intentionality is relative to a stance we adopt toward a system, the "intentional stance." Adopting the intentional stance serves to reveal "real patterns" in the system's behaviour, but the explanation of how these patterns are generated resides at the design level, i.e., the level at which a complex system is a syntactic engine. For Searle, however, physical systems can be syntactic engines only in relation to the intentional agents who can assign syn-

tactical interpretations to physical phenomena. For Dennett's position, see 'Intentional Systems', in *Brainstorms* (Cambridge, Massachusetts: The MIT Press/A Bradford Book, 1978), pp. 3–22, *The Intentional Stance* (Cambridge, Massachusetts: The MIT Press/A Bradford Book, 1987), and 'Real Patterns', *Journal of Philosophy* (1991): 27–51.

[14] See John Haugeland's discussion of this point in his *Artificial Intelligence: The Very Idea* (Cambridge, Massachusetts: The MIT Press/A Bradford Book, 1985), pp. 96, 119–123.

[15] Jerry Fodor and Zenon Pylyshyn, 'Connectionism and Cognitive Architecture: A Critical Review', *Cognition* **28** (1988): 3–71. Page references are to the article as reprinted in Steven Pinker and Jacques Mehler, eds., *Connections and Symbols* (Cambridge, Massachusetts: The MIT Press/A Bradford Book, 1989).

[16] *Ibid.*, pp. 13–14.

[17] Zenon Pylyshyn, *Computation and Cognition. Toward a Foundation for Cognitive Science* (Cambridge, Massachusetts: The MIT Press/A Bradford Book, 1984), pp. 54–59.

[18] See *ibid.*, pp. 57–58, and Jerry Fodor, *Representations. Philosophical Essays on the Foundations of Cognitive Science* (Cambridge, Massachusetts: The MIT Press/A Bradford Book, 1980), pp. 22–23.

[19] *Computation and Cognition, op. cit.*, p. 56.

[20] *Ibid.*, p. 57.

[21] *The Rediscovery of the Mind, op. cit.*, p. 210, first full paragraph.

[22] Ray Jackendoff, *Consciousness and the Computational Mind* (Cambridge, Massachusetts: The MIT Press/A Bradford Book, 1987), pp. 29–36.

[23] *Ibid.*, p. 30.

[24] W. S. McCulloch and W. Pitts, 'A Logical Calculus of the Ideas Immanent in Nervous Activity', reprinted in W. S. McCulloch, *Embodiments of Mind* (Cambridge, Massachusetts: The MIT Press, 1965).

[25] Searle considers this line of argument (see the passage cited in note 21), but dismisses it by simply repeating his assertion that "syntax is not intrinsic to physics," which of course begs the question at issue.

[26] For different statements of this point, see Humberto R. Maturana and Francisco J. Varela, *Autopoiesis and Cognition: The Realization of the Living. Boston Studies in the Philosophy of Science, Volume 43* (Dordrecht and Boston: D. Reidel, 1980), pp. 19, 124–134; Walter Freeman and Christine Skarda, 'Spatial EEG Patterns, Nonlinear Dynamics, and Perception: The Neo-Sherringtonian View', *Brain Research Reviews* **10** (1985): 145–175; Gerald Edelman, *Neural Darwinism* (New York: Basic Books, 1987); and Andreas K. Engel, Peter König, Andreas K. Kreiter, Thomas B. Schillen, and Wolf Singer, 'Temporal Coding in the Visual Cortex: New Vistas on Integration in the Nervous System', *Trends in Neuroscience* **15**, No. 6 (1992): 218–226.

[27] See 'Intentional Systems', reprinted in *Brainstorms, op. cit.*

[28] Paul Smolensky, 'On the Proper Treatment of Connectionism', *Behavioral and Brain Sciences* **11** (1988): 1–74.

[29] See Fodor and Pylyshyn, 'Connectionism and Cognitive Architecture', *op. cit.*, p. 10, n. 4.

[30] *Ibid.*, p. 13. See also note 9 on the same page.

[31] See Stephanie Forrest, ed., *Emergent Computation, op. cit.*

[32] Stephanie Forrest, 'Emergent Computation: Self-Organizing, Collective, and Cooperative Phenomena in Natural and Artificial Computing Networks. Introduction to the Proceedings of the Ninth Annual CNLS Conference', in *ibid.*

[33] See Martin Gardner, 'Mathematical Games: The Fantastic Combinations of John Conway's New Solitaire Game of "Life" ', *Scientific American* **224**, No. 4 (1970): 120–123.

[34] John Maynard Smith, *The Problems of Biology* (Oxford: Oxford University Press, 1986), p. 19.

[35] *Ibid.*, p. 21.

[36] H. H. Pattee, 'Dynamic and Linguistic Modes of Complex Systems', *International Journal of General Systems Theory* **3** (1977): 259–266.

[37] John Von Neumann, *The Theory of Self-Reproducing Automata*, ed. A. Burks (Urbana, Illinois: University of Illinois Press, 1966).

[38] 'Dynamic and Linguistic Modes of Complex Systems', *op. cit.*, p. 262.

[39] *Ibid.*, p. 263.

[40] For further discussion of this important point see Susan Oyama, *The Ontogeny of Information: Developmental Systems and Evolution* (Cambridge: Cambridge University Press, 1985).

[41] See F. J. Varela, H. Maturana, and R. Uribe, 'Autopoiesis: The Organization of Living Systems, its Characterization and a Model', *Biosystems* **5** (1974): 187–195, and Humberto R. Maturana and Francisco J. Varela, *Autopoiesis and Cognition, op. cit.*

[42] Francisco J. Varela, *Principles of Biological Autonomy* (New Jersey: Elsevier North Holland, 1979), p. 75.

[43] 'Artificial Life', *op. cit.*, p. 11.

[44] The clearest statement of this idea is still to be found in Humberto R. Maturana and Francisco J. Varela, *Autopoiesis and Cognition, op. cit.*, pp. 77–84.

[45] See Ned Block, 'Introduction: What is Functionalism?', in Ned Block, ed., *Readings in the Philosophy of Psychology. Volume One* (Cambridge, Massachusetts: Harvard University Press, 1980), pp. 171–184.

[46] There is a difference in the type of computational approach taken in each case, however. In computation-representation functionalism, mental processes are recursively *decomposed* in a *top-down* manner, whereas in AL biological processes are recursively *generated* in a *bottom-up* manner. This difference does not affect the point being made, however.

[47] Claus Emmeche, 'Life as an Abstract Phenomenon. Is Artificial Life Possible?', in Francisco J. Varela and Paul Bourgine, eds., *Toward a Practice of Autonomous Systems. Proceedings of the First European Conference on Artificial Life* (Cambridge, Massachusetts: The MIT Press/ A Bradford Book, 1992), pp. 466–474.

[48] H. H. Pattee, 'Dynamic and Linguistic Modes of Complex Systems', *op. cit.*, p. 264.

[49] Christopher G. Langton, 'Artificial Life', *op. cit.*, p. 41.

[50] Susan Oyama, *The Ontogeny of Information, op. cit.*, p. 22.

[51] *Op. cit.*

[52] See Paul Smolensky, 'On the Proper Treatment of Connectionism', *op. cit.*

[53] J. Doyne Farmer, 'A Rosetta Stone for Connectionism', in Stephanie Farmer, *Emergent Computation, op. cit.*, pp. 153–187.

[54] See Francisco J. Varela and Paul Bourgine, eds., *Toward a Practice of Autonomous Systems, op. cit.*, and Francisco J. Varela, *Principles of Biological Autonomy, op. cit.*

[55] See David E. Rummelhart, 'Brain Style Computation: Learning and Generalization', in Steven F. Zornetzer, Joel L. Davis, and Clifford Lau, eds., *An Introduction to Neural and Electronic Networks* (San Diego: Academic Press, 1990), pp. 405–420.

[56] See Christoph von der Malsburg, 'Network Self-Organization', and Gail A. Carpenter and Stephen Grossberg, 'Self-Organizing Neural Network Architectures for Real-Time Adaptive Pattern Recognition', both in *ibid.*, pp. 421–432, and pp. 455–478.

[57] See the article by Carpenter and Grossberg cited in the previous note. In addition see Stephen Grossberg, 'How Does a Brain Build a Cognitive Code?', *Psychological Review* **87** (1980): 1–55, and 'Competitive Learning: From Interactive Activation to Adaptive Resonance', *Cognitive Science* **11** (1987): 23–63.

[58] Gail A. Carpenter and Stephen Grossberg, 'Integrating Symbolic and Neural Processing in a Self-Organizing Architecture for Pattern Recognition and Prediction', *Technical Report* CAS/CNS-93-002, Boston University Center for Adaptive Systems and Department of Cognitive and Neural Systems. To appear in V. Honavar and L. Uhr, eds., *Symbol Processors and Connectionist Networks in Artificial Intelligence and Cognitive Modelling: Steps Toward Principled Integration* (San Diego: Academic Press, forthcoming).

[59] Alan Newell and Herbert A. Simon, 'Computer Science as Empirical Inquiry: Symbols and Search', reprinted in John Haugeland, ed., *Mind Design* (Cambridge, Massachusetts: The MIT Press/A Bradford Book, 1981).

JAMES A. McGILVRAY

MAKING COLORED OBJECTS

The many discussions of color in recent philosophical literature have tended to focus on whether the subjectivist or the objectivist is right: on whether colors are "out there" as properties of things we see, or "in here" as features of either minds or neural events – perhaps as ways of seeing. The dispute has sometimes focussed on the nature of color – how colors are ordered or related to one another, and whether this way of individuating them makes them candidates for identification with certain physical properties, for example. It has also focused on standard epistemic issues, such as our access to colors and their roles in description and explanation. I speak to some of these arguments in this paper, but my primary intention is to look at the dispute between subjectivist and objectivist by asking what *sorts of things can in the final analysis bear color properties*.

I. COLORED THINGS: AN OVERVIEW

Current philosophical literature offers several different candidates for the status of color bearer. There are at least four non-relational candidates: (1) entities as described in everyday affairs ('common sense objects') (2) entities as described in relevant physical theories ('scientific objects'), (3) qualia and sense fields, and (4) entities defined over neural objects and processes ('neural events'). There are some relational candidates for color bearers too; I mention them briefly later.

Before discussing these options, let me define terms. To *bear* a property is to have it, all things considered, where "having a property, all things considered", amounts to saying that a thing's having a property gives it a no-holds-barred explanatory status. Having a property in this sense allows for complete explanations – for answers to everything that one can legitimately ask with regard to some relevant domain. In this case, it provides for answers to what it is to be and have a color, even to the extent of explaining why matters might appear to be different than they are. This strong notion of bearing a property is defined over the weaker notion of having a property; some thing *x has* a property within a theory of framework, not "no holds barred". This amounts to something like "It is correct in theory T/framework F to attribute property P to *x*'s. For purposes of discussing colors, a useful condition on some thing having a property is that it *track* it. The idea is that colors that belong to some thing move with their object and remain approximately the same during the relevant form of movement, transformation, or deformation, if any. The color of the falling leaf, for instance, remains the same as it moves in position and deforms under air resistance. What counts as movement and what sameness

81

M. Marion and R. S. Cohen (eds.), Québec Studies in the Philosophy of Science II, 81–97.
© 1996 *Kluwer Academic Publishers.*

depends on the kinds of object one deals with. The red of a common sense
object like a mailbox moves as the mailbox moves in space – as the mailbox
is carried from one spot to another – and the object is the same through these
changes. Significantly, the object remains the same when painted a different
color, but note that its color is changed only through special effort and outside
influence. The red and green (say) color patterns in the visual field of the
migraine sufferer, however, track no common sense thing like a mailbox; it
is not obvious that these patterns are in space in the same way that mailboxes are; and tracking tends to be very short-lived, for in fact a change in
color can constitute a change in "object". Nor – to take a common case –
does the blue "of" the sky track anything like a common sense object. The
sky's and migraine sufferer's colors can, however, be said to track (sometimes short-lived) parts of a visual field. Furthermore, should such "parts"
of a visual field prove identifiable with certain neural events, they could be
said to track these neural events, which are located in space – specifically in
an organism's brain. In any case, I shall often speak, not of having a color,
but of a candidate for a color-bearer tracking a color.

1. The common sense object is the object of common sense belief – the thing
that answers to the descriptions and explanations we routinely employ in
everyday life. These beliefs, and thus the natures of these objects, are artic-
ulately expressed by Wittgenstein (following, in some respects, Goethe) in
Remarks on Colour. Extensions, applications, and implications of this view
of the colored object are found in the work of many (e.g., Westphal, 1987).
These objects are the objects of Sellars' "manifest image" and they figure
prominently in our folk theories and folk psychological descriptions and explanations. Colors for common sense belief are persisting properties of (surfaces
of) those familiar persisting objects "out there" that we act with and upon,
such as people, mailboxes, and stones.

The colors and things of the common sense framework have a central and
virtually ineradicable epistemic role, for their function in innumerable cognitive affairs – including our full-fledged scientific enterprises – is more than
just a matter of convenience. Colors are central to our efforts to classify,
identify, and re-identify things like birds and minerals. Colors of objects attract
and hold attention. Colors heighten or diminish certain activities, such as
heart rate. Colors figure prominently in instrument and test design (PET scan
displays, litmus paper). And so on. The strength of the claim color-related
common sense beliefs have on us and on our intuitions is undeniable. They
virtually dictate what might be called *the* logic of color. Because of this, any
full-fledged theory of colors and colored objects must be able to explain and
underwrite the role of colors in the common sense framework.

I mention two settled beliefs that can be stated as principles of the common
sense logic of colors and colored objects. The first is that colors track common
sense objects or "substances" in a stable, observer-independent way, so that
if the surface of some object outside us has such-and-such a color, it tends

to keep it. Call this the external/stable belief. The second belief – autonomous order – is that colors are ordered in several ways, including 'betweenness'. For instance, orange is between red and yellow. It is important that this and other ways of ordering colors seem to have nothing to do with the natures of the common sense objects that bear colors, for there is no correlation: while orange is between red and yellow, orange things are in no sense (except for their colors) between red things and yellow things. Colors seem, then, to be autonomously ordered in the common sense framework.

I mention these two principles of the logic of color and colored objects not only because they are epistemically necessary conditions that any account of color must explain, but to emphasize that it is not easy to adequately explain both. I shall suggest that the second belief – autonomous color order – is best explained by identifying colors with properties of neural events, making them subjective and not properties of things "out there". The external/stable belief, however, is most easily explained by identifying colors with surface spectral reflectance (SSR) values, placing colors "outside". There is no obvious, direct way to say that colors are both inside and outside in the same way, and without qualification. It is because of this that relational accounts of color that try to place colors both in the head and out there might prove tempting. But if the aim is to decide what counts as a color bearer, relational accounts are *ad hoc*, for they try to say colors are "both in and out" without being able to say why in a principled way. Non-relational alternatives are to be preferred, if at all possible. Unfortunately, there are few straightforward strategies for non-relational accounts to pursue. The non-relational objectivist really has only one strategy available: s/he has to make color order "merely subjective", without somehow also locating color in the head. The non-relational neural subjectivist, on the other hand, must make colored objects "out there" an appearance, and explain how colors get routinely projected by neural machinery (not mysterious minds) onto external things.

2. Scientific objects that bear "surface spectral reflectance" properties appear virtually to be designed to accommodate the first principle of the logic of common sense belief. The objectivist who identifies colors with surface spectral reflectances (SSRs) (Hilbert, 1987; Matthen, 1988) makes colors into observer-independent, persistent dispositional properties of things outside ourselves and thus explains the first epistemic condition in a straightforward way.[1] Colors seem to be persistent, observer-independent properties of surfaces of things outside us because colors *are* SSRs, which are respectable properties of physical objects. I speak to the efforts of the SSR objectivist to deal with autonomous color order and other apparently subjective features of color below.

3. The two objectivist candidates for color-trackers just discussed – the common sense view and the surface spectral reflectance view – contrast with subjectivist views of color and color bearers. One form of contemporary sub-

jectism is little different from a phenomenalist tradition that is centuries old. This isolationist phenomenalism holds that colors somehow track objects "inside" us – perhaps not mental objects, but certainly neither common sense objects that we see outside ourselves nor external scientific objects. The view is that colors are properties of a special class of sensory entities called sensa, qualia, or sensations. Those who now work in this tradition tend actually to say very little about colors as such and only slightly more about what sorts of entities are colored; because of this, they are not well positioned even to deal with color order, not to mention stability and externality. Their interest is elsewhere, focussed on isolating some aspect of experience or knowledge of color from "objective" investigation (e.g., Jackson, 1982, 1986; Jackson and Pargetter, 1987; Nagel, 1974, 1986). Regrettably, this focus inhibits the development of a serious phenomenalist account of color, and seems wrongheaded to boot.

Virtually defying the isolationist's focus, an extraordinary amount has already been learned about color and bearers of color by psychometricians who employ techniques of "objective" investigation. In all important respects, these techniques not only capture the internal features of sensory qualities (including their order), but provide as fine-grained an account as possible of what colors and other sensory qualities look like "from inside". Indeed – these objective phenomenalists claim – virtually all that there is to be known about the nature of color as such (as opposed to what kinds of physical events are the distal causes of color sensations, for example) is *now* captured by scaling techniques that provide a coordinate system and metric for the color spaces of individual organisms and groups of organisms, human and otherwise.[2] Perhaps statistically based scaling sciences do not tell us everything about "what it is to actually undergo" a color-experience, for they obviously do not actually make one undergo an experience at a particular time. But if this is all that the fuss over "essential subjectivity" in isolationist phenomenalism is about, it is difficult to understand why the concept attracts so much attention.

4. The fourth view of color-trackers is that colors are (identical with) kind properties of neural events. The neural color subjectivist holds that colors are ways of classifying certain neural events, and that this form of classification not only provides an account of what it is to have a color, but explains both principles of the logic of color and colored objects. The point is to make neural events the sole color-bearers. Classification here is not just functional classification that says that there are neural events that perform a particular information-displaying job; it is a classification that specifies for a particular class of organisms and a particular "color experience" which sets of neurons are firing, their firing rates, and the relevant patterns of relationships between relevant neurons and their firing rates (it specifies neural locations, thresholds, wave patterns, coordination, feedback, and the like). This is, I believe, a defensible form of color subjectivism and a plausible view of

what bears color properties. According to it, colors are kind properties that classify neural events, and being a kind of event is a way of being a property of an event. To "track" such an event is just for the event to occur.

Neural subjectivism differs from isolationist phenomenalism in two ways. First, it accepts cheerfully the results – based on statistically based scaling techniques – of the objectivist phenomenalist and is equally willing to call itself "objectivist" in this regard. Second, it explains the fundamental features of the ordering found by these scaling techniques – particularly color opponency, mixture, and degree of difference between colors – by identifying colors with kinds of neural events.[3] It is thus positioned not only to articulate color order, but explain it. So neural subjectivism deals handily with the order principle of the logic of color and colored objects. Can it also deal with the first? It can if it can explain how colors somehow appear to us as properties of things outside. I suggest how it can do this below.

5. Finally, current literature also offers various relational views of color. Dispositional accounts of color, for example, are still ably defended (e.g., Johnston, 1992), but they tend as they always have to collapse into one or more of the above views about colors and colored objects. I also mention a recent enigmatic variation on a relational view of color, offered by Thompson, Palacios, and Varela (1992). It seems to make colors out to be properties of pairs of classes of organisms and classes of ecologically specified external things: for each relevantly different class of organism, a set of objects "for" that class of organisms is colored with that class of organism's colors. This view looks interesting, but it provides no clear answer to the question of what is colored. We must, I think, await further details.

This brief survey introduced some useful notions (color bearer, color tracker) and two epistemically necessary conditions on theories of color. It also suggests that the serious contenders at the moment for answering the question of what things can properly be said to bear colors are the scientific object with its SSRs and the neural event. These are the serious contenders because they purport to explain the logic of colors by identifying colors with features of scientific objects. Nevertheless, as we have seen, each seems to be particularly well suited to explain either the first or the second of the principles of this logic, not both at once. Until one or the other adequately explains both, it cannot properly claim to tell us what bears colors. I start by examining the claims of the SSR objectivist.

Note that the issue of what bears colors has here come to whether colors are borne by scientifically described objects outside us, or by scientifically described objects inside us. If they are properties of so-described objects inside us, of course, colors will turn out to be objective in one sense, but not in that sense the color objectivists have in mind. For even if objective in this sense, they are not properties of things "out there".

II. SSR COLOR OBJECTIVISM

The SSR color objectivist I describe is represented by no single philosopher; it is a philosopher-type that happens to employ three arguments I outline below to defend the view that colors are surface spectral reflectance values of physical objects, or SSRs. (Hilbert occasionally resembles a philosopher who offers the first.)

The three arguments in outline follow. (1) The first argument the idealized SSR color objectivist offers begins by taking the first of the two principles of the logic of common sense color belief – that colors track things we see outside ourselves – at face value. This is a good initial strategy, for no one seriously denies that this principle needs to be explained. As Hilbert puts it in (1987), "[a]ny view that denies that any external things are colored flies in the face of the way we conceive and talk about colors." Then the SSR objectivist points out that if the surfaces of the things we in our epistemically primitive common sense way say we see were identical with the surfaces of things described by the colorimetry specialist as having certain SSRs, the fact that objects outside ourselves appear to track colors in an observer-independent and persistent way is explained, for SSRs have just the second-order properties (observer-independence, persistence) that principle says colors have. So, one should identify colors with SSRs. (2) The second argument is this: because it is the case that of all known properties of surfaces of scientifically described objects, SSRs provably most nearly correlate with colors as experienced, SSRs should be identified with colors. (3) The third argument is a functional-computational one: there is strong evidence that color vision fulfills the function of "solving" for surface spectral reflectance values, so we should identify SSRs with colors. The third argument gets the most play in current literature, perhaps because it can legitimately appeal to the success of several relatively new computationalist models of color vision.

In more detail now, there is little doubt that we commonsensical folk expect the colors of objects we see to remain the same even when we are not looking at something, and that this expectation is what is expressed by the first principle of the logic of color. The tree is green in the dark, and in the light when no one is looking at it. Changing colors are not the norm, and require some special effort, for normally colors are (said or thought to be) stable and to move with the object even when it undergoes deformation – in the case of the green tree, this effort consists of shutting down its photosynthetic chemistry; in the case of a mail box, it consists of our painting it or letting it rust. As a rule, the familiar things of "our" world reliably keep their colors for long times and they take them with them. Exceptions, such as the chameleon, provoke surprise. So it must be granted that an ontology of colored objects consistent with the first principle of the logic of common sense belief must make colors out to be properties of things out there, properties that they have independently of anyone's seeing them. In offering the first argument, the SSR objectivist not only appeals to this naive color objectivism, but shows how to make it sophisticated. S/he points out that there happens to be a property of the surfaces of things as described by the physical sciences, a surface's spectral reflectance, which suits the logic of common sense color objectivism. A thing's SSRs are properties that something carries around with it and that it has independently of illumination and/or whether some organism is looking at it. The parallel between these features of SSRs and features of our entrenched beliefs about colors provides a *prima facie* case in favor of the idea that colors are SSRs. And the proposal has the virtue of simplicity: there is no easier way to account for the first principle of the logic of color.

Second, my idealized SSR objectivist argues that SSRs are the only properties of external objects that qualify at all as candidates for colors, for if *any* recognized properties of surfaces as described within the relevant sciences correlate with the colors we experience, SSRs do. It must be admitted that the correlation is far from one-to-one: the phenomenon of metamerism shows that (specifically, it demonstrates that surfaces with wildly different SSR curves can still look the same under the same lighting conditions). It must also be admitted that some surfaces are light *emitters*, rather than reflectors, so that these surfaces cannot have SSR values assigned to them (unless one allows values greater than 100% reflectance, which is a zealous but *ad hoc* way to guarantee that all surfaces get SSR assignments). Further, it must be admitted that not all colors are "surface colors", for not only are there illumination colors like those just mentioned, but film or transmission colors (the colors of surrounding gases and tinted glasses) and even other causally different colors that do not fit peacefully into this classification (phosphorescence, Moiré patterns, . . .). And finally and most importantly, it must be admitted that the ways in which SSRs are ordered with respect to each other (represented as curves on a reflectance scale (0–100%) for each wavelength where radiant energy reflects at all) do not in any way look like the ways in which colors in experience are ordered with respect to each other. Nevertheless, if what is at issue is "the colors of objects that we (typically) see", there is for the colors of objects as experienced like these no more closely correlating physical property than the scientifically described objects' SSR(s).

The third argument combines several considerations to try to convince us, first, that the function of color vision is to detect or solve for SSRs and, second, that in presenting something to us as colored, color vision is representing the SSRs of objects we look at. A good part of the impetus for this argument came from Marr's (1983) effort to bond Gibsonian realism (which emphasized the role of constancies in perception and based realism on this) to a computational theory of vision; Marr's view suggested that a computational account of vision's task made vision out to be an effort to represent features of the things of the world, including their colors. Marr's Gibsonian views concerning color were buttressed by Land's efforts to model human color constancy phenomena in his "retinex" theory, developed thirty years ago, and more recently by increasingly successful computational algorithms. Land himself was a subjectivist on color, rather than an objectivist,[4] but many computationally minded perceptual psychologists who used and improved on Land's algorithms (for instance, Maloney and Wandell (1986)) were objectivists and took their efforts to produce an algorithm for computing the SSRs of things to be not just a description of the function of color vision, but a model of how an organism like ourselves (with three separate cone systems) detected what are always spoken of as the *colors* of its world. Computationalists think of color vision as solving for SSRs, and inevitably think of SSRs as colors.

However, the only way to take the claim that vision (normally) solves for SSRs to the view that SSRs are colors is to adopt realist assumptions about the nature of representation. Thus, another strain in the third argument takes representational (descriptive, depictive, . . .) success and combines it with a realist assumption to yield the view that SSRs are colors. The realist assumes that if a sentence, description, depiction, or other representation is true or correct, it describes things as they are; more to the point, if a color-representation is correct, it represents the colors of things as they are. So if there is evidence that color-vision operates successfully – and there is millions of years of evidence in the success of creatures with color vision thriving in their world to suggest there is – we have reason to think that color vision that has the function of computing for SSRs does indeed represent the colors of things in correctly depicting their SSRs. The strength of this effort to show that color vision in representing SSRs is representing colors is muted by the observation that colors that are found in experience do not map in anything like a one-to-one fashion onto SSRs; metamerism demonstrates this by showing that a single color as experienced can map onto a large group of very different SSRs that by no means form a kind, natural or otherwise. Furthermore, it is muted by the observation that the order of colors as experienced in no way resembles the order of SSRs: a particular shade of green experienced on an occasion could not only be correlated with many SSRs, but these many SSRs bear no systematic relationship either to each other or to the SSRs that correlate with an experience of red on the same occasion. (Certainly there is no way to say that the "red" SSRs are *opposed* to the "green" SSRs as red is opposed to green. So there is evidence that color vision does not completely or fully adequately represent SSRs. Nevertheless – the argument goes – this does not show that color vision does not represent colors in representing SSRs. The criterion of success here is the success of a species at getting on in its world, not full-fledged articulate and well-confirmed sentential truths. And if this borders on realists' appealing to the bugbear of partial truths, it soothes to speak instead of the visual system as gathering "information" and then to say, as Hilbert does, that this information is "partial" or even, as Matthen says, that it is "systematically in error". This way of speaking assumes from the start that truth and error are truth and error about colors as SSRs, and – the subjectivist suggests – masks the real issue. It makes color order not just "anthropocentric", as Hilbert says, but essentially inexplicable and subjective, for there is nothing in SSRs to explain it. Nevertheless, ignoring this fact, this argument with strains from Gibsonian (and Aristotelian) realism, from successful modelling of human color vision, and from a realist assumption about correct representation, seems to argue in favor of identifying SSRs with colors. When these considerations are combined with the two arguments just rehearsed, moreover, the claim that SSRs are colors can appear very compelling. On the whole, SSR color objectivism is well situated to deal with the common sense belief that colors are stable properties of things outside ourselves. On color order, however, it offers little or nothing.

III. COLOR SUBJECTIVISM RESPONDS

The color subjectivist answers these three objectivist arguments best by beginning with a response to the third; answering the third is easy, and shows how to answer the second. Answering the first argument is in some respects the most difficult task, for while the objectivist argument is faulty, a worry remains. When all is said and done, the subjectivist still has to contend with our very well-entrenched common sense belief that colors are properties of things "out there" – that colors are properly speaking features of common sense objects.

As for the SSR objectivist's third argument, most of the objectivists' observations, and much of their argument from computational success and the function of the visual system, are simply accepted by the objective phenomenalist and the neural subjectivist. Yes, color vision's normal function (at least usually in the case of human beings) is to "compute" for SSRs. It would be difficult to explain the evolutionary development of color vision and its role in recognition[5] – particularly as it is linked with color constancy – in any other way. And yes, color in the visual system does represent (or better: 'present') differences between SSR values where it concerns itself with (pays attention to) a physical object at all; furthermore, it does so by presenting these differences as differences in colors. The only serious problem with the objectivists' account of the function of color vision is that SSRs are not colors. For everything the objectivist claims about the typical function of color vision being to solve for SSRs is correct, but no one need hold that in presenting SSRs in a colored way, or "coloredly", the visual system is representing the colors of things outside ourselves.

The crucial move the subjectivist makes in rejecting the objectivists' third argument is this; s/he denies, at least in the case of colored-object representations, that the objectivist may legitimately appeal to the realist principle that a correct representation describes things as they are. The subjectivist does not reject the principle out of hand. In fact, the neural subjectivist is a straightforward scientific realist, and adopts the principle where it is warranted. But it is not warranted here. It is not warranted, in general, wherever there is a better account of the matter that is consistent with the best theories and that explains why things appear as they do. The subjectivist has such an account, and it begins with the fact that color vision (re)presents SSRs, but that these are not colors.

Locke said something like this when he claimed that certain properties are secondary as opposed to primary, and that while these secondary properties contribute essentially to the way things outside appear to us, nevertheless these properties do not have "resemblances" in things. We are far beyond Locke in understanding the natures of things and of colors; we can in fact prove that colors do not have resemblances in things. We know, for example, that they do not correlate with spectral frequencies – no one today would defend this Newtonian position. And while in the case of vision directed to physical objects they correlate most closely with SSRs, they do not correlate at all in their second-order properties (the ways they are arranged with respect to one another) – they do not correlate sufficiently to merit speaking even of resemblance, not to mention identity. This is because colors are ordered with respect to each other as they are in the color sphere that has yellow opposite blue and green opposite red (for these are "opponent" colors); it is because green, yellow, blue, and red are "unmixed" while all other hues are mixed, and mixed in a particular order; it is because hue, brightness (or lightness) and saturation taken together uniquely specify a color, and it is because there are more discriminable colors between some colors than there are between others.[6] None of this is true of SSRs: they simply do not stand to each other in these ways, nor are there unique SSRs, nor an order of SSRs like the order of colors, nor does hue, brightness, and saturation uniquely specify an SSR, nor do SSRs tell us about degree of discriminability. Nevertheless, there is nothing wrong with saying that colors (re)present SSRs, so long as it is understood that this does not involve resemblances, only normally reliable guides to SSRs. This relatively thin thesis should be acceptable to everyone, for the functional argument for (re)presentation that the objectivist and subjectivist both accept does not rely at all on resemblances.

This is not the whole story, of course, because in order to explain how color experiences can seem to have distal content, one must show that they do refer in some way to "things outside". This requirement poses no additional hurdles, however: if the functional argument for (re)presentation goes through, it is not hard to hang an appropriate concept of reference on it. To be sure, the relevant concept of reference cannot rely on color experiences' being "descriptively correct" of scientific objects with SSRs, but for this purpose all one need insist on is an appropriate form of non-descriptive reference. Some story about causal flow of information like Dretske's (1981) will do.

In addition to showing that colors do not resemble SSRs, the argument that SSRs cannot be identified with colors shows that there is no force to the objectivists' second argument. It is correct that (surface) colors as experienced most closely correlate with SSRs. And given what the subjectivist is willing to accept from the functional argument, it is permissible to talk of color experiences providing partial information and perhaps even systematically incorrect information concerning SSRs. But there is nothing in this to support the view that SSRs *are* colors.

What, then, of the objectivists' first argument? This argument for identifying colors and SSRs rests on the entrenched common sense belief that colors are stable properties of the things we see "outside ourselves", and claims that the relevant features of the first principle of the logic of color talk in our everyday affairs – stability, externality, and independence from observers – are duplicated by features that all acknowledge are characteristic of SSRs. It must be conceded that they are. To show that physical objects with SSRs are color-bearers, however, the argument must also show that *all* relevant features of color are similarly explained by identification, not just these. But order (specifically: opponency, mixture, array, and a variable metric), are not so explainable. It is not enough in this regard to say that these are merely subjective features of colors, reflecting our inability to capture all differences between SSRs, for that simply says that there are features of color that are inexplicable on this view. So the argument fails, quite spectacularly.

This response to the SSR objectivist constitutes a negative, and by no means sufficient argument in favor of some form of color subjectivism. A positive argument must first show that subjectivism can explain color order. This is done in detail in Hardin (1988) and Clark (1993), and I do not repeat it here. Second, it must show how it is that we could have come by the entrenched belief that colors are stable, observer-independent properties of things outside ourselves. A full defence of neural color subjectivism, incidentally, not only explains the autonomous order principle and the stable/external principle, but offers considerably more. It offers a full explanation of color "illusions"; provides for a comparative study of the color vision and color spaces of other species; deals with colors with all causal etiologies, including dreams; and in general (as Sellars hoped it could) explains why everything (not just common sense objects) appear in the way they do to creatures like us. I cannot pursue any of these matters either. I will show how neural subjectivism can deal with the first principle of the logic of color.

IV. EXPLAINING APPEARANCES

Let me clarify two things at the beginning. First, to say that colors are properties of neural events is not to suppose that neural events *appear green*. Neurons of a conscious brain seen through a trepanned skull are not green even if the organism is looking at something green. The neurons in such a case constitute just another reflecting surface. (In fact, since blood and nervous tissue are all that one can usually see through a trepanned skull, they are not likely even to appear green.) Saying colors are borne by neural events is saying that colors are classifications of kinds of neural events. Second, taking seriously the considerations raised earlier about how color vision is a way of (re)presenting SSRs, we can say that functionally speaking, colors typically appear to us when and as our brains undergo the neural events that (re)present SSRs. This is the key to the explanation of how colors appear as properties of things outside ourselves.

Let us recall once again that our epistemically most fundamental knowledge of colors is knowledge of colors *as* properties of surfaces of common sense things; our words and actions show that we identify the surfaces of which colors are properties with those we think of as surfaces of common sense objects. The surfaces are also the ones that we touch and perhaps distort. They are those that get hotter more quickly in the sun if they have darker colors and they are those that can be painted. They certainly appear to be colored: when we put our hands up in front of our eyes, the color *of the object* – as we say – is occluded. But these surfaces, if they are outside ourselves, are not colored, for no color can really be a property of something outside of the brain. If there is something that appears to be outside and (genuinely) colored, then, it must be a fiction. Explaining how colors appear as properties of things outside ourselves must be in part a matter of explaining how we could possibly have developed such complete and settled agreement in our beliefs concerning the colors of fictions, not to mention epistemic dependence upon them.[7]

The phenomenon of color constancy must play a role in any story about how colored external objects come to be constructed. Color constancy – or more accurately, approximate color constancy – is an often-experienced fact: within wide limits of variation in illumination intensity and chromaticity, an observed surface appears the same in color to the human eye and – there is strong evidence for this – appears the same (for that species) to the eyes of a large number of other organisms. Relatively simply devices like cameras using color films adjust to illumination intensity; most modern cameras automatically adjust their f-stops. But they cannot adjust to chromaticity in the illuminant, so they render a surface in very different colors when encountering chromaticity that falls outside of the designed chromatic range of the film. Photographic prints made from daylight film will when rendering a white surface outdoors in full sun depict it as white; when the same film is used on the same surface indoors under low color temperature incandescent light, prints made from it will depict that surface as yellow. More complex devices can compensate. Sophisticated circuitry is used in television cameras to afford approximate color constancy for those watching a screen so that the surfaces of television screens that render images of things the camera scans duplicate our expectations with regard to things we see and describe as colored. The interesting point here is that the circuitry is obviously needed to meet our expectations. Our expectations, described in part by the principle that colors are stable properties of things outside ourselves, crucially depend for the subjectivist upon the fact that our visual systems are – like color TV cameras – "designed" to yield constant results with variable input. It is only because of this, the neural subjectivist says, that we are able to develop the belief that colors are stable properties of things outside of us at all. We "make" constant-colored objects. Looked at from this subjectivist perspective, the idea that our visual systems function to solve for SSRs is properly described, not as the objectivist hopes in terms of an effort on our part to represent the *colors* of things out there, but as an effort on the part of our visual systems to yield a reliable means of getting around in our world by providing reliable colored presentations of things. These presentations permit reliable identification and reidentification, and so on, *via* colors.

If the aim is to construct common sense objects with their colors, however, this cannot be just a matter of remarking the raw fact that our visual systems yield constant colors when directed to certain surfaces with SSRs. Among other things, it is important to explain how (genuine) colors get (or appear to get) projected "outside",[8] how they seem to us to remain the same when we are not looking, and how they seem to be publicly colored. I suggest that to make sense of what construction of stably-colored observer-independent objects and projection amounts to here, we settle on the "units" of construction and the techniques. The phenomenon of color constancy is a useful guide: we can begin with colors that are occurrent properties of neural events (they track neural events) and end with colors that are stable properties of observer-independent (albeit fictional) things.[9]

Since what is at issue is the construction of persisting external observer-independent common sense objects with genuinely colored surfaces (not SSRs) from internal resources, start with something that is a genuine colored surface. This colored surface must have sufficient continuity and stability to count as object-like at all (for a color to "belong" to it or to track it), so assume that it persists for at least a brief time. This genuinely colored surface must be in the head (think of them as "coloring-surfacings" or as "hue-brightness-saturation-azimuth-altitude-depth locations" in a color-surface phenomenal space (Clark, 1993) to avoid thinking of the head as full of little colored and shaped things that we see with an inner eye), for that is where all colors are. These surface need not, of course, be colored surfaces that I also touch with my hands, although it must be possible to tell a story about constructing common sense objects that allows that tactile properties be joined to visual. The only plausible candidates for "internal" colored surfaces are neurally definable analogues to the phenomenalists' colored patches; call them, then, *'phenomenal objects'*.[10] Unlike the phenomenalists' colored patches, however, these colored patches are identical with certain neural events. Like the phenomenalists' colored patches, they are typically short-lived, which suits the status of colors as "occurrent properties"; but they have sufficient stability to count as "units" of construction. Like the phenomenalists' patches again, they are "in experience", although not typically experienced; for they are "undergone" by an organism to the extent that they are in the history of an organism's sensory states, although not (typically) something to which the organism "pays attention". Unlike the phenomenalists' colored patches, they are not private except in the sense that a particular instance belongs in the (neural) sense history of one organism as opposed to another, and are presumably only "undergone" by that one organism. They can be type-identical with those of other organisms that have and that exercise the same neural machinery too. Notice that as introduced, color and shape constancy are constitutive of a particular phenomenal object (neural event); as either changes, a new object begins. That is, as either changes, the color ceases to "track" the object that it partially determines, and it no longer "belongs" to it.

Clearly, the phenomenal object is far too episodic to us "as is" in the construction of a common sense object with stably colored surfaces. So as a step in the right direction, I introduce an intermediate object that maintains color constancy but allows for changes in shape and position. The dark brown and white patches that I experience as my-dog-presentations remain dark brown and white as Murphy changes shape and moves from rest to greeting me. It is plausible to regard the sequential record of patches that maintain color constancy as a single *sense object*. This sequence of patches is still genuinely colored; it is clearly also still only in my head in the same way as the episodic patch is. It lasts longer than the episodic patch but it is not by any means something that exists independently of perceivers. Colors remain constitutive of these objects, and constant colors "track" sense objects because they effectively define them. And yet these sense objects are still too episodic, they are "private" (certainly not-observer-independent), and they are still inside the head. They too will not do "as is" in the construction of the colored common sense object.

The next step is to introduce a "perceptual object". There is good reason to hold that when any of a group of organisms of the same species with color vision look at (attend to) Murphy thought of as a scientific object with SSRs, they have sense objects that are – when perspectival differences are taken into account – approximately the same, or "similar". It is surely legitimate to appeal to something like "normal observer position" in defining such an object. Thus, I do not need to insist that the perceptual object be definable over all conceivable observer positions, such as looking at Murphy cross-eyed from underneath from a distance of one thousand and one point thirty three meters. With the aid of normal observer position, the concept of sense objects for members of the same species, and the concept of attention, I define a *perceptual object*. It is defined as a similarity class of shape-size-color sense objects at a particular time, and is got by appeal to a hypothetical: if any member of a class of organisms attends to some particular scientific object with its SSRs, it will experience approximately the same sense object as others in the class. The similarity is explained and grounded by virtue of stable scientific objects (with stable SSRs), perceptual machinery designed to yield constant results in varying circumstances which is possessed in common by each member of a class of organisms, and by non-descriptive reference based on causally-provided information. The result is a public object. It is also a fiction. Unlike phenomenal objects and sense objects, which are identical with neural events, these perceptual objects no more exist than the standard perceiver or average taxpayer does.[11]

Unfortunately, perceptual objects still need to be supplemented to get persisting things with persisting colored surfaces, for while public they are defined over a physical object with SSRs at some time. So let us introduce the *perceptual object over time*. This can be defined by using counterfactuals and perceptual objects: it is what would have been seen, had there been. . . . There should be nothing controversial about this move.

The common sense object can now be reached by introducing beliefs. Common sense objects as thought of by human beings are repositories of an extraordinary number of beliefs that assign an object all sorts of perceptual and theory-laden properties, including dispositional properties. Not all beliefs play a role. The important beliefs here are those that are used to explain recognition by sight and color; these need not be linguistically expressed. The organism that recognizes something that it has seen before, and does so at least in part because of "its" constant colors, can be said to believe that that object has those colors during times when it is not looking at it. So there is good reason to introduce colored objects that are independent of perceivers, perceptually accessible, and that have color properties relatively independently of time. These objects are genuinely colored; we project the colors we experience into them by constructing them. But because they purport to be outside ourselves *and* genuinely colored, these things do not exist. That does not stop them serving as reliable perceptual guides to the things that do exist – scientific objects with SSRs.

Thus, the neural subjectivist can explain the first principle of the logic of color and colored objects and have genuine colors (not just SSRs) track common sense objects, so long as it is understood that there really are no common sense objects, and that the colors that they appear to have "out there" are really borne by neural events in the head.

McGill University

NOTES

[1] To say that SSRs are dispositions is not to say that they are dispositions to produce color experiences. They are dispositions of surfaces (with various structural properties) to reflect radiant energy of certain wavelengths in various ways. A particular physical surface might, for instance, reflect 73% of light at 473 nm., 22% at 501 nm., and so on for each wavelength at which it reflects at all. This property is instrumentally and objectively determinable. It has nothing to do with observers.

Incidentally, while I do not discuss dispositional accounts of color here, they, like the SSR account, preserve the common sense view that colors are persisting properties of things outside ourselves. But while dispositional accounts come in various forms, they all have at least one liability: they specify the dispositional property of the thing outside that has the disposition in terms of the perceived color, rather than (as with the SSR approach) by identifying it with some instrumentally determinable, genuine property of the object.

[2] Austen Clark's *Sensory Qualities* (1993) explains how this is done and offers an interesting defence of the idea that scaling sensory qualities and constructing sensory spaces provide a kind of "objective phenomenology" that answers Nagel's challenge to find a way of describing alien sensory spaces, such as those of bats.

[3] Neural subjectivism is not particularly well-advanced yet because there is still so much to learn about the brain, even in the comparatively well-developed domain of vision. Nothing that has already been discovered, however, is inconsistent with the view of neural subjectivism that I present, and opponency *does* appear to be well on the way to being explained.

[4] I read him this way in 'To Color' (1983). Land claimed that the immediate locus of color vision is in what he called "lightness *sensations*" in (1977). Those works of Land's that appeared after Marr's did not change their orientation in this regard.

[5] Color vision can and does serve several cognitive functions. Recognition is one of the most prominent. See in this regard Hardin, 1992. The role of color in recognition is displayed in a quite spectacular way in recent work that has found that there is highly organized visually-directed columnar organization in the inferotemporal cortex, part of the higher associative cortex (Fujita, Tanaka, Ito, Cheng, 1992); here colors, detailed shapes, and relative positions of features all combine to aid a creature's capacity to distinguish, etc.

[6] These features of colors (experienced by human beings) are largely accepted by the scientific community that deals with colors. There is considerable experimental evidence from neuropsychology, psychophysics, and perhaps even Berlin and Kay's analysis of color terms to support them. Moreover, they are readily demonstrable to and by anyone with a set of color samples and/or relevantly controllable monochromatic light sources.

It is difficult to find the right word to describe these features of colors as experienced by us. I suggest that they be thought of as conceptually necessary intrinsic properties.

[7] While I argue that common sense objects are fictions, I do not argue as some phenomenalists have done that they are epistemically secondary. They are epistemically prior, although they are constructed. The construction of the common sense object I outline below does not depend on anything like "sense data".

[8] Boghossian and Velleman (1989) defend a projectivist account of color that is "potentially neutral on the metaphysics of mind". I confess to finding it difficult to see how an account of color can be neutral in this way. And unlike B and V, I hold that one can truly say that common sense objects are colored (although this is not strictly true). In any case, what follows is an effort to explain projection assuming that neural subjectivism is right. A more detailed account is found in my 'Constant Colors in the Head' (1994). An early effort that ignores the realist assumptions of most computationalists is found in 'To Color' (1983).

[9] Constructing objects is not just a philosophical exercise. It is something that the neural system does. Cognitive psychologists offer various models that purport to explain how this construction proceeds. An interesting version of one that deals with the role of colors in particular in object construction is found in (Davidoff, 1991). Davidoff is primarily concerned to show that color "processing" is not totally modular (informationally encapsulated), but his argument also amounts to an effort to show how color contributes to the construction of object-(re)presentations.

[10] I assume that it will eventually be possible to specify, for a particular class of organisms, and in principle for a single organism, the neural event that is identical with a given case of a phenomenally characterized object. "Token identity" specification is of course possible in principle now. But there is no reason – given the way in which the sciences in question work – that we should exclude useful forms of event type descriptions that will specify an event type as identical with, for example, a relatively fine-grained coloring-surfacing (for a class of normally functioning individuals of the relevant species). When we have these type identities, it will no doubt turn out that in practice a particular individual's event tokens prove interesting only if they appear to be aberrant.

[11] I know of no reason to think that only humans are capable of the kind of "reasoning" involved in the construction of perceptual objects. Non-descriptive reference is assumed so long as it makes sense for the functional argument to go through at all, and the hypothetical "reasoning" involved is only that which is sufficient to yield an assumption that organism x sees what another organism y does. Murphy, not known for his technical philosophical skills, can evidently do that: he has no difficulty realizing that Maggie is watching and has designs.

REFERENCES

Boghossian, P. A. and Velleman, J. D., 1989, 'Colour as a Secondary Quality', *Mind* **98**, 81–103.

Brainerd, D. H. and Wandell, B. A., 1986, 'An Analysis of the Retinex Theory of Color', *Journal of the Optical Society of America A* **3**, 1651–1661.

Davidoff, J., 1991, *Cognition Through Color*, MIT Press, Cambridge.

Dretske, F. I., 1981, *Knowledge and the Flow of Information*, MIT Press, Cambridge, MA.

D'Zmura, M. and Lennie, P., 1986, 'Mechanisms of Color Constance', *Journal of the Optical Society of America A* **3**, 1662–1672.

Fujita, I., Tanaka, K., Ito, M., and Cheng K., 1992, 'Columns for Visual Features of Objects in Monkey Inferotemporal Cortex', *Nature* **360**, 343–347.

Hardin, C. L., 1988, *Color for Philosophers: Unweaving the Rainbow*, Hackett Publishing Co., Indianapolis.

Hilbert, D., 1987, *Color and Color Perception: A Study in Anthropocentric Realism*, Center for the Study of Language and Information, Stanford.

Hilbert, D., 1992, 'What Is Color Vision', *Philosophical Studies* **68**, 351–370.

Jackson, F., 1982, 'Epiphenomenal Qualia', *Philosophical Quarterly* **32**, 127–136.

Jackson, F., 1986, 'What Mary Didn't Know', *Journal of Philosophy* **83**, 291–295.

Jackson, F. and Pargetter, R., 1987, 'An Objectivist's Guide to Subjectivism about Colour', *Revue Internationale de Philosophie* **160**, 127–141.

Johnston, M., 1992, 'How to Speak of the Colors', *Philosophical Studies* **68**, 221–263.

Land, E. H., 1977, 'The Retinex Theory of Color Vision', *Scientific American* **237**, 108–128.

Land, E. H., 1983, 'Recent Advances in Retinex Theory', in Ottoson and Zeki (eds.), *Central and Peripheral Mechanisms of Colour Vision*, Macmillan, London.

Land, E. H., 1986, 'An Alternative Technique for the Computation of the Designator in the Retinex Theory of Color Vision', *Proceedings of the National Academy of Sciences U.S.A.* **83**, 3078–3080.

Maloney, L. T. and Wandell, B. A., 1986, 'Color Constancy: A Method for Recovering Surface Spectral Reflectance', *Journal of the Optical Society of America A* **3**, 29–33.

Marr, D., 1983, *Vision*, Freeman, San Francisco.

Matthen, M., 1988, 'Biological Functions and Perceptual Content', *Journal of Philosophy* **5**(1), 5–27.

McGilvray, J., 1983, 'To Color', *Synthese* **54**, 37–70.

McGilvray, J., 1992, 'Colors Really Are Only in the Head' (peer commentary on Thompson *et al.* (1992)), *Behavioral and Brain Sciences* **15**, 48–49.

McGilvray, J., 1994, 'Constant Colors in the Head', *Synthese* **100**, 197–239.

Nagel, T., 1974, 'What Is It Like to be a Bat?', *Philosophical Review* **83**, 435–450.

Nagel, T., 1986, *The View from Nowhere*, Oxford University Press, Oxford.

Sellars, W., 1960, 'Philosophy and the Scientific Image of Man', in *Science, Perception, and Reality*, Humanities Press.

Thompson, E., 1992, 'Novel Colours', *Philosophical Studies* **68**, 321–349.

Thompson, E., Palacios, A., and Varela, F., 1992, 'Ways of Coloring: Comparative Color Vision as a Case Study for Cognitive Science', *Behavioral and Brain Sciences* **15**(1), 1–26.

Westphal, J., 1987, *Colour: Some Philosophical Problems from Wittgenstein*, Basil Blackwell, Oxford.

Wittgenstein, L., 1977, *Remarks on Colour* (ed. Anscombe, tr. McAlister and Schättle), University of California Press, Berkeley and Los Angeles.

PAUL BERNIER

WHY MARR'S THEORY OF VISION
IS NOT ANTI-INDIVIDUALIST

INTRODUCTION

Tyler Burge has argued that Marr's theory of vision is anti-individualist.[1] If correct, his argument would show that visual states, as they are characterized in that theory, do not supervene locally on inner non-intentional states of the subject and, thus, that individualism is false in Marr's theory. As it is well known, according to mental local supervenience, if two psychological subjects share all their non-intentional inner physical properties – that is if they are so-called physical twins – then they also share all their mental properties; or, in other words, if two subjects are identical in all inner physical respects, then they cannot fail to be identical in all mental respects. Burge's argument rests essentially on two general assumptions. First, it is assumed that on Marr's theory the physical environment imposes constraints on the individuation of representational primitives used in what is called the theory of computation.[2] Secondly, it must be possible to produce a thought experiment showing that physical twins, or *doppelgängers*, inhabiting relevantly different physical environments would have visual states carrying different informational content. It is the second of these two general assumptions that I want to question.

Two crucial premises of Burge's argument rest on these two assumptions. My purpose is to outline a conflicting tension in these two premises. My main claim will be that even if we accept the first of these two general assumptions, anti-individualism does not follow. Roughly stated, my argument will be as follows. On the one hand, although it is possible to imagine a thought experiment – namely a standard Twin-earth case – involving different physical conditions, this difference in the external physical conditions does not entail a difference in the visual content of the physical twins, insofar as we take into account the general assumptions of Marr's theory of vision concerning the role of the physical environment in constraining the theory of computation. On the other hand, I will argue that if we are to imagine a counterfactual situation in which the physical conditions are relevantly different, namely a situation which would induce a difference in the visual content of a human-like inhabitant of that world, then it would be very implausible that a *doppelgänger* could inhabit such a world. In other words, my general claim is that Burge's argument is confronted with a dilemma: Either (i) we can imagine a counterfactual environment regularly causing the same non-intentional physical regularities, but then this counterfactual situation would not be visually different from ours, since vision in that environment would be constrained by the same general assumptions as those that constrain the theory of vision on earth, or

99

M. Marion and R. S. Cohen (eds.), Québec Studies in the Philosophy of Science II, 99–111.
© 1996 *Kluwer Academic Publishers.*

(ii) if we are to imagine a counterfactual situation that is different from ours, in such a way that a visual system in that environment would have to operate on different general assumptions about the physical environment, then this would go against the hypothesis of the existence of a *doppelgänger* in that environment.

That is, schematically, how my argument will proceed. However, before I can formulate it in detail it is important to clarify the role of the physical environment in constraining the theory of computation, since this is obviously what motivates an anti-individualist interpretation of Marr's theory.

1. PHYSICAL ASSUMPTIONS IN THE THEORY OF COMPUTATION

First, we must acknowledge that, on Marr's theory, it is quite clear that the visual structure of the world does impose important constraints on the theory of computation.[3] Thus, it comes as no surprise that one may be tempted to argue for an anti-individualist interpretation of that theory, as Burge has done. It is this claim about the role of the visual structure of the world that constitutes the first general assumption on which Burge's argument rests. Generally speaking, it is quite clear that Marr's theory is *success oriented*, in the sense that it aims at describing what is *correct* vision for an organism that is visually adapted to his or her environment. This very general constraint seems quite relevant to the theory of computation, which aims at solving the following computational problem: given some gray arrays (or matrix of intensity values) as (proximal) input, what are the computational processes that will eventually produce three-dimensional descriptions of the distal scene, as output? Given the poverty of the proximal stimulus, if we want to understand how the visual system can produce 3 D representations which, under normal conditions, are successful representations of the distal scene (i.e. the external visual environment), the theorist must first say in which representational format the retinal image is to be represented at the early stage of the computational process. Then he must say what are the representations that are to be obtained at the next stage and how they are to be obtained – that is in accord with which processes. In other words, solving the computational problem consists in saying how the image is to be represented and processed, at different stages, in order to eventually produce a 3 D representation of the distal scene. The hypothesized representations and computational processes defined over these representations must be such that at each stage, they gradually recover more and more information about the distal scene. Marr notes:

In the theory of visual processes, the underlying task is to reliably derive properties of the world from images of it; the business of isolating constraints that are both powerful enough to allow a process to be defined and generally true of the world is a central theme of our inquiry.[4]

The purpose of these representations is to provide useful descriptions of aspects of the real world. The structure of the real world therefore plays an important role in determining both the nature of the representations that are used and the nature of the processes that derive and maintain them.[5]

For the purpose of the present discussion, we don't need to go very deeply into the details of the theory of computation, it suffices only to recall that it divides in four stages, which are also divided in further sub-stages. The first stage corresponds to the formation of the retinal image. The second consists in the construction of the primal sketch from the retinal image. It is at that stage that the theorist is first faced with the crucial task of saying how the retinal image is to be represented, at the most basic level of representation. The third stage is the construction of what is called a $2^1/_2$ D sketch, and the last is the construction of a 3 D sketch.[6] The important point to note is that at each stage the visual representations and the processes must be construed in such a way that they recover some visual information about the distal scene, and in such a way that once this information is recovered it can be used at the next stage. For instance, the representational primitives and the processes must be such that the representations they produce contain (and carry) information about the geometry of the scene, the reflectances of surfaces, the illumination of the scene and the fact that the scene is viewer-centered.[7] This makes it clear that if the system is to produce 3 D representations of the distal scene which, under normal conditions, constitute correct vision, the theorist must make what Marr calls physical assumptions.[8] In other words, given the poverty of the proximal stimulus, the theorist must make some assumptions about the visual properties of the "real world", which ought to be represented at each stage. As Marr notes: "An important part of the theoretical analysis is to make explicit the physical constraints and assumptions that have been used in the design of the representations and processes."[9] And, since these constraints are essential to the individuation of visual states, we understand how one can be led to endorse an anti-individualist interpretation of that theory. As I noted, the fact that the theory of computation must be constrained by such physical assumptions is crucial to Burge's argument. However, I will argue that it does not support his anti-individualist conclusion.

We may recall some physical assumptions, suggested by Marr as imposing constraints on the representations and processes involved in early vision.[10] For example: (i) "the visual world can be regarded as being composed of smooth surfaces having reflectance functions",[11] (ii) "the spatial organization of a surface's reflectance function is often generated by a number of different processes, each operating at a different scale",[12] (iii) "the items generated on a given surface by a reflectance generating process acting at a given scale tend to be more similar to one another in their size, local contrast, color, and spatial organization, than to other items on that surface."[13] Burge illustrates one of these assumptions, namely the constraint of spatial localization which is crucial in the representation of some features of the image as edges. If the various inputs fed from different channels are to be processed to produce the visual representation of an edge, we must assume that "things in the world that give rise to intensity changes in the image, such as changes of illumination [. . .] or changes in surface reflectance [. . .] are spatially localized, not scattered and not made up of waves."[14]

Thus, it seems quite clear that such physical assumptions do impose some constraints on the individuation of visual states. And since these physical assumptions refer explicitly to some external visual properties, such as spatial localization of objects or the fact that distal scenes are composed of smooth surfaces, it seems that we must accept the first general assumption of Burge's argument, viz. that the physical environment does impose some constraints on the individuation of representational primitives in the theory of computation. Thus I agree with Burge when he notes that "the method is [. . .] to identify general physical conditions to motivate constraints on the form of the process that, when satisfied, will allow the process to be interpreted as providing reliable representations of the physical environment."[15] The fact that representational content on Marr's theory is so closely linked to such physical assumptions is particularly relevant to the discussion of Burge's argument, because the argument, as we are about to see, rests essentially on the assumption that different physical conditions may produce the same internal neurophysiological states non-intentionally described, which would realize visual states with different representational content, given the difference in the physical conditions. But this second general assumption of Burge's argument raises the following question: In which sense must the physical conditions be different in order to produce visual states that have different informational content, on Marr's theory?

2. WHY BURGE'S ARGUMENT MUST BE REJECTED

Burge's argument is as follows:

(1)   The theory is intentional.

(2)   The intentional primitives of the theory and the information they carry are individuated by reference to contingently existing physical items or conditions by which they are normally caused and to which they normally apply.

(3)   So if these physical conditions and, possibly, attendant physical laws were regularly different, the information conveyed to the subject and the intentional content of his or her visual representations would be different.

(4)   It is not incoherent to conceive of relevantly different physical conditions and perhaps relevantly different (say, optical) laws regularly causing the same non-intentionally, individualistically individuated physical regularities in the subject's eyes and nervous system. It is enough if the differences are small; they need not be wholesale.

(5)   In such a case (by (3)) the individual's visual representations would carry different information and have different representational content, though the person's whole non-intentional physical history (at least up to a certain time) might remain the same.

(6)   Assuming that some perceptual states are identified in the theory
        in terms of their informational or intentional content, it follows that
        individualism is not true for the theory of vision.[16]

My discussion will focus on premises (3) and (4) which are crucial to the
argument.[17] Premise (3) rests quite clearly on some considerations that I have
outlined in the previous section. The following will be central to my objection: If "physical conditions were regularly different" so as to "convey different
information to the subject" as it is stated in premise (3), this would have to
be because these different conditions would be such that a human-like visual
system operating in such conditions would have to be described by making
different physical assumptions about the visual features of the environment.
For instance, given the assumption of the poverty of proximal stimulus, and
the fact that the theory is success oriented, the content of visual representations depends essentially on the general visual properties which are involved
in the physical assumptions made by the theorist to constrain the different
stages of the theory of computation. Thus, if a human-like creature would
inhabit a world in which some of the physical assumptions about our environment would be clearly false, then if we were to make a Marrian theory
of that creature's visual system, that theory should be constrained by assumptions which are different from the ones imposed on the theory of human
vision. We can thus understand what it means to hold that if the physical
conditions were different, the visual states of the subject would carry different information. For example, if the counterfactual environment was
such that things giving rise to intensity changes in the retinal image were
not spatially localized, contrary to a physical assumption we make in our environment in characterizing some representational primitives, then the theory
of computation would clearly have to be constrained differently than the theory
of *human* vision. If the physical conditions were different enough, in that sense,
we could clearly accept that such a difference in physical conditions entails
a difference in visual content. But it seems that it is only in that sense that
one can claim that a difference in the physical conditions would entail a
difference in visual content.

Thus, it is not my contention to show that premise (3) is necessarily false
– and neither that premise (4) is. What I just noted indicates in which sense
premise (3) could be true. However, I will argue that these two crucial premises
are not necessarily true, and that it seems implausible that they can be jointly
true, for the following reason. On the one hand, a standard Twin-earth case,
which is a likely example to support premise (4), would make premise (3) false
since the different physical conditions would not make any difference in the
physical assumptions constraining the theory of computation, and thus these
differences in physical conditions would not entail a difference in visual
content, on Marr's theory. On the other hand, an interpretation on which
(3) is true, with respect to visual content in Marr's theory, makes premise
(4) rather implausible, because it becomes very hard to imagine a coherent

thought experiment to support premise (4), because the latter must be inter-preted relatively to physical conditions that would entail a difference in visual content, on Marr's theory, namely physical conditions that would violate some of the physical assumptions made to construe the representational primitives, and it seems rather implausible that such conditions could cause "the same non-intentionally, individualistically individuated physical regu-larities in the subject's eyes and nervous system."

Premise (4) makes it quite clear that Burge's argument must rest on a thought experiment similar to the various ones we have been accustomed to in recent philosophy of mind.[18] But it seems that a thought experiment that would meet the ends of the argument would be very implausible, since the counter-factual situation would have to be quite different from the actual one, and it is not clear that it makes sense to make the metaphysical assumption that a *doppelgänger* could inhabit such a world. I will try to substantiate this claim, but before let us first consider a standard Twin-earth case, for which it is indeed plausible to imagine that a difference in the physical conditions would cause the same non-intentionally, individualistically individuated physical regulari-ties in the subject's eye and nervous system. As it will turn out, these conditions are not different in the required sense, since they would be perfectly compatible with the physical assumptions constraining the theory of computation on earth and thus such a difference in the physical conditions would not entail a difference in visual content, on Marr's theory, contrary to what is expressed by premise (3). Although that is not sufficient by itself to refute Burge's argument, it is useful since it underscores some general conditions that must be met by a thought experiment that would give support to premise (4).

First, we assume that Oscar is located on earth and that he has a visual state caused by the distal scene of a pond of water (namely a pond of H_2O). Next, we assume that there exists a planet (Twin-earth) which is identical to earth in all respects, except for the fact that the stuff which is very much like water (call it "twater") – it fills the ocean of Twin-earth, it pours down on Twin-earth from clouds, etc. – is not in fact H_2O but a complex chemical (abbreviated as XYZ). Moreover, we suppose that on Twin-earth there exists a *doppelgänger* of Oscar (call him Twin-Oscar), who is a molecule-for-molecule twin of Oscar and who has a visual state caused by the distal scene of a pond of twater (namely a pond of XYZ). We may further assume that the visual system of Twin-Oscar and its visual states would be correctly char-acterized within the general approach of Marr's theory of vision.

Given the hypothesis of neurophysiological identity between Oscar and his twin, their respective proximal stimuli (or gray arrays) can be described as being of the same physical type. Now, the question is whether we should conclude that their visual states have a different content, given the differ-ence between their respective environments. But it is quite clear that to give a positive answer, the difference in content would need to show up in the fact that the twins' visual systems – more specifically the theory of compu-tation of these systems – would have to be constrained by different physical assumptions. But clearly there is no reason at all for that. Earth and Twin-earth being so similar and particularly visually similar, the general assumptions about the structure of our environment would certainly be true of Twin-earth, simply because the physical assumptions used to describe Oscar's visual system are very general assumptions about the visual *structure* of our environment, and not about the "deep nature" of objects in the environment. In other words, the physical assumptions are about general visual properties of the world and not about the nature of particular physical items instantiating these properties. And, *ex hypothesis*, the objects on earth and Twin-earth – particularly water and twater – do share all the same visual properties, even though they are different natural kinds. This shows that there is a plausible interpretation on which the physical conditions would be different, but on which the information conveyed to the subject, and the intentional content of his or her visual representations, would not be different, namely an interpretation on which premise (3) is false. It is important to note that in the foregoing argument, I have used a quite standard and quite plausible thought experiment, in which the differences between the actual and counterfactual situations are minimal. Keeping the differences as minimal as possible is quite important, because that is precisely what gives some plausibility to the hypothesis of the physical identity of the twins. And as far as I can tell it is doubtful that this hypothesis would remain plausible if the physical conditions in the counterfactual situation were different enough to entail a difference in visual content, on Marr's theory.

As I have noted, premise (4) assumes that it is possible to imagine a plausible thought experiment. But a standard Twin-earth case will not do, since the difference between earth and Twin-earth would not bring about a difference in visual content.[19] We are now in a position to see which kind of thought experiment – that is, which kind of differences in the physical environments – is required for Burge's argument. The required thought experiment would need to invoke a counterfactual situation which is much more different from the actual one, than Twin-earth is. The counterfactual situation must be different enough, so that the visual system of a human-like organism, in that environment, would have to be constrained by different physical assumptions. The important point is that premise (4) cannot be assumed to be simply a quite general metaphysical truth, stating that different conditions can cause the same non-intentionally described physical states. The problem lies in the fact that the different physical conditions must be *relevantly* different, in a sense that is constrained by how the representational primitives get their informational content, in Marr's theory, namely by the physical assumptions the theorist makes in positing them. This point is very important because it shows that the interpretation of premise (4) is constrained by Marr's theory of vision, which after all is an empirical theory.

Let us now consider in more details what would make premise (3) true, on Marr's theory, since it is not necessarily true, as it can readily be seen from the standard Twin-earth case. For physical conditions to be relevantly different we must assume, of course, that these differences would bring about a difference in intentional content. But what does that mean on Marr's theory? Well, it simply means that the theory of computation of a human-like organism in such an environment would have to be characterized by imposing some constraints which are different from the constraints imposed on Oscar's system, for example. In other words, the physical conditions in the counterfactual situation must be such that the very general physical assumptions made by the theorist in his construction of the theory of computation for human vision, would not be correct if he were to describe adequate vision of a human-like organism in such a counterfactual environment. For example, we could imagine that in such an environment, objects are scattered and that the assumption of localization in space should be replaced to describe vision for the human-like creature. But, if it is so, then we see that the counterfactual environment must be quite different from ours, given that Marr's physical assumptions are very general assumptions about the visual *structure* of our world.

Now, my main claim is that if we are to imagine such a counterfactual situation, instantiating visual properties different from the general visual properties of the objects on earth, then it becomes quite implausible to assume that there could be a molecule-for-molecule duplicate of Oscar in such a world. Generally stated, my claim is simply that if the counterfactual environment must be different enough from ours, then it is likely that these differences would affect the physical (non-intentional) properties of the human-like creatures inhabiting such a world. For example, if the assumption of spatial localization does not hold in that world – that is, if objects were scattered – then it seems quite obvious that the putative *doppelgänger* would also have to satisfy such a constraint, namely he himself would be scattered. But then it becomes rather unclear how we could still hold that he is a duplicate of Oscar. Moreover, if we assume that the visual system of the human-like creature is constrained by different physical assumptions, because that creature has adapted to a physical environment having a visual structure different from the earth's visual structure, then it is very likely that these adaptive differences would show up in his physical make up, more precisely in his genetic make up. And that, once again, goes against the hypothesis that such a creature would be a duplicate of a human being (say, Oscar). Thus, for these reasons, it seems that a counterfactual situation that would satisfy premise (3) would make premise (4) very dubious.

The preceding remarks seem sufficient to reject Burge's fourth premise, but the following could be added as additional ground. First, we may recall that if a human-like creature (a putative twin of Oscar in a counterfactual situation), is to have a visual state with a content different from the content of Oscar's visual state, then we need to assume that the theory of computation and the representational primitives would be characterized by invoking different physical assumptions. Now, even if it seems implausible to hold that there could exist a duplicate of Oscar in such an environment, for the reasons I have just noted, we could still grant – for the sake of argument –

that Oscar and his putative twin would have the same retinal image at a certain time. But then, if we want to make sense of the claim that their visual states have different visual contents, we must assume that these physically identical gray arrays would be represented and transformed (stage by stage) in such a way that eventually they would produce different 3 D representations. And it is quite clear that to do that, each system's representational primitives and computational processes would have to be constrained by some different assumptions about the structure of the environment. But then we may wonder how these different processes could be realized by exactly the same (non-intentional) neurophysiological processes and exactly the same (non-intentional) algorithmic processes. In other words, if these processes are indeed constrained by different assumptions, this difference would have to show up, at some point, in the neurophysiological or algorithmic processes implementing them.

My point here is simply to insist that it would be rather surprising that two visual processes described computationally by invoking quite different general assumptions about the visual structure of the environment could be physically (and algorithmically) implemented in exactly the same way. The main reason is that the physical assumptions the theorist makes do not depend only on our pre-theoretical or commonsense intuitions about the visual structure of the world. These assumptions also depend on what we know about the psychophysics of the human visual system. If the theorist were to discover that the same neurophysiological processes which realize Oscar's visual processes also realize correct vision for a system located in an environment where quite different physical assumptions seem to hold, then it is hard to tell what he would do. But it is quite plausible that the theorist would want to revise the physical assumptions he made in the first place to constrain Oscar's visual system. Accepting the claim that the same neurophysiological (and algorithmic) states and processes could realize representational primitives and computational processes constrained by different physical assumptions, is like accepting that the physical assumptions in Marr's theory – for example, the ones describing Oscar – are not specific enough to say in which sense the neurophysiological processes are achieving the task of visually relating the system adequately to its environment. It seems to me that if our visual system, as it is described at the neurophysiological level, could also be visually adapted to an environment which violates some of the physical assumptions made in Marr's theory, that would simply indicate that the theorist was wrong in the first place in assuming that these particular physical assumptions are the ones that must be invoked to constrain the representational primitives and the computational processes in the theory.

For these different reasons, it seems that an interpretation that would make premise (3) true, on Marr's theory, would entail that premise (4) is at best dubious, but most probably false, because what would constitute *relevantly* different physical conditions, cannot be interpreted by abstracting away from how intentional primitives are characterized on Marr's theory.

Now, it could be objected that the second part of my argument – which suggests the implausibility of premise (4) on the assumption that (3) is correct – gives support to a form of externalism.[20] This may well be true, since what I have been arguing seems to suggest that, in some cases, differences in physical conditions would entail differences in intentional content. But this must be qualified, since I have also noted that a difference in the physical environment does not necessarily entail a difference in visual content, as can be seen from the standard Twin-earth case. Yet, I agree that my argument does entail a form of externalism, but that form of externalism seems innocuous with respect to individualism, since it would be a rather weak form of externalism that is compatible with the thesis of local supervenience of mental states on internal non-intentional physical and functional states.

3. A POSSIBLE REJOINDER

One may agree with the first part of my argument and concede that a standard Twin-earth case will not meet Burge's need, since it is a case where premise (3) is false. But one may try to resist the second part of my argument, by insisting that there is indeed a reading of premise (4) which is consistent with (i) the assumption of the identity of the twins and (ii) a correct interpretation of "relevantly different physical conditions" in that premise. To do that we could imagine, for instance, that there is a counterfactual situation (say Earth-3) where at least one of the general physical assumptions constraining the theory of human vision (on earth) does not hold. (To illustrate this point in a simple way, we can think that objects are scattered on Earth-3.) Then, we would only have to imagine that Oscar is transported to that bizarre environment. In such a case, we could hold that Oscar's visual system is not visually adapted to Earth-3. We could even suppose that Oscar could not survive very long there. All we would need to suppose is that Oscar can be there long enough to have an internal physical state non-intentionally individualistically individuated, which is identical to his visual state in front of a pond of water, back on earth. Suppose that this state of Oscar occurs when he is confronted with some strange phenomenon on Earth-3. In such a case, which one might be tempted to describe as a case of visual misrepresentation, one could insist that Oscar's visual states, in the two situations, do not have the same visual content. Though I find such a thought experiment hard to swallow, it seems that even if we were to accept it, we would still be left unable to say that Oscar's visual state on Earth-3 has a different content from his visual state back on earth, unless we are willing to beg the question at stake.

What is peculiar about this case is that although the visual structure of Earth-3 is such that the physical assumptions made to characterize Oscar's system (on earth) would not be adequate to constrain a (human-like) visual system adapted to Earth-3, yet we must assume that Oscar's visual system is still constrained, on Earth-3, by the same physical assumptions. This is precisely why Oscar would not be adapted to Earth-3. But then how could we hold that Oscar's visual state (on Earth-3) has a different content from Oscar's? This possibility seems to be blocked. One might want to say that it is different since it has been caused by some weird stuff that has visual properties different from the visual properties shared by water and twater. But that would beg the very question at stake, since to give that answer we must assume that the content of Oscar's visual state is determined by its particular distal cause (on Earth-3), which is precisely what is at stake.

The problem is that in such a case we assume that Oscar's visual system is still operating by respecting constraints and physical assumptions which are established relatively to the visual structure of earth. Moreover, we assume that Oscar has the same non-intentionally described internal states, in both situations. But then it seems that there is nothing left on which the difference in content could rest, on Marr's theory. And, as I have noted, if one insists that the difference must rest on the fact that Oscar's state, on Earth-3, is not caused by its normal distal cause, this would simply beg the question. It seems that the only way to resist my present objection, would be to insist on the fact that Oscar's visual state while it is visually adapted to earth is not adapted to Earth-3. But to me, this only indicates that such a case is what can be called a limiting case showing, in another way, why it is not very plausible to imagine that a twin could inhabit a counterfactual world that meets the needs of premise (3). What enables us to have the intuition that Oscar's visual state would have a different content, on Earth-3, is either that Oscar would perish, in the long run, or that he would adapt to the environment. But in either of these cases, Oscar could hardly be considered to be physically identical to himself, back on earth. Thus, it seems that even if we were to accept that thought experiment, it would not suffice to defend the conjunction of premises (3) and (4).

Let me recapitulate the main points of my discussion. Burge's argument needs to rest on a thought experiment. In the first part of my discussion I have argued that on a standard Twin-earth case, premise (3) of Burge's argument must be rejected. In the second part of my discussion, I have also argued that on an interpretation that would make premise (3) true, according to Marr's theory, premise (4) loses its plausibility and there is no reason why it should be endorsed. Though the latter assumes that differences in the environments may bring about differences in the contents of visual states, this hardly supports the anti-individualist conclusion, since it is most likely that such differences would also bring about differences in the relevant internal properties of the visual subject. For these reasons, it seems that we must admit that Marr's theory of vision is not anti-individualist. On the contrary, it seems to be compatible with the thesis of individualism, insofar as the latter is interpreted as the thesis of local supervenience.[21]

City University of New York

NOTES

[1] Cf. 'Individualism and Psychology', *Philosophical Review* 95 (1986), 3–45. As Burge notes, what he calls "Marr's theory of vision" is the result of many diverse contributions (see note 14, p. 27). For a general account of that theory, see D. Marr, *Vision*, New York, Freeman (1982).

[2] The theory of computation is one of the three levels of explanation in Marr's general approach (see Marr, *op. cit.*, pp. 22–25). That level aims at explaining *what* the visual system does and *why*, namely to derive three-dimensional representations of the distal scene from the retinal image. As Marr notes: "Here lay a way to formulate [the] purpose [of human vision] – building a description of the shapes and positions of things from images." (*Op. cit.*, p. 36) The other levels of explanation are the following: (a) the theory of algorithm, which aims at explaining how the theory of computation can be formally described: "[W]hat is the representation for the input and the output and what is the algorithm for the transformation" (*op. cit.*, p. 24). That level is concerned with selecting the adequate formalism to represent the computational processes described at the upper level, and with characterizing the algorithms that implement these processes; (b) the theory of the neurophysiological realization of the representations and processes: "[A]t the other extreme are the details of how the algorithm and representation are realized physically." (*Ibid.*)

[3] On that score I disagree with Frances Egan's objection to Burge's argument according to which Marr's theory is not intentional. See F. Egan 'Must Psychology be Individualistic', *Philosophical Review* 100 (1991), 179–203, and 'Individualism, Computation and Perceptual Content', *Mind* 101 (1992), 443–459. Egan denies that visual states are individuated essentially in relation to some objective visual properties. My main reason to disagree with that view is that Marr insists over and over on the importance of imposing physical assumptions and constraints when facing the task of saying how the image is to be represented, and these physical assumptions make explicit reference to objective visual properties of the distal scene. See note 17, for more on Egan's view.

[4] *Op. cit.*, p. 23.

[5] *Op. cit.*, p. 43.

[6] *Op. cit.*, pp. 36–38.

[7] *Op. cit.*, pp. 41–42.

[8] *Op. cit.*, pp. 44–51.

[9] *Op. cit.*, p. 43.

[10] *Op. cit.*, pp. 44–51.

[11] *Op. cit.*, p. 44.

[12] *Op. cit.*, p. 46.

[13] *Op. cit.*, p. 47.

[14] *Op. cit.*, p. 30.

[15] *Op. cit.*, p. 32.

[16] Cf. Burge, *op. cit.*, p. 34. G. Segal has objected to this argument, by insisting on the fact that visual states must be individuated in relation to the discriminative abilities of the subject. On his view, the visual states of twins in relevantly different physical conditions would have the same content, since they would have the same discriminative abilities. Cf. 'Seeing What is Not There', *Philosophical Review* 48 (1989), 189–214. My own objection is different. It focuses on the role of the so-called physical assumptions in determining visual content.

[17] Frances Egan (*op. cit.* (1991), pp. 195–202) has objected to premise (1). She claims that it is not the theory itself that is intentional, but only its models, and that the relations to our environment in individuating these states are not constitutive of these states, but only have a heuristic value. This is not the place to discuss Egan's view in detail. But let me only point out that although I think Egan is probably right to resist Burge's claim that the first premise is "sufficiently evident" since "the top levels of the theory [namely the theory of computation] are formulated in intentional terms" (*ibid.*, p. 196), insofar as Burge means that these primitives are individuated with respect to objective visual properties such as edges, bars, blobs,

terminations and the like. Nevertheless, I disagree with Egan if what she claims is that no objective visual properties, whatsoever, are involved in the individuation of visual states and of the representational primitives. For it seems clear that some general visual properties such as the fact that distal scenes are composed of spatially localized objects and of smooth surfaces, are essential to the individuation of visual states. Concerning the second premise, we can accept it as uncontentious, insofar as we do not consider "the reference to contingently existing physical items or conditions by which [intentional primitives] are normally caused" as constitutive of the intentional primitives. In other words, we can accept it as long as this reference is only epistemic and it does not entail that visual states have a deep individuative relation to "contingently existing physical items", otherwise this premise would obviously be question begging, as it has been noted by L. Shapiro in 'Content, Kinds, and Individualism in Marr's Theory of Vision', *Philosophical Review* **102**, 489–513.

[18]   For the more widely discussed of these thought experiments, see H. Putnam, 'The Meaning of "Meaning" ', in *Mind, Language and Reality. Philosophical Papers, Vol. 2*, Cambridge University Press (1975), 215–271.

[19]   At this stage one may be tempted to object that Burge did present a thought experiment that would do the trick, namely his crack/shadow thought experiment (see Burge, *op. cit.*, pp. 41–43). I will not discuss this thought experiment here, since I think that the conclusion of my discussion of the standard Twin-earth case would also apply to that thought-experiment. Let me only mention that this thought-experiment involves a counterfactual situation that differs from the actual one only in the fact that there are no shadows in that environment, but there are only cracks. Where in the actual situation there are shadows and where in the actual situation there are cracks, in the counterfactual one there are only cracks. But there seems to be no reason at all to claim that different physical assumptions hold in that counterfactual situation. It is implicit in Burge's example that the cracks the putative twin would perceive, in the counterfactual situation, instantiate the very same visual properties as the cracks in the actual situation. Thus it's hard to see how one could insist that different physical assumptions hold in such an environment.

[20]   The following point addresses a problem briefly suggested by Burge. He notes that "to reject this step [premise (4)] would be self-defeating for the individualist." (*Op. cit.*, p. 35) But even though what I have argued with respect to premise (4) may entail a weak form of externalism, it would still be compatible with individualism, insofar as the latter is simply interpreted as the claim that mental states do supervene locally on non-relational physical properties of the subject. Thus denying premise (4) is not self-defeating for the individualist, as long as he restricts his claim to local supervenience. See L. Shapiro, *op. cit.*, for a discussion of a different interpretation of individualism.

[21]   This paper was made possible by a postdoctoral fellowship from the Social Sciences and Humanities Research Council of Canada (no. 756-93-0039). I have had many useful comments on earlier drafts. Special thanks to Frances Egan, Daniel Laurier, David Rosenthal, Stephen Schiffer and Michel Seymour.

MICHEL SEYMOUR

"THREE THOUGHT EXPERIMENTS REVISITED"

I. INTRODUCTION

I would like to reexamine the thought experiments that are very often invoked by those who argue for anti-individualism. I will try to reevaluate those experiments by interpreting the statements that they purport to elucidate within a general semantical framework which accounts for the meaning of sentences in terms of assertability conditions.

First, it is important to show that there are "three grades" of anti-individualistic involvement. There are experiments that can be used to illustrate what is very often labelled as "externalism". With these, philosophers try to prove that the physical environment determines the nature of mental content. This is very often seen as the minimal version of anti-individualism. A second grade is arrived at as soon as an experiment involves an attempt to show that the sociolinguistic environment can also be described as playing a major role in the determination of mental content. It is argued that a change in the linguistic conventions of a community will entail changes in the contents of thought entertained by the members of the community. Finally, there is a final grade involved in experiments invoked by Ludwig Wittgenstein in his sceptical considerations on rule-following. These can be used in the course of defending a community view of language.

What are we to say about these different anti-individualistic conclusions, once it is noticed that they all rely on such thought experiments?[1] One might wonder whether the thought experiments are by themselves sufficient to support the arguments that philosophers want to elaborate on their basis. I would like to show that the adoption of a semantics of assertability conditions must allow for many different uses of propositional attitude sentences[2] and that this forces us to relativize in different ways the results obtained by each version of anti-individualism.

I would be inclined to admit that there are uses of expressions such as "water", "arthritis", "sofa", etc., that enable us to draw very different conclusions from those that were reached by anti-individualist philosophers. When we recognize the variety of these uses, we are led to results that are quite different and that seem to lend a little more plausibility to the opposite view. This is especially the case for the first grade of anti-individualism. Of course, these "conflicting results" do not refute those obtained by anti-individualist philosophers, but we now have to relativize them in an important way.

Concerning the third grade of anti-individualism, the one involving a community view of language, it must also be admitted that the thought experiment on which it is based is precisely what provides justification for the adoption

113

M. Marion and R. S. Cohen (eds.), Québec Studies in the Philosophy of Science II, 113–138.
© 1996 Kluwer Academic Publishers.

of a semantics of assertability conditions. Now since I am invoking precisely such a semantic framework, it may look as though I should relativize the very claims that I am making. If we are to relativize the results of the thought experiments, shouldn't we relativize the result of the experiment that leads to the adoption of a semantics of assertability conditions? But I will conclude by showing that there are independent justifications for the framework that do not rest merely on *a priori* considerations. Ruth Millikan's work in bio-semantics is an attempt to cast such a semantic theory into an empiricist and naturalistic mold, and I will end the discussion by briefly considering her main theoretical assumptions. The community view of language and the semantical framework that goes with it must surely be amended in an important way if we are to follow Millikan's project, but its main assumptions remain valid. And the same kind of remarks apply to anti-individualism itself.

II. THREE GRADES OF ANTI-INDIVIDUALISTIC INVOLVEMENT

Let us first distinguish between those three grades of anti-individualism. The first thought experiments have been associated with the names of Hilary Putnam and Tyler Burge.[3] Putnam's thought experiment concerning brains in a vat and Burge's adaptation of Putnam's twin earth examples are good instances of the first kind of experiments. We cannot count the initial work of Putnam in 'The Meaning of Meaning' as providing a good example since, as it is well known, Putnam only meant to show that "meaning" was not in the head and his argument did not clearly entail anything concerning the environmental determination of mental states in general.[4] But it is not as though he had nothing to say concerning intentional states, quite the contrary. He assumed at the time that intentional states, as narrowly construed, could be individualistically individuated and construed as internal functional states. There is, according to the early Putnam, a way to characterize mental states in terms of narrow functional states only, and it was precisely this fact that enabled him to claim that meaning was not in the head. In the thought experiment, the physical environment is changing while the mental states are fixed.

It is only with the 'Brains in a Vat' chapter of *Reason Truth and History* that Putnam arrives at a claim concerning the nature of mental states in general. It now seems that mental states should always be understood as involving relational properties and that they can no longer be narrowly construed. Or at least it no longer seems plausible to postulate narrow contents that would not ultimately be individuated in terms of the relational properties that an agent entertains in a physical environment. Does it make sense to suppose that, contrary to what we think, our brains are not inextricably tied to the world? Putnam uses his causal theory of natural kind terms to show that the very formulation of that possibility is self refuting. We have to formulate the doubt by referring, as it were, from the outside to the postulated brain since causal connections are at the basis of the reference relation. This proves that if we are to make sense of the claim according to which we are brains in a vat, it must be false because it can only be formulated if it is assumed that we are able to entertain a causal relation with an external object.[5]

Burge arrived at similar conclusions through thought experiments that were developed along the lines of the earlier Putnam. We were invited to reconsider the twin earth examples and to wonder whether it made sense to fix the mental states of an agent while the external physical environment was changing. Burge argued against this claim. He also showed that we could fix the internal physical properties of the agent and allow for a change in her mental states. Since the distal properties of the world determine content and since content serves to individuate the very nature of the mental states, a difference in the physical environment could lead to a difference in the mental states themselves even if the internal physical states of the agents remained the same. Mental properties do not supervene upon the internal physical properties of the agents. According to Burge, this is true of propositional attitudes and of perceptual states in general.

So a first kind of anti-individualism is revealed by experiments that serve the purpose of showing that the individual's mental states or perceptual states are partly individuated in terms of the *physical* environment. With such a definition, anti-individualism amounts to externalism. Even if intentional states are not in themselves relations to the environment, they are at least in part individuated by reference to items in the physical environment. For example, a *de dicto* thought involving water is not to be analysed as involving only a direct or *de re* relation to water, whether it is direct acquaintance or another kind of direct relation. But one would never be in a *de dicto* state involving a thought concerning water if there weren't water in the physical environment.

Burge's initial contribution to the subject of the individuation of mental states was different however. In his famous first paper on the subject, he argued for the determination of the sociolinguistic environment upon mental contents.[6] His thought experiments purported to show that when the sociolinguistic community stipulates a change in the semantic rules of some of the words of the language, this has immediate consequences upon our characterizations of mental states. Different linguistic resources within the sociolinguistic community determine distinct characterizations of the mental contents themselves. This is at least true in the case of higher order propositional attitudes. The conceptual resources with which agents articulate their thoughts, beliefs, judgments, sayings and propositional knowledge are different when the linguistic conventions of the community are undergoing a change and, for this reason, mental contents themselves differ.

This is already a second kind of anti-individualism because we are no longer requiring any changes to occur in the environment apart from those that take place within the linguistic community. Semantic stipulations are what makes the difference. For our present purposes, it does not matter whether this thought experiment is understood as an independent result or as a particular application of the first. We could for instance interpret it as providing evidence only for a particular instance of externalism. According to this interpretation, mental contents can be determined by the physical environment in general, whether by "physical environment", we mean the distal properties of the physical objects or the linguistic behaviour of the community. In this sense, the first thought experiment would be more basic. This is the interpretation that Burge himself seems to favour.[7]

But we could also interpret his original thought experiment by resorting to phenomenological notions of the linguistic community, of linguistic conventions and semantical rules. In a phenomenological perspective, these notions have no ontological import and we only rely on common sense. When we interpret it that way, this second thought experiment becomes an independent, autonomous, result and can even be understood as more basic than the first.[8]

These diverging interpretations will not affect the arguments put forward in the present paper. We need not interpret the first grade as a minimal thesis on which it would then be possible to argue for a stronger thesis such as the one that postulates an influence on the part of the linguistic community. And neither do we need to understand the second grade as an independent result that would require just phenomenology and serve as a basis for externalism. It is only important to remember that it is in principle possible to defend these two views independently of one another. Indeed, the only thing important for my argument is that externalism and anti-individualism (in the second sense) *could* be argued for independently of one another.

Externalism is in principle compatible with a rejection of anti-individualism in the second sense. Davidson, for instance, is both a defender of externalism and a critic of anti-individualism. His individualistic inclination is an inescapable consequence that follows from his arguing simultaneously for a theory of radical interpretation and semantical holism, while understanding the principle of charity as an *a priori* constraint on radical interpretation. For Davidson, understanding amounts to interpreting and the latter involves maximizing agreement with others. The rough picture that emerges from Davidson's theorizing is that the radical interpreter is all by herself and cannot ever encounter divergence, error or even true intersubjective agreement. He can only map his already existing "conceptual scheme" in order to make sense of the beliefs, actions and meanings of others. This is what explains his individualism. And since he also rejects the third dogma of empiricism and refuses to distinguish between conceptual scheme and content (or if one prefers, between theory and experience), there is no room for a distinction between distal and proximal objects of experience. There is no such thing as a proximal object of experience. This is what explains his externalism.

Now from an opposite perspective, the admission of a determination of the linguistic community on content is as we shall see compatible with the existence of narrow contents, and therefore with a certain form of internalism. This could for instance be claimed concerning perceptual states. The very linguistic practices adopted by our community may justify categorizations in which our perceptual mental states are internally individuated. But since these categorizations are only relative to our communal practices, they are in some way determined by the sociolinguistic environment. Quine illustrates this second approach. Contrary to Davidson, he accepts the distinction between scheme and content and distinguishes between proximal and distal objects of experience. The mental states that are related to these proximal objects can therefore be categorized as involving narrow contents only. This is what explains Quine's internalism. But at the same time, the subordinate clauses used in the social practice of attitudinal ascriptions play for Quine an essential role in the individuation of mental states, as long as we think of mental states in the context of folk psychology. This thesis enables him to show the irreducible character of the intentional idiom and it is an important step in his argument for eliminationism. Whether or not we accept Quine's eliminationism is of no importance in the context of the present discussion. What is important is his claim that the resources of the linguistic community determine mental content. This is what illustrates Quine's anti-individualism.[9]

It may perhaps sound controversial to describe Quine's approach to folk psychology as anti-individualistic. But this impression may be attenuated if we think of anti-individualism as a doctrine which stipulates that if intentional states exist, then they are not individuated individualistically. It is true that, in a way, he does not subscribe to an anti-individualistic characterization of mental states as described in folk psychology since he denies the very existence of such entities. But he would surely accept the counterfactual conditional thesis according to which if these mental states had existed, then they would not have been individuated individualistically.

So one reaches a second grade of anti-individualism as soon as one argues that the community or social environment plays a major role in the individuation of content. It does not have to be interpreted as a particular instance of the first brand of anti-individualism. One important remark is that only these last thought experiments, and not the first, imply a rejection of methodological individualism. The first grade of anti-individualism is still compatible with the idea that all pertinent social phenomena can ultimately be reduced to properties of the individual psychology. It only implies that thought contents must be understood as relational properties, i.e. properties that involve a reference to items in the physical environment. The second brand of anti-individualism, however, implies that the sociolinguistic environment plays a major role in the individuation of one's thought and, in general, it is claimed that the sociolinguistic properties of the environment do not supervene upon the relational properties of the individual agent. They could supervene upon the physical environment itself, but not upon the properties of the individual, whether these are internal or relational properties. It could not then simultaneously be argued without a vicious circularity that all pertinent sociolinguistic phenomena can ultimately be analysed in terms of psychological properties of the individual.

A third grade of anti-individualism is reached as soon as one incorporates a view in which language is understood as essentially relative to communal practices. We are at this point no longer merely holding that sociolinguistic practices determine intentional content, for this is still compatible with the view that intentionality and language are two interdependent notions, and compatible with "intentionalism".[10] We are rather moving a step further and arguing for a view according to which semantic properties can be explained without recourse to intentionality. This third grade makes it even possible to explain intentionality partly in terms of language.

It is important to realize that one could argue for anti-individualism in the second sense while refusing to make this additional move. Margaret Gilbert, for instance, argues both in favour of methodological holism and intentionalism. She holds the view that certain collectivity concepts are irreducible notions, such as the notion of a "plural subject", and this is enough to reject methodological individualism. But she also thinks that social phenomena, whether it is language or even "plural subjects", do not exist independently of speaker's intentions. In the case of Davidson, the idea is that meaning and belief are two interdependent notions.

We shall see that Burge could not afford to argue both for intentionalism and the sociolinguistic determination over mental content. Or at least he could not do so in the attempt to argue for anti-individualism in the full blooded sense of the term. This is however not because the two theses are logically related. The so-called determination of the "linguistic community" could after all have been framed in the context of an interaction between intentional agents having primitive "meaning intentions". It did not need to rest upon a view in which language was presupposed as essentially community relative. But a properly understood anti-individualism requires such an additional claim.[11] Burge himself seems to be willing to argue for a community view of language and to take this additional step.[12] The idea is that language can be accounted for without resorting to notions such as speaker's intentions, mental representations or *intentional* states. Even if many social notions are at least in part subjective, this subjective component need not be characterized as "intentional".

According to this view, it is wrong to suppose that there has to be a strong connection between meaning and belief or between language and theory. This assumption is often made by philosophers with a holistic bias, but there are not many clear arguments in its favour available in the literature.[13] As far as I can see, it is founded mainly upon the rejection of building block theories, i.e. atomistic approaches to the problem of meaning.[14] But from the rejection of atomism, it is wrong to infer the truth of semantic holism. There are alternative solutions which require only the truth of molecularism. As a matter of fact, the most important arguments for the community view of language stem from approaches to meaning that are molecularist in spirit. (Wittgenstein, Dummett, Kripke and Millikan) In a way, this is not surprising. A fundamental premise in all the arguments for semantic holism is precisely that meaning is interdefinable with belief (or mental representations, or speaker's intentions).

It is therefore only natural to expect a genuine community view of language to resist this assumption.[15]

Molecularism, in the very vague sense of the term, is perhaps itself a certain form of "holism". But here the word "holism" means only "not-atomistic". It is also true that a molecularist semantics of assertability conditions is intimately related to methodological holism and is compatible with epistemological holism. But if such a semantical framework is to do more than just pay lip service to the community view of language, it must be opposed to philosophers that try to establish a strong link between meaning and belief. And since this claim plays the role of a fundamental premise in the arguments for semantic holism, it is only natural for a molecularist philosopher to entertain doubts toward semantic holism.

Semantic holists share a view of language in which idiolects occupy a central position. This too is not very surprising. Since meaning is, according to the holist, dependent upon the intricated web of beliefs of particular agents, and since these belief systems differ from one agent to another, it is only natural to suppose that the "communal" language is nothing but the intersection of different idiolects. For the partisan of the community view, this is just another reason for resisting holistic temptations in matters of semantics. As soon as we account for natural languages without the help of intentional notions, and are able to discard Gricean reductive approaches or those that appeal to the interdefinability of belief and meaning (as in Davidson, Quine or Putnam), we are even in a position to explain the intentionality of mental states in terms of the "intentionality" of language. More precisely, the intentionality of the mental could then be explained in terms of the semantic properties of natural languages. But I shall not consider this final move as an integral component of the third and final grade of anti-individualism. What's important is only that our account of language does not presuppose intentionality.[16]

III. THINKING ABOUT THE EXPERIMENTS

Before moving on, I would like to say a few words concerning thought experiments in general. I am not among those who denigrate these kinds of strategies in philosophy. There are reasons to believe that they form a good part of what philosophy is all about. Philosophy is to a very large extent an *a priori* field of research and thought experiments contribute to its advancement. By assessing the soundness and validity of philosophical arguments, by a proper use of conceptual analysis and by the development of thought experiments, analytic philosophers have contributed to the development of that portion of our philosophical thinking. I therefore see nothing wrong with thought experiments *per se*, as long as they are seen as merely providing "intuition pumps" to use Daniel Dennett's happy phrase, and as long as they are not the only grounds for accepting a particular thesis.

Since Frege, we generally agree that *a priori* truths are truths that *can* be established without the recourse of sense experience and not truths that *must* be so established. A sound management of the relations between philosophy and sciences requires that parallel researches be done on common topics. The work done by the first ones constrains, and imposes limits upon, the work done by the others. The theories developed by one group fix the room to manoeuvre within which the others will be entitled to work. Of course this cannot always be true, but the point is that philosophy cannot pretend to have a privileged access to a certain class of truths.

Conversely, one cannot always expect scientific results to bring definitive answers to our philosophical perplexities. There are loads of examples confirming that these results have to be "interpreted" and, very often, philosophers disagree on the appropriate interpretation. This is true, for instance, of the debate between the simulation theory and the "theory theory" approaches to mental attribution. Marr's theory of vision also offers an edifying example. If some philosophers saw in Marr's theory a confirmation of their own internalist view,[17] Burge saw it quite differently and thought that Marr's theory confirmed the truth of externalism.[18] So one cannot always expect to get clear cut scientific answers to our philosophical queries. In any case, it is on the philosophical arena that I will now find myself reevaluating the results of the thought experiments of Putnam, Burge and Kripke.

These thought experiments rely on intuition and it is there that they gain their persuasive force. But it seems to me that these intuitions are not entirely satisfactory. Let's take, for instance, Burge's experiment. Let's consider an agent confronted with a perceptual experience that prompts her to say: here's water. Does that agent think about the same thing when confronted with an indiscernible liquid substance having a different chemical composition? We are tempted to answer that it depends on the use that we make of the term "water". According to one of those uses, it serves the purpose of referring to a certain substance, understood as a substratum of essential properties. If this is the use of the term "water", we must admit that the agent does not think about water on twin earth, but rather about "twater". The use that we are considering forces us to say that the agent does not mean the same thing on twin earth even though the substance has the same phenomenal properties. The same kind of remarks could be made concerning Putnam's experiment involving brains in vat. If we use the expression "brain in vat" as a natural kind term and the reference relation is at least partly causal, then perhaps the claim that we are brains in a vat is self refuted.

But there can be other uses made of the same terms. We could, for instance, use the term "water" in the context of an observational statement. There is surely no such thing as an observational term in the absolute, but some sentences are available for observational *uses*. Couldn't we think about water intending only to refer to a bunch of phenomenal properties? An odorless, colorless, transparent liquid that quenches our thirst and that runs in the rivers is maybe the thing that could be meant by the word. Even if there are always theoretical properties associated with any given word, it could very well be that these turn out not to have any role to play in certain statements whose sole purpose is to attract the hearer's attention to a phenomenal experience.[19] A sentence like "Look there is water!", used in a certain appropriate context, can be meant to describe a particular phenomenal experience without any regard for the chemical structure of the substance.

It is true that nowadays the average English speaker happens to know that water is made out of molecules of hydrogen and oxygen, but this may be a contingent fact about her semantic competence. English speakers in the eighteenth century would have used the same word and would have ignored this fact about water. The same kind of remarks apply nowadays to a child learning to use the word. Moreover, even if there might very well be a distinction made by contemporary speakers between what the substance looks like and what it is really, that would not suffice to justify us in claiming that all speakers who make use of the term always refer to the substance, as distinct from its empirical properties.

It is important to note also that the use we are alluding to does not presuppose the truth of phenomenalism. Not only is it compatible with the naturalization of epistemology, it should even be interpreted as ontologically neutral. We very often refer to objects without invoking any specific claims concerning their nature. Like Quine, we can acknowledge the epistemological role played by certain observational uses of sentence, without reifying these phenomena into *sense data*. Of course, the anti-individualist could be willing to accept the suggestion, while insisting that the essential point in the anti-individualist doctrine concerns "individuation" and not "categorization", and therefore ontology and not epistemology. But this is not satisfactory because, as we shall see, there are reasons to believe that from an ontological perspective, the distal properties of substances that serve to individuate perceptual contents may be properties that supervene upon narrow contents.

In any case let us grant that there are at least certain uses of a sentence like "Look there is water" in which the term is used only to designate a perceptual experience. If we were to make this assumption, we would arrive at a very different conclusion in the thought experiment. It seems that Oscar might be thinking about the same thing when he finds himself on twin earth saying "here's water". My conjecture is that in these cases, it is only the proximal properties of the objects that count in determining the content of the thought. In other words, Oscar and twin Oscar have the same thoughts in both worlds. This does not rule out the result of Burge's own thought experiment. It only means that the result is relative to a particular use of the term, one in which it is used to refer to the distal microphysical properties of the substance. The fact that expressions may have different contributions to the assertability conditions of a sentence, i.e. different uses, is what explains that it can express different thoughts in different contexts, and provide evidence for a certain form of internalism as well as externalism.

It is important to clarify the sense in which it can be said that the mental state could, under a certain use of the term "water", be "internally" individuated. As I said, this concerns only a certain use of the term. It involves only the categorization that we are inclined to adopt, relative to particular epistemological interests and specific explanatory purposes. And since it is epistemologically motivated, it does not, initially at least, have any consequence for the ontology of the states. Saying this, however, does not mean that the appropriate ontology of perceptual states is such as described by Burge and Putnam. If, for example, we are to rely on Marr's theory for an appropriate characterization of our visual experiences, we might have to accept the idea that only some general distal properties such as shape and colour play a role in the individuation of visual experiences and not the physical micro-structure itself.

Moreover, there might be a supervenience relation that holds between those general distal properties and some intrinsic internal properties of the visual apparatus. It would then perhaps still be true to suggest that the physical environment serves to individuate perceptual contents, but this would now only amount to a weak form of externalism, for externalism is now compatible with supervenience upon the narrow psychological properties of the agent. And if there still were a use that confirmed the experiments performed by Burge and Putnam, it would no longer have any ontological bearing. It would now be wrong to suggest that the microstructure of physical objects in the environment plays a role in the individuation of perceptual states.

The arguments for externalism have very often been formulated in terms of experiments purporting to show that the microstructure of physical objects serves to individuate mental states and the way to establish this has been to deny that the states supervene upon the internal physical constitution of the organisms. If externalism is understood in this way, it is perhaps false at least in the case of perceptual states such as vision. It could very well be that, in the case of vision, the only distal properties that serve to individuate the states are precisely properties that supervene upon the agent's physical constitution. A change in these environmental properties would normally lead to a change in the agent's physical constitution.

If externalism is understood only in the weaker sense of a theory which purports to establish a determination of some general properties of the physical environment upon the mental states of the agent, then it should perhaps not imply a denial of supervenience. The fact that some properties supervene upon the physical constitution of the agent is by no means a proof that they do not play a role in the individuation of her physical constitution. But in this weaker sense of externalism, it would fail to vindicate the intuition behind Burge's thought experiment. Once again, it may very well be that the microphysical structure of objects has very little to do with the nature of our visual experiences. But then again, a weaker form of externalism requires only a theory like Marr's and nothing else.

This does not show that Burge's experiment is entirely ill-conceived, for there could very well be a *use* of the term "water" according to which a change in the microstructure of the objects would involve a change in the mental state.

And if we allow for such a use, the properties of mental states would no longer supervene upon internal properties of the agents. But the justification for this use is not to be found in Marr's theory or in some other ontological argument, but rather in our communal linguistic practices which allow for a directly referential use of the term "water". In short, there might be a use that confirms Burge's experiment, but we can no longer rely on it to draw an ontological conclusion.

There is also a use that coincides with an internalist categorization. It is the one according to which we use the term "water" in order to refer to phenomenal properties on which the distal properties of the environment supervene. And from the point of view of ontology, the view would now be compatible with what I have labelled as "weak externalism". Indeed, the two claims that there are narrow contents and that some general properties of the physical environment serve to individuate these contents are now just like the two sides of a coin and they reinforce each other.

The same kind of remarks can be made regarding the thought experiment in which it is claimed that mental states are determined by the sociolinguistic environment. We have to recognize that a term like "arthritis" might have a use which makes it a term in Oscar's idiolect in addition to a use in which it is understood as part of the dialect of the community. It is perhaps not necessary to represent the situation in which Oscar finds himself as one in which he has an incomplete mastery of the term. Oscar might have a complete understanding of the dialectal use, but might prefer a broader use of the term according to which it serves to refer to a wide range of pains, including those that one could experience in the thighs. The gulf that might subsist between an individual's use of a term and the one made by the community of speakers need not always be explained in terms of an imperfect mastery of the language. We can of course admit a distinction between an idiolectal and a dialectal use of the term, can't we?

So let us suppose that we do. Oscar behaves exactly as initially in the story. He believes that his father suffers from arthritis, that he has been spared from such a disease until now, that it is better to have arthritis than cancer, etc. And he also believes that he has arthritis in his thighs. When the term is used as belonging to Oscar's idiolect, these beliefs conform to the semantic rules of his language. Oscar thinks in perfect accordance with his own idiolect and does not exhibit an incomplete understanding of the communal language. And in that sense of the word, Oscar's last belief is true.

Now, let us consider what happens if, on twin earth or in another possible world, the community's dialect contains an expression "arthritis" whose meaning now coincides with Oscar's idiolectal meaning. Are the variations in the semantic rules of the community going to affect Oscar's mental contents? This time, the answer seems to be negative. Oscar needs not have different thoughts, because he is still using the term as part of his own idiolect. There is certainly a use of the term that confirms Burge's initial thought experiment, but there also seems to be a perfectly acceptable use that falsifies it.

It is true that Burge only supposes that Oscar *could* have an incomplete understanding of the word "arthritis". And when he discusses the possibility of generalizing his findings to many other types of expressions in the language, he seems to be implicitly acknowledging that there could be exceptions. What about beliefs in mathematical or logical propositions? Are these not individuated individualistically? Whether or not they are, we are now in a position to uncover a more substantial list of exceptions. These concern all the cases where a term belonging to the idiolect of the agent is considered.

This conclusion immediately requires qualifications however. And here we can draw on an anology with the thought experiment that was discussed earlier. When Burge argues for externalism, for instance, he always insists that the thought experiments must apply to *de dicto* and not to *de re* thoughts as such. This is not only because it would be trivially true to argue that a *de re* thought is partly individuated in terms of the relation entertained with distal objects of experience. It is also because it is important to distinguish externalism from the thesis according to which our thoughts should be understood as identical or reducible to relations to the physical environment. He is willing to admit the autonomy of *de dicto* thoughts and the claim that our thoughts are partly individuated by reference to the physical environment must not be confused with the claim that our thoughts can be *reduced* to relations to the physical environment. An agent can have complete access to a certain concept of water while being ignorant of the chemical composition of water and this is so even if the micro-structure of the liquid is responsible for our having the thought. According to Burge, the inner composition of the liquid is causally responsible for the occurrence of the concept. It is my concept of water, but I would not have that concept if I weren't in an environment in which there is H_2O.

Now I have expressed reservations concerning that idea as it applied to the case where our thoughts are directed upon perceptual contents. It may very well be that perceptual contents are individuated partly in terms of general distal properties belonging to the physical objects themselves, but this would not be sufficient to show that the micro-physical structure of the object, namely that it is H_2O, has anything to do with the individuation of our thoughts. I want to grant all of this but claim that the situation is different as regards the relationship between idiolects and dialects. In that particular case, we can all at once claim the autonomy of *de dicto* thoughts, admit the possibility of idiolects and argue that the agent could not have such a thought if it weren't determined by the sociolinguistic environment. Burge need not argue that our thoughts in general are *reducible* to direct relations that we entertain towards the conceptual resources of our linguistic community. It is claimed only that, in general, they are *partly* individuated in terms of the conceptual resources of the community.

Understood in this way, the general conclusion that our thoughts are individuated in terms of the linguistic resources available in the community of speakers remains unaffected by the counterexample that I have just been discussing. There prevails, broadly speaking, a global supervenience of our thoughts over the linguistic conventions adopted by our fellow speakers.

In that sense, Burge's thought experiment is compatible with the admission of idiolects. It is just that we can only make sense of them if they are individuated in terms of concepts expressed in the dialect of the community. However, it also shows that the experiment is not by itself sufficient to draw a general conclusion concerning the social character of our thoughts. An additional premise is needed, one that establishes the primacy of the community's dialect. The community view of language must be accepted if we do not want the counterexamples to threaten anti-individualism.

Anti-individualism remains unaffected by the counterexamples we have been discussing. What we have been able to establish is rather that it is all a question of use. It may very well be that there are no reasons to go beyond these uses. If we are right, it would mean that the debate between internalists and externalists can partly be resolved by saying that they all are correct accounts of particular uses of the sentences of the language, but that neither are universally true. It all depends on the assertability conditions. From an ontological point of view, Marr's theory would only confirm a weak version of externalism. And as far as the experiment involving Oscar's use of the term "arthritis" is concerned, the fact that the word could be part of his idiolect shows that the arguments favourable to anti-individualism in the second sense may still be true, but that they require an additional premise.

IV. THE COMMUNITY VIEW OF LANGUAGE

We must now inquire into what was earlier described as a third grade of anti-individualism. Burge admits the possibility of idiolects in 'Wherein is Language Social?',[20] but it seems that he did not seriously consider enough this possibility when he wrote 'Individualism and the Mental'[21] and that this could lead to a relativization of the results of his own experiment. We have reached such a conclusion by adopting a semantics of assertability conditions, and we were able to conclude that, without an additional premise, Burge's experiment would no longer be in itself sufficient to conclude that our thoughts are individuated in terms of the relations that we entertain with the socio-linguistic environment. We have shown that the experiment would fail in the case where Oscar articulates his thought in his own idiolect, and the question remains open whether this constitutes a serious counterexample for Burge's main conclusion.

In order to secure the initial results of the experiment, we need to show that there cannot be private idiolects. We should try to establish that there is no such thing as a language that can, in principle, be understood by no one except the speaker. In order to show this, we have to recognize the priority of the dialectal over the idiolectal use. An immediate consequence of doing so will be that private languages are impossible. In effect, we need to establish much less than the rejection of the logical possibility of private idiolects.

We only need to show that idiolects are individuated in terms of the dialect of the community. But by doing so, we happen to rule out the very possibility of there being private idiolects.

One must not confuse the notion of an idiolect with the notion of private language. But without the primacy of the communal language over the idiolects, we would be unable to refute the logical possibility of private languages. If languages were idiolects, it would be logically possible to suppose the existence of a private one. If so, arguing against the possibility of primitive idiolects amounts to no more and no less than an argument for the community view of language.

Now the community view of language is precisely a view that lies behind a semantical framework such as the one that I have been assuming all along in the present discussion. Indeed a semantical approach founded upon the notion of assertability conditions must partly be justified by an argument that rules out the possibility of private idiolects and it is irrevocably linked with a community view of language. So what about this view? Shall we have to relativize the results of the experiment that leads to this conception of language?

Before answering this question, it is worth noticing that the communal approach to language is compatible with the admission of a distinction between the observational and theoretical uses of natural kind terms such as "water", "lion", "planet", etc. Even if there are uses according to which the term "water" only serves to describe a certain phenomenal experience, these do not entail private idiolects. The reason is that there is no such thing as a purely observational vocabulary serving to describe an object belonging to a private experience. There are only observational *uses* of certain terms and they are uses that are accepted by the members of the community.

Within such a semantical framework, we are also in a position to accept a distinction between idiolectal and dialectal expressions, but this distinction is made possible only because idiolects are parasitic upon dialects in general. We could describe the situation as one involving global supervenience over the dialect of the community. In order to make sense of Oscar's linguistic behaviour, we must perhaps represent him as forging his own notion of arthritis with the help of the vocabulary already available within the community. Specifically, the definition of his idiolectal expression must be cast in a vocabulary that we understand. Words like "disease", "pain", "joint", "muscular", "thigh", etc. must be understood as they are understood in the vocabulary of the community. The community view of language prescribes that the only accepted notion of idiolect is one that is definable in terms of the dialect of the community.[22]

Of course, one could instead adopt the following approach. It could be claimed that we cannot make sense of someone's idiolect if we are unable to translate it in our own. So it is true that an individual linguistic behavior can only make sense relative to an interpretation, but the latter need not be couched in a dialect, for it can be made in the interpreter's idiolect. This is indeed the line taken by Davidson.[23] But we must ask: under what conditions are we going to be able to translate the idiolect of the speaker into our own? It seems that there must be a core meaning shared by the two speakers for each word, and it is natural to represent this core meaning as a system of meaning postulates similar to the ones found in dictionaries. This suggests once again the existence of a communal vocabulary shared by the two speakers, but the Davidsonian insists in construing this core meaning as the overlap of the idiolects.

For philosophers such as Davidson, Quine or Putnam, languages are certainly public in nature, but this is explained in terms of an overlap between the idiolects. Now their view concerning the primacy of idiolects is motivated by their holism and it is certainly not within the confines of this paper that we are going to deal with such an important issue. Let me just note that holism only follows if, in addition to the inscrutability of reference, one also accepts the suggestion that meaning is information or cognitive content (whether it is to be construed in terms of truth conditions or verification procedures). But within the framework of a semantics based on the notion of assertability conditions, there are at least two essential components to the meaning of a sentence: the locutionary or cognitive component and the illocutionary component. And one can discriminate, among all the sentences uttered by agents in a given community, those that behave as meaning postulates, given that agents have a propensity to stipulate them, i.e., perform declarative illocutionary acts on them, and treat them as rules governing their linguistic behavior.

But the community view of language cannot be taken for granted. We cannot discard in principle the possibility of private idiolects. And for our present purposes, we must question the thought experiment that leads us into thinking that private languages are impossible. So what are we to say concerning the sceptical argument discussed by Kripke? Aren't we able to relativize the result of that experiment also, just as we were able to do for the two previous forms of arguments leading to a certain form of anti-individualism?

So let us now consider the thought experiment suggested by Kripke. It takes the form of a sceptical paradox and it presupposes that language is a system of semantic rules. Now it has been claimed by Davidson that we should not view languages as systems of semantical rules.[24] This view of language is however very minimal. If Davidson feels like rejecting it, it is perhaps because he wrongly assumes that it is intimately linked with the notions of linguistic conventions, institutions and social customs. But we are not assuming that in the argument. The social character of language is not taken for granted. It is rather a conclusion we want to reach.

Another criticism of the idea that language is to be understood as a semantical system comes from those who interpret the claim as an implicit endorsement of semantical atomism. According to such a view, semantical rules seem to hang in the air in Plato's Heaven and we should want to deny that. And if it is not a good approach to view language in terms of semantical rules, why should we bother about Kripke's sceptical paradox? This would certainly be one way to dissolve the sceptical paradox and it has been at the center of some of the replies formulated by Hacker and Baker.[25] But these reactions are also out of place. The view of language we are assuming at the beginning is neutral between atomism, molecularism or holism. Under an atomist account, semantic rules specify the meanings attached to the primitive vocabulary of the language. Within a holistic perspective, the semantical rules are nothing over and above a photographic picture that describes the state of the language practice at a given moment. These semantical rules may have in addition a normative force, but it emerges from the practice of language. Under such an account, the essence of language is located in the linguistic practice itself and not in the system of semantical rules, but there is no need to deny the existence of such a semantical system. Similarly, under a molecularist approach, there is also room for accepting the existence of semantic rules, but these are now understood as sentences having special assertability conditions. The experts in the community have a propensity to stipulate these sentences as if they provided guide lines in the interpretations of the speech acts performed by their fellow speakers.

And so understanding language as a practice governed by semantical rules involves no particular prejudice concerning the form a theory of meaning should take. Saying that the linguistic practice is to be partly explained in terms of semantical rules is just one other way of saying that language is compositional. It surely is a constraint on any acceptable theory of meaning, and a very weak one indeed. Compositionality is not a condition that must be respected only by building block theories of meaning. It is also one that can be met by holistic and molecularist approaches.

But if language is a system of semantical rules, it is going to be hard to show that the normative force of these rules could be captured and realized in the mental life of the agents and difficult to imagine how the potentially infinite applications of the rules could be kept, given the finite number of dispositions that one can have. How could it be that, by contemplating these semantical rules, I am in effect anticipating all their applications in advance? And how could I be sure that, by behaving in the way I do, I am in effect behaving in accordance with the normativity of the rule? It seems difficult to show that there is a *fact of the matter* about meaning and understanding. Whether these are dispositions or mental capacities, it seems that we cannot postulate the realization of a semantic competence on the part of the individual without the disappearance of the normative force attached to the rules and without losing the potentially infinite applications of the rule. Indeed, the problem is that in concrete cases, I always follow the rule blindly and antici-pate at best only a small number of applications of the rules. Since our brains have a finite capacity, they cannot by themselves produce a substitute for the rules governing our linguistic behaviour and they cannot capture their normative force.

The problem we are alluding to is not one that can be circumvented by a holistic account of meaning. It is true that under such an account, the normative force of the semantical system is entirely parasitic upon the language practice of the agent, taken as a whole. The normative force of the rules emerges, as it were, from the language practice itself. But this does not suffice to solve or dissolve the paradox. Whether the source of the normative source is to be located in the system of semantical rules itself (as in atomistic or molecularst accounts) or ultimately in the language practice from which it emerges (as in holistic accounts), it remains that it is the semantical system as such that must somehow be known by the speaker and integrated as propositional knowledge.

Together with the assumption that language is a system of semantical rules, we also have to assume that understanding a language is at least in part a matter of propositional knowledge and it cannot be accounted for only in terms of practical abilities. As a matter of fact, this second assumption is an inescapable conclusion that follows from our assuming that language is a practice governed by semantical rules. According to this view of language, we have two essential components involved in language. We have the notion of a practice and the notion of a system of semantical rules. Therefore it is only natural to assume on this basis that understanding a language will also involve two components. It will involve a practical ability and a propositional knowledge of the semantical rules. The semantically competent speaker is one that knows the semantical rules of the language and that is able to use linguistic items in accordance with the semantical rules.

But how is this system of semantical rules realized in the mind of the semantically competent agent? As soon as it is granted that semantic competence involves both propositional and practical knowledges, the question can be asked: how do I know that I am now behaving in accordance with the semantical rules? The question is one that remains even if the system of semantical rules somehow originates from language practice. For we are not asking anything concerning the gap that occurs between the language practice and the semantical rules, but rather between my practical ability and my propositional knowledge of the semantical rules. There always seems to be room for doubting that I am behaving in accordance with my propositional knowledge of the system of semantical rules. And since I do not know for sure that my language practice connects with my propositional knowledge, my epistemological doubt paves the way to an ontological doubt. We cannot claim that there are for sure such things as meaningful linguistic behaviors. The conclusion seems to be that there are not facts corresponding to the understanding of language, a paradoxical conclusion indeed. But this is precisely the conclusion that is reached by the sceptical paradox.

The solution is to diagnose that the source of the sceptical paradox resides in our assuming that truth conditional semantics is the appropriate semantical framework for the interpretation of semantic statements. If we assume such a framework and consider the problem raised by the sceptic, then it seems that we are inevitably confronted with a sceptical paradox. But the paradox is "solved" as soon as it is discovered that the meaningfulness of semantic statements does not depend upon expressing objective truth conditions. We wrongly suppose that in order to mean anything, semantic statements must themselves express entirely objective truth conditions and we then find ourselves unable to specify the facts that exemplify such a competence in the real world. But the meaning of an indicative sentence does not always rest in its expressing objective truth conditions. The essential ingredient of meaning for a sentence is rather to be located in its assertability conditions. These assertability conditions associated with a particular sentence may allow for a variety of uses and vary from time to time and from community to community. This is all that is required for a sentence to acquire meaning. It is not always essential that it expresses objective truth conditions. We are able in this way to understand why semantic statements could be meaningful even if they did not express entirely objective truth conditions.[26]

This, in rough outline, is the essence of the sceptical solution. It leaves the sceptical doubt intact and does not attempt to provide a direct answer to the question whether there are entirely objective facts of the matter concerning meaning and understanding. It is not an attempt to dissolve one of the premisses in the sceptical argument. It is at best an attempt to deny an implicit premise in the argument, namely the one according to which the only appropriate semantics is truth conditional semantics. If that claim were explicitly built into the sceptical argument, then it would perhaps be appropriate to describe Wittgenstein's answer as a dissolution of the paradox. But it is better to understand it as a presupposition made by the sceptic. Our answer to the sceptic is not motivated by a desire to describe an objective fact of the matter concerning meaning, nor to dissolve one of the premisses in the argument. The solution is a sceptical solution, one that leaves the sceptical paradox intact, but one that tries to justify our language practices without invoking entirely objective semantical facts.

The sceptical solution must not be interpreted as a claim that meaning and understanding pertain to *objective facts* concerning the community and not to facts regarding the individual. This solution would be a positive answer to the question raised by the sceptic and would not be a sceptical solution. In addition, it would not involve a rejection of truth conditional semantics. The objective truth conditions would be discovered in the practices of the community instead of being confined to the individual. But this is not Wittgenstein's solution, if Kripke is right. We have to change our semantical framework and think about the role played in our lives by semantical statements. These statements are not always used by us in order to describe a prevailing objective fact. They are very often used in the course of making stipulations based on the satisfaction of an objective criterion that has been conventionally adopted as indicating semantic competence.

I do not wish to discuss issues related to Kripke's interpretation of Wittgenstein's text. The problem as I said is that the argument is essentially based on *a priori* considerations and one might wonder whether the conclusions carry enough weight because of that. We should be able to provide independent arguments for the community view of language. Thought experiments are useful tools for philosophers but they are not reliable enough to justify the adoption of philosophical theories.

V. MILLIKAN'S APPROACH

I think that biosemantics does provide such an independent justification. Ruth Garrett Millikan's work on the subject can be interpreted as a research program with a sound heuristics that could empirically be fruitful and that would be our desired independent confirmation of Wittgenstein's views.[27] The application of Millikan's theory to natural languages is a molecularist semantical theory that takes into consideration the locutionary and illocutionary aspects of meaning, that allows expressions to have many different uses or functions, and it is cast within a community view of language. For these reasons, it is interesting to take it as an illustration of what Wittgenstein's conception of language would amount to if it were to take the form of an empirical theory.

I do not wish to enter into the detailed discussion of her work in biosemantics. This task would carry us too far away from the limited goal of this paper. It is only important to describe the general lines of the program. I shall conclude by showing how Millikan's theory forces us to qualify Wittgenstein's claims concerning the impossibility of a private language and by addressing some of the difficulties that are raised in the attempt to reappropriate the results of her theory in an anti-realistic framework such as the one involved in the semantics of assertability conditions.

Millikan's semantic theory, it will be remembered, is an attempt at redefining the semantic notions within a conceptual apparatus similar to the one developed in evolutionary biology. The concept of proper function is at the center of this conceptual reduction. Meanings are accounted for in terms of proper functions. This is not to say that semantic facts are biologically realized in individual organisms. Millikan would strongly be opposed to the idea of reducing semantic competence to an innate mental capacity in the agent and she criticizes Chomsky's work on this score in various places. What is probably innate, as far as semantic competence is concerned, is at best a capacity on the part of the agent to adapt herself to the linguistic behaviour of her linguistic community. It would therefore be more appropriate to describe her program as teleosemantics instead of biosemantics.

Millikan criticizes the rationalist account of language according to which meanings would be known *a priori*. This is, according to her, the reason why it has often been held that intensions were the most important components of meaning. Indeed, if it is assumed that knowledge of meaning has to be *a priori*, it must be admitted that we do not know *a priori* whether or not a given term has a denotation. And since it might even turn out also that a given expression is devoid of denotation, this latter component cannot be part of what we understand when we say that we understand the meaning of a word. But the only thing left seems to be the intension. So intension must either be identical to meaning or must determine the meaning of an expression. Now

since intensions are psychologically realized entities, it could be argued
that psychologism is almost inevitably implied by the Cartesian model
according to which meaning is known *a priori*. . . .[28] So by attacking as she
does the rationalist view of language, her standpoint seems clearly to be anti-
psychologist and anti-individualist.

Three basic ingredients may be involved in the meaning of an expression.
Millikan distinguishes between the stabilizing function, the "Fregean sense"
and the intension of an expression.[29] The stabilizing function is the essential
ingredient since it is involved in all the meaningful expressions in the
language.[30] It is according to Millikan essentially social in character.[31] Sense
must also be social because it requires the existence of previous occurrences
of a term that successfully referred to objects in the world.[32] And among all
intensions, there are explicit and implicit intensions.[33] The first ones are
either part of the dialect or part of the idiolect of the agent, but they presuppose the notion of sense.[34] Implicit intensions may be understood as internal
properties of devices belonging to a language of thought, but it seems clear
that Millikan would in general treat these inner features as less important
than the public features of the language.[35] Implicit intensions may have a
private character but they do not seem to constitute enough to form a language
by themselves. This is however just an *a posteriori* claim that could be refuted
by experience and that does not rule out the logical possibility of human private
idiolects.

Millikan does not rule out in principle the possibility of innate mental
representations, but they are to be accounted for also in terms of their exem-
plifying proper functions. And since in the case of natural languages the
latter apply only once the whole complex history of relationships between
the members of the community has been taken into consideration, the notion
of proper function itself also ultimately has a social character.[36]

In short, by insisting on the ethological character of the notion of a proper
function, by analysing all the ingredients of the meaning of natural human languages in terms of the notion of proper function and by introducing a hierarchy
in the structuration of these ingredients of meaning which gives priority to
the notion of the stabilizing function, Millikan offers a naturalistic approach
to the problem of meaning which is not mediated by psychologism. In order
to argue for naturalism, it is not necessary to be a Gricean in semantics.[37]

Millikan's theory enables us to draw the broad features of an account that
could provide an empirical basis for a semantics of assertability conditions,
one that would be molecularist and community relative. She assimilates the
notion of meaning with that of proper function, and she is in this way able
to show how the meaning of a sentence would be related to its use or assertability conditions. She formulates an account of meaning that divorces it from
belief, and she explains the meaning of an expression in terms of the different occurrences of the same expressions independently of the other
expressions of the language. She is in this way able to establish molecularism.[38]
And finally, by a structural hierarchy of the different components of meaning

in which the notion of stabilizing function occupies a central place, she is able to confirm the social character of meaning for natural languages.

It could however be replied that the analogy I am trying to establish between Wittgensteinian views and those of Millikan suffers from an irremediable flaw. Wittgenstein's account is anti-realistic in spirit while Millikan's theory must be understood as an attempt to give new foundations for realism. Realism should no longer be, according to Millikan, grounded on a truth conditional theory but rather on a teleological functionalist theory. By characterizing the sense of a referring expression in functional terms and by showing that an expression has sense if it is supposed to refer, Millikan finds a way to reintroduce reference in the semantic picture. The same kind of remarks apply to individual sentences which are after all the most basic semantic units. Sentences are meaningful because they have a stabilizing function, but also because they are supposed to represent actual states of affairs. It is their purpose to represent what is the case. And actually, if words ever exemplify their referring function, it is precisely because they enter into sentences whose function is precisely to represent actual states of affairs.

Wittgenstein's approach seems to be miles away from Millikan's. There is a strong tendency in favour of anti-realism in Wittgenstein's philosophy and it is not motivated solely by his *penchant* for finitism in the philosophy of mathematics. It is to be traced back to the very heart of his philosophy of language. In the *Investigations*, Wittgenstein not only criticizes his earlier Tractarian views, but he replaces them with an account inspired by a rejection of metaphysical realism. The anti-realism does not have its source simply in the equation between meaning and assertability conditions, for there could be instances of statements whose assertability conditions amount to asserting the existence of an objective states of affairs. Nor could it be located in the relativization to the community, because such a relativization could have taken place in the context of a realist approach in which meaning would have been identified with facts concerning the community.

No, it is rather in the partly subjective character of meaning and understanding that Wittgenstein's anti-realism reveals itself. Words have the meanings that they have because of semantic stipulations made by some of the members of the community, i.e. the experts whose role it is to legislate in accordance with the principle of the division of linguistic labour. And speakers are said to understand these semantical rules as long as they behave in accordance with them and as long as the other speakers of the language are willing to stipulate that they have such a semantic competence.

Now how could this be made compatible with Millikan's views? Millikan was the first to think that her approach could provide a direct answer to the challenge raised by the sceptic.[39] There are facts of the matter about meaning and understanding as long as we follow the history of the use of an expression back to its original occurrences in the complex relations that characterize the members of the community. As long as we look for these very broad facts that involve the other members of the group and look at the history of these relations, we are going to be able to prove the factual character of meaning. Understanding, for a human organism, would then be located in her adaptative capacity to conform with the already complex set of existing customs that are associated with each and every sentence of the language.

I do not wish to deny entirely Millikan's claims, but I do feel that she has not sufficiently acknowledged the importance and role of semantic stipulations in the community of speakers. In order to have a meaning in the full sense of the term, it is not enough for an expression to acquire a stabilizing function, a sense and an intension. It must also have been stipulated by the members of the community that this or that particular exemplified proper function is to count as its meaning. Millikan is willing to recognize that words may have many different proper functions but she does not seem to recognize how big a problem this could be for her account. Here I am not merely alluding to the plurality of uses but rather to the unstable situation that could originate from such a plurality of uses. The normative force of a fixed choice of basic semantical rules for each expressions of the language may be essential for the stability of the relationships between the members of a given linguistic community. This is why grammars and dictionaries are so indispensable.

Notice, however, that this account refutes Millikan's realistic approach. If languages are semantical systems and semantical rules have a normative force to be explained in terms of subjective factors, it seems that no objective facts of the matter will ever capture this normativity and this is true even for Millikan's teleological semantics. From the fact that meaning and understanding are at least in part subjective notions, it does not follow that semantics statements have no application. It does not follow that we are going to have to abandon our talk about meaning and understanding, and the reason is that the appropriate semantic framework is the one based on assertability conditions. The idea is that a sentence may express assertability conditions and can therefore have a use in a community even if it does not express entirely objective truth conditions. Millikan has perhaps come close to recognizing assertability conditions, but she has wrongly been led by her realism to equate them with entirely objective facts.

With such an amendment made to Millikan's original approach, it becomes once again possible to use Millikan's theory to acknowledge the phenomenon of meaning. It could be argued that a sentence acquires a meaning if and only if it exemplifies a stabilizing function, a sense and an intension, *and* if it generates on the part of some of the members of the community (i.e. the experts) a propensity to stipulate that it has this proper function as part of its meaning. We could also claim that a speaker understands a word if there are objective facts in her mental states or in her behaviour that reveals an aptitude to use the word in accordance with its meaning, *and* if a critical mass of speakers in the community have a propensity to stipulate that she has such an aptitude.

Having said this, we must conclude against a strong form of realism in the theory of meaning and understanding. If there are facts of the matter about meaning and understanding, they are not entirely objective. And at the same time, the conclusion entails a refutation of nihilism since there is after all a difference between a meaningful behaviour and no meaningful behaviour at all. That difference is to be traced back into a mixture of objective and subjective features of the physical states and behaviours of the agents belonging to a given community.

Returning now to our main theme, we could very briefly conclude our essay by qualifying the result obtained by an *a priori* argument to the effect that there are no private idiolects. Let us suppose that an account such as the one developed by Millikan is correct. What are we to say of Wittgenstein's claim? Well in the first place it no longer appears as an *a priori* philosophical argument and it is available for refutation. It is like all empirical theses a fallible statement. Secondly, it no longer implies the idea that private languages are metaphysically impossible. We have to admit the logical possibility of private idiolects. And since we have turned molecularism, the semantics of assertability conditions and the community view of language into empirical hypotheses, the conceivability of private languages is no longer what is at stake. The communal view of language becomes at best a claim about human natural languages and the way they happen to be individuated in the actual world. In an account such as the one developed by Millikan, we can at most claim that human languages happen to be a social phenomenon, irreducible to the relational or internal properties of the individual psychology. Furthermore, we are now also in a position to acknowledge a distinction between idiolects and dialects. Finally, nothing in principle can help to prove *a priori* that none of our concepts are innate instead of being acquired. All of these claims are important amendments to the Wittgensteinian view of language, but they do not threaten the hypotheses made by anti-individualist philosophers. As far as I can see, their main conclusions remain essentially right.

Université de Montréal

NOTES

[1] Of course, thought experiments are not the only arguments available to the anti-individualistically inclined philosopher. There are general epistemological arguments to be invoked as well, and there are arguments that stem from the study of particular theoretical approaches. See for instance Tyler Burge's 'Individualism and Psychology', *Philosophical Review* **95**, 1986, 3–45.

[2] For an illustration of this view, see my 'A Sentential Theory of Propositional Attitudes', *Journal of Philosophy* **LXXXIX**(4), April 1992, 181–201.

[3] See for instance Putnam's 'Brains in a Vat', Chapter 1 of *Reason, Truth and History*, Cambridge, Mass., Cambridge University Press, 1981, and Burge's 'Other Bodies' in Andrew Woodfield (ed.), *Thought and Object*, Oxford, Clarendon Press, 1982, 97–120, as well as 'Cartesian Error and the Objectivity of Perception' in McDowell and Pettit (eds.), *Subject, Belief and Context*, Oxford, Oxford U.P., 1986, 117–136.

[4] Putnam, 'The Meaning of Meaning', in *Philosophical Papers, Vol. 2, Mind, Language And Reality*, Cambridge, Cambridge University Press, 1975, 215–271.

[5] Actually, Putnam's externalist view of the mental was already looming in the background even at the time when he wrote 'The Meaning of Meaning'. He equated intensions with intentional states and argued that, in a sense, one could agree that intensions determine extensions, as long as extensions are themselves involved in the individuation of intensions. The same kind of remarks could be made concerning mental states themselves. They too could be seen as determining extensions as long as the physical environment played an important role in the individuation of those states.

[6] Burge, 'Individualism and the Mental', in P. A. French et al. (eds.), Midwest Studies in Philosophy, Vol. 4. Studies in Metaphysics, Minneapolis, University of Minnesota Press, 1979, 73–121; see also 'Intellectual Norms and the Foundation of Mind', Journal of Philosophy LXXXIII(12), 697–720.

[7] Burge, 'Individuation and Causation in Psychology', Pacific Philosophical Quarterly 70, 303–322.

[8] I have argued for this alternative interpretation in 'L'expérience de Burge et les contenus de pensée', Dialectica 46(1), 1992, 21–39.

[9] Anti-individualism needs not be interpretated as involving a realist commitment to the attitudes. Burge himself claimed that his anti-individualist thesis in 'Individualism and the Mental' is neutral as to the truth of realism, anti-realism, instrumentalism or eliminationism. See op. cit., p. 113. It is also the same anti-individualistic assumptions that are used by Quine in the course of defending his eliminationism. See Word and Object, pp. 219–222.

[10] Intentionalism is the view that "according to our everyday collectivity concepts, individual human beings must see themselves in a particular way in order to constitute a collectivity. In other words intentions (broadly construed) are logically prior to collectivities". (M. Gilbert, On Social Facts, Princeton, Princeton University Press, 1992, p. 12.)

[11] That claim is well presented in Saul Kripke, Wittgenstein on Rules and Private Languages, Cambridge, Harvard University Press, 1982.

[12] Burge 'Wherein is Language Social?' in Alexander George (ed.), Reflections on Chomsky, Oxford, Basil Blackwell, 1992, 175–191.

[13] Many holist philosophers simply assume that belief and meaning must be related. In the case of Quine, the connection takes the particular form of a linkage between meaning and verification. See for instance 'Two Dogmas of Empiricism', in From a Logical Point of View, Cambridge, Mass., Harvard University Press, 1953. Davidson seems also to take holism for granted as the only possible alternative to building block theories of meaning. See 'Thought and Talk', in Samuel Guttenplan (ed.), Mind and Language, Oxford, Clarendon Press, 1975, and reprinted in Donald Davidson, Inquiries into Truth and Interpretation, Oxford, Clarendon Press, 155–170. Putnam himself makes a similar claim. See for instance Representation and Reality, Cambridge, Mass., MIT Press, 1988, p. 119.

[14] The only argument for semantic holism seems to be the indeterminacy thesis. But there are many different arguments for the indeterminacy thesis and many different sorts of indeterminacy. See Quine's 'Three Indeterminacies', in Robert Barrett and Roger Gibson (eds.), Perspectives on Quine, Oxford, Basil Blackwell, 1990, pp. 1–16. The most famous argument for indeterminacy, the one involved in Chapter Two of Word and Object, now appears to Quine to be an argument for the inscrutability of reference, and it does not imply semantic holism. There is indeed an argument for indeterminacy which is closely linked with semantic holism, but it is an argument based on the assumption that semantic holism is true. In this argument, Quine uses semantic holism in order to prove the indeterminacy itself.

[15] It is true that holist philosophers will most of the time want to acknowledge the fundamentally social character of language, but it is very often not substantiated or derivable from their most important philosophical claims. We have seen above how it is questionable in the case of Davidson. He does not accept the connection between meaning and communal agreement but rather between meaning and the maximization of agreement with one's own conceptual scheme. As regards Quine's repeated statements favourable to the communal character of language, they seem to rest more or less on wishful thinking since a Quinean communal language stems from many overlapping idiolects. As far as Putnam is concerned, we must grant that his views

concerning the division of linguistic labor are certainly indicative of a community view of language and play a central role in his philosophy of language, but they affect mostly the reference of our terms. Linguistic meaning, as opposed to reference, seems according to Putnam to be, for a large part, subjective. Of course there are also many atomistic approaches that have attempted to establish a close connection between meaning and attitudinal states (e.g. Gricean reductive characterizations in terms of speaker's meaning intentions), but I am not aware of any molecularist account that would follow that line. I am also not claiming that a molecularist semantics of assertability conditions is inconsistent with a denial of the community view of language. I am just suggesting that it is very congenial with such a view.

[16] There are those who, like Millikan and Dennett, believes that intentionality comes "by degree" and who uses a minimal notion of intentionality which amounts to 'aboutness'. I have no objections to this and would agree that, in this minimal sense, intentionality can partly be used in the explanation of language. My point however is that a rich notion of intentionality (involving intensionality, reflexivity and first person authority) could not have any explanatory value in our understanding of language, but is itself rather to be explained in terms of language.

[17] G. Segal, 'The Return of the Individual', *Mind* **98**, 1989, 39–57.

[18] Burge, 'Marr's Theory of Vision', in Jay L. Garfield (ed.), *Modularity in Knowledge Representation and Natural-Language Understanding*, Cambridge, Mass., MIT Press, 1989.

[19] Unfortunately, very few anti-individualists have been ready to accept this intuition. Apart from Putnam, we could mention Patricia Kitcher's 'Narrow Taxonomy and Wide Functionalism', *Philosophy of Science* **52**, 1985, 78–97.

[20] Burge 'Wherein is Language Social', pp. 186–187.

[21] In 'Individualism and the Mental', pp. 100–102, Burge correctly points out that in order to refute the result of his thought experiment, one would have to show that a belief involving an incomplete understanding of a concept should *always* be explained in terms of beliefs involving notions that one completely understands. And this is surely not forthcoming from the mere invocation of counterexamples such as those under consideration. But if idiolects are not harmful to the account, it is because they should be countenanced within a general picture of language which makes it community relative. And this was not made explicit by Burge when he wrote 'Individualism and the Mental'.

[22] The above remarks serve to illustrate what is otherwise described in the theory of radical interpretation as the principle of charity. However, it does not have the same consequences as in Davidson's theory for we do not accept semantic holism and we do not agree with the idea that a theory of understanding must take the form of a theory of radical interpretation. With those additional premises, Davidson is able to argue against the relativity of conceptual schemes and indeed against the meaningfulness of the very idea of a conceptual scheme. If on the other hand the semantics is molecularist, we are in a position to distinguish a special class of statements that have the property of functioning within the community as semantic rules. We are then able to accept the distinction between language and theory or between meaning and belief, or linguistic meaning and conceptual role. This is what allows us to distinguish between different conceptual schemes and therefore allows us to make sense of divergent conceptual schemes within our language. The principle of charity needs to be applied only to our primitive vocabulary. In the interpretation of someone else's discourse, we maximize agreement with the hearer on the linguistic meaning of our primitive vocabulary.

[23] Davidson, 'The Second Person', in Peter A. French *et al.* (eds.), *Midwest Studies in Philosophy, Volume XVII, The Wittgenstein Legacy*, Notre Dame, University of Notre-Dame Press, 255–267.

[24] Davidson, 'A Nice Derangement of Epitaphs', in Ernest Lepore (eds.), *Truth and Interpretation*, Oxford, Basil Blackwell, 1986, 433–446.

[25] In a nutshell, Kripke thinks that, according to Wittgenstein, language is to be understood as a system of such rules in a way that recalls the view of Rudolf Carnap. But this, according to Hacker and Baker, is precisely the mistake that should not be made. By insisting on the idea that language is an institution, that linguistic practices are customs and that understanding is at least in part a practical ability, Wittgenstein is in effect dissolving what could look like a

sceptical paradox. The paradox occurs only if we assume the existence of rules that have an independent existence and a normative force, and inquire subsequently about the connection with linguistic practices. They claim on the contrary that, by attracting our attention to language games, Wittgenstein meant to argue in favour of the view according to which there is an internal connection between applications of linguistic expressions and their semantical rules. Hacker, P. M. S. and G. Baker (1984), *Scepticism, Rules, and Languages*, Oxford, Basil Blackwell. It is not my purpose to discuss such a matter and to enter into the debates between the Kripkensteinians and ordinary language philosophers, even if my sympathies clearly go to the former group. My concern is more with the *a priori* character of the argument. We cannot take it for granted that language can best be viewed as a communal system of semantical rules.

[26]   A semantics of assertability conditions takes into consideration the illocutionary conditions as much as the locutionary conditions of uses of the sentences and semantic statements have at least one use that allows them to be uttered in order to perform a declarative speech act. These statements are semantically assertible in part because of objective conditions prevailing in the real world, such as the behaviour of the agents and their mental dispositions, but we very often have to take into consideration also subjective factors such as the propensity on the part of the other members of the community to perform declarative speech acts on them.

[27]   Millikan, Ruth (1984), *Language, Thought and Other Biological Categories. New Foundations for Realism*, Cambridge (Mass.), MIT Press, Bradford Books.

[28]   *Op. cit.*, p. 157.

[29]   *Op. cit.*, p. 147.

[30]   *Op. cit.*, pp. 5, 30.

[31]   *Op. cit.*, pp. 77, 82.

[32]   *Op. cit.*, p. 133.

[33]   *Op. cit.*, p. 148.

[34]   *Op. cit.*, p. 135.

[35]   *Op. cit.*, p. 149, where Millikan writes: "Thus, Quine's remarks about the "diversity of connections between words and experience" among individuals that yet yield "uniformity where it matters socially" aptly express the relation between private intensions and public *sense* . . .". See also pp. 157–158.

[36]   *Op. cit.*, pp. 27–30.

[37]   *Op. cit.*, p. 52.

[38]   *Op. cit.*, pp. 80–81, 104, 106–109.

[39]   Millikan (1990), 'Truth Rules, Hoverflies and the Kripke-Wittgenstein Paradox', *Philosophical Review* **99**, 323–353.

DENIS FISETTE

DAVIDSON ON NORMS AND THE
EXPLANATION OF BEHAVIOR*

In 'Three Varieties of Knowledge', D. Davidson distinguishes three types of knowledge: knowledge of the self, of others' thoughts, and knowledge of the world. He notes that the Cartesian tradition privileged the first type of knowledge believing that the other two could be derived from it. Against Cartesianism and logical positivism, Davidson maintains that these three modes of knowledge are irreducible, although complementary. I am particularly interested here in one of the arguments brought up by Davidson against the reduction of any one of these modes of thoughts. I will begin my analysis with his views on the characterization and explanation of intentional behavior. According to Davidson, the normative character of rationality and of the concepts used in the characterization and explanation of behavior constitute an inevitable obstacle to their reduction to an explanation using nonnormative concepts, and more particularly, to strict laws. The thesis of the irreducibility of explanations by rationalizations of the explanatory type found in physics but also in biology and neurophysiology, refers, of course, to his theory of identity, and more particularly, to his third principle: the anomalism of the mental. As we all know, this doctrine has been the object of numerous debates in contemporary philosophy. However, it is not my intention to add to this debate. My goal is to examine if the normative character of rationality, which is also an argument in support of the anomalism of the mental, can support the thesis of the irreducibility of explanations through reasons.

THE CARTESIAN THEATER

In a series of articles that have recently appeared and more particularly, in 'The Myth of the Subjective', D. Davidson (1989: 163) claims to have introduced a revolution in philosophy by proposing a perspective concerning thought/world relations that is radically different from the one normally expressed by classical metaphysics and, more precisely, by the Cartesian tradition. Davidson's target, or what he calls the "Cartesian theater", is described in the following manner:

... the mind is a theater in which the conscious self watches a passing show (the shadows on the wall). The show consists of ‹‹appearances››, sense data, qualia, what is given in experience. What appear on the stage are not the ordinary objects of the world that the outer eye registers and the heart loves, but their purported representatives. Whatever we know about the world outside depends on what we can glean from the inner clues. (1986: 453)

---

* Translated by Lina Di Blasio.

M. Marion and R. S. Cohen (eds.), Québec Studies in the Philosophy of Science II, 139–157.
© 1996 Kluwer Academic Publishers.

This image evokes an ancient myth, but it is primarily directed to a tenet of traditional philosophy, the tenet being that "to have a thought is to have an object before the mind." (1986: 455)

A more contemporary version of the Cartesian conception of thought is given by the Brentanian concept of intentionality and his doctrine of intentional inexistence. In fact, in a famous passage of his *Psychology from an Empirical Standpoint* Brentano introduced his concept of intentionality in order to dissociate the nature of mental phenomena from that of physical phenomena. Mental phenomena are characterized by the idea of directedness (*sich richten auf*), the idea that all representations are representations of something or are directed towards something. Brentano coupled his thesis of intentionality as directedness with his doctrine of "intentional inexistence", a doctrine according to which each act "includes something as object within itself", something which "inexists" intentionally in the act. Brentano's mental phenomena, derived from his doctrine of *intentional inexistence*, are defined as those phenomena "which contain an object intentionally within themselves" (p. 89).

As we all know, this definition of intentionality in terms of intentional objects is fraught with numerous problems. These problems affect both the psychological thesis,. the very way in which the concept of intentionality is characterized, and the ontological thesis, the doctrine of intentional inexistence. These problems also have some bearing on those who conceive the terms of an intentional relation in terms of propositional attitudes, Fregean "*Sinn*", "sense data", mental representations, or internal objects. In Brentano's terminology, and by taking into account the ambiguity of what he calls "intentional objects", one wonders if intentional objects are to be distinguished from real objects and how they are supposed to account for thought/world relations. As Husserl demonstrated and as Brentano recognized it himself in a letter to his disciple Kraus,[1] this problem, which Davidson (1989: 18) accurately called "Brentano's problem", is constitutive of what is called a «*Gegenstandstheorie*» to the extent to which the latter conceptualizes intentionality in terms of objects rather than in terms of relations. We must abandon the Cartesian myth and the idea that thought is comparable to a small closed box, to a monad which never has anything to do with things other than its own mental representations or internal objects.[2] Davidson thus suggests that thought and mental states ("intentions", beliefs, and desires) be conceived as relational concepts.

An analogy will be helpful here in illustrating both Davidson's solution to Brentano's problem and his alternative. The analogy is concerned with the measurement of temperature and of an object's weight whereby numbers are assigned to objects. When using grams or ounces to measure the weight of an object, we do not have to presuppose that the grams or ounces belong to the ontology of the object being measured. Likewise, the intentional objects, which we attribute to an agent to explain his behavior, do not indicate entities or intentional objects that supposedly inexist in the agent's head. Davidson applies the Quinian doctrine of the indeterminacy of translation to thought by pointing out that there is no "fact of the matter" as to the choice of one unit of measurement over another. For there are many different but equally acceptable ways of interpreting the verbal and nonverbal behavior of an agent.

This seemingly isolated criticism of the conception of thought in terms of intentional objects has consequences that go above and beyond the ontological status of these objects of thought. The first consequence is concerned with the foundational claim of the cogito in the Cartesian tradition. The evidence of the cogito that Descartes conceived in terms of privileged access, absolute certainty, and adequacy is meaningful only if we presuppose that the ego maintains a privileged and intimate connection with its thought contents. Thus, by showing that human thought is not to be conceived in terms of absolute certainty or adequacy, Davidson concludes that there is nothing that justifies the privilege of this mode of knowledge over the knowledge that we have of the world and of others. In 'Knowing one's own mind', Davidson proceeds to a reversal of Cartesianism:

Our beliefs about the world are mostly true, but we may easily be wrong about what we think. (1986: 446)

If we recognize the merits of Davidson's damaging criticism, what will become of the intentional idiom? Does Davidson's revolution also carry with it the entire intentional discourse on objects? Certainly not, for like Quine, Davidson recognizes the irreducibility of the intentional idiom. He thus places us before a dilemma that is similar to the one Quine (1960: 221) was confronted with in his *Word and Object*, between two opposing and apparently irreconcilable attitudes right where this irreducibility is situated. The irreducibility means either that the intentional is an empty idiom or that it is indispensable to and important for a science of the intentional. Quine, as we all know, adopts the first alternative by pointing out that Brentano's thesis goes hand in hand with the indeterminacy of translation and that it therefore has no place in science. For different reasons, Davidson recognizes the indeterminacy. However, he reduces its extent by imposing more constraints on the interpretation of verbal and nonverbal behavior than does Quine. Thus constrained, the indeterminacy would not be more troublesome "than the fact that weight may be measured in grams or in ounces" (1980: 6). In this manner, numbers are used to establish certain relations between objects and their weights. The use of these numbers does not in any way commit one to include ounces and grams in one's ontology. All that is needed here are objects with a certain weight and numbers. Likewise, the attribution of beliefs, desires, and intentions to an agent in order to measure his or her behavior simply requires a "collection of entities related in ways that will allow us to keep track of the relevant properties of the various psychological states" (1989: 11). This analogy would be trivial if it did not highlight two of Davidson's claims

regarding the intentional idiom. First, there is the ontological range of the indeterminacy which, when applied to thought, signifies that there are no such
things as "inexistent objects" which, Brentano claimed, characterize thought.
More importantly however, this analogy highlights the relational character
of psychological concepts. In this manner, the intentional contents attributed
to an agent on the basis of his or her behavior are not "real properties" of
objects to which they are attributed. They are concepts which characterize
the relation that takes place between two events described in intentional terms
– the agent, or one of his or her states, and the objects and events of the
immediate world. The effectiveness of this relational concept of intentionality is measured by its capacity to identify and to explain intentional behavior.
Several remarks on the characterization of action will give us the opportunity to unveil some of its properties.

THE CHARACTERIZATION OF INTENTIONAL BEHAVIOR

As we all know, the theory of action is a discipline that developed toward
the end of the 1960s under the influence of the second Wittgenstein and as a
reaction against positivism. It grew out of the problem of how action should
be explained. From the outset, this problem set the adherents of a causal
approach to action against the protagonists of an "hermeneutic" or teleological approach. The former, such as Hempel (1962), recommended the reduction
of what is supposedly specific to the domain of action to instances particular to the kind of standard explanation endemic to the natural sciences. On
the other hand, the "neo-Wittgensteinians", such as Anscombe (1957), claimed
to have been capable of dispensing with the concepts of causality and that
of law in their characterization and explanation of action. In his 1963 article
'Action, Reasons, and Causes', Davidson traced a path between positivism and
the neo-Wittgensteinians by reestablishing the concept of causality but by
dissociating it from that of law.

The first question that must be dealt with in the theory of action is concerned with the characterization of behavior. Wittgenstein formulated it in
the following manner. "What must be added to my arm going up to make it
my raising my arm?" He considered the following three statements: The
arm's muscles contract; he lifts his arm; the agent lifts his arm with the intention of looking at his watch. The first statement simply indicates an event,
whereas the second and third describe an action; the second by naming it
and the third by stating the reasons behind the action, reasons which are
supposed to explain it. There is thus a difference between lifting an arm and
the fact that an arm lifts. However, in response to Wittgenstein's question,
nothing is added to the event itself for these three statements relate to the
same event. Here is where the concept of "description", which Davidson
borrowed from Anscombe (1957), comes in to characterize intentional

behavior. A single event or a given behavior, for example, can just as well be described using the vocabulary of physics as it can be described using an intentional vocabulary. Furthermore, according to Davidson,

an event is an action if and only if it can be described in a way that makes it intentional. (1980: 229)

Thus, what makes a given behavior an action is the possibility of describing it in a vocabulary that makes it intentional. The same event can equally be redescribed using the vocabulary of physics, and as we will see, it can also instantiate laws. Consequently, the difference brought up by Wittgenstein's question is not concerned with the event, the descriptum, but with the description that is given to it.

Given what we now know, what are the ingredients which figure in the description of intentional behavior? The first problem is concerned with the very concept of intentionality. In fact, in addition to the numerous theories which are currently circulating in the literature on action theory,[3] Davidson, in his preface to *Essays on Action and Events* (p. XIII), does recognize that of the three uses of this concept enumerated by Anscombe (1957: 1),[4] he had initially privileged the syncategorematic usage, to subsequently adopt the substantive concept of pure intention. This seemingly isolated change has significant consequences especially with regard to his views on the explanation of action. Aristotle's useful syllogism, frequently used as a schema to explain action, will allow us to evaluate not only the breadth of these changes but also and especially the effectiveness of the concept of pure intention. Let us return to Wittgenstein's example which we could schematize in the following manner:

(1)   A wants to know the time.
(2)   A believes that the appropriate way to know the time would be to look at his watch.
_____
(3)   A looks at his watch.

The premises of the syllogism express the agent's reasons, that is, his desires and beliefs or the means he judges to be appropriate to obtain the desired object. The conclusion expresses the agent's action, which explains the desires and beliefs present in the premises. The desires and beliefs that are attributed to an agent as we seek to explain and predict his behavior correspond to the "reasons" which represent the conditions of intelligibility of a meaningful act.

The characterization of intentional behavior in terms of reasons, which Davidson advocated in his 1963 article, poses numerous problems. As discussed in his 'How is Weakness of the Will Possible?', the cases of incontinent behavior of akrasia forced him to modify his conception of intentionality and practical reasoning. In fact, the entire philosophical tradition from Aristotle on, maintained that to act intentionally, an agent had to act in a way which he judged to be the very best. Akratic behaviors, which are nonetheless intentional behaviors, are cases whereby the agent behaves against his better judgment, against what he judges to be the best for him, and thus against his will. The tradition claimed to have resolved this problem by pointing out that these behaviors, since they are not voluntary, cannot be considered as actions. But, asks Davidson (1980: 30): Can an action be both incontinent and intentional? It all depends, of course, on what one means by "intentional action".

As Davidson has shown several times,[5] his solution to the problem of incontinence rests essentially on the distinction made between two types of evaluative judgments: conditional or *prima facie* judgments and unconditional or "all out" judgments. The first correspond to Kant's categorical imperative and imply an obligation: "it is desirable that . . .". From this point of view, all actions would be carried out in order to fulfill an obligation. Furthermore, we would say that an agent acted with a given intention because he had certain desires and beliefs, the desirability of the action being simply deduced from those premises. It is precisely this type of reasoning that akratic agents violate. For if, as Davidson thought (1963), acting with an intention is simply acting for certain reasons, then an agent who carries out action "a" while believing that action "b" is just as appropriate, and all things considered, is the best action, does not act for reasons and thus does not act intentionally. In other words, if the action must be deduced from desires and beliefs, from which it is supposed to be derived, and if, as cases of akrasia show, the agent who nonetheless acts intentionally does not arrive at this conclusion, then we must revise our conception of practical reasoning. In fact, for the deductive conception of practical reasoning, the problem of knowing how an agent can conclude, on the basis of a belief and a wish, that "this" action is desirable and more desirable than any other possible alternatives for the same circumstances, remains to be resolved. For, from this point of view, all actions would be desirable. According to Davidson, more is needed than simply reasons to explain why an agent acts in a certain way rather than in another. Practical reasoning, which leads to an intentional act and at once to an intention to act, must be restored by presupposing a new judgment on the agent's part, a judgment concerning the value of an action. Davidson qualifies this judgment as unconditional in the following way: "this action is desirable". In this manner, an agent can value a certain state of affairs and believe that a given action will promote it. But the end does not justify the means. We must therefore presuppose a new judgment on the agent's part, an inductive judgment concerned with the value of one of the agent's options: all things considered, this option is better or more desirable than another. But the emergence of an intention, specifies Davidson (1987: 40), demands, in addition, that the agent cross that stage where he can commit himself to act in the future. We can schematize those steps leading to the emergence of an intention to act and to practical reasoning in the following manner:

(A)   A values a certain state of affairs or a certain end.
       A desires (health)
(B)   A evaluates various means, alternatives which take into account
       his beliefs and limitations. (e.g., exercising, living in the country,
       travelling . . .)
(C)   A judges that, all things considered, a given action is more desir-
       able than another. (He believes that exercising is an appropriate
       means to attaining the state of affairs he desires.)

(D)   The agent intends to act in the future. (He commits himself to
       exercising.)

The thesis that Davidson (1980: 99) defends in 'Intending', is that pure intention is quite simply a particular case of an unconditional judgment:

A present intention with respect to the future is in itself like an interim report: given what I
now know and believe, here is my judgment of what kind of action is desirable. (1980: 100)

In a nutshell, the reversal that takes place in the Davidsonian conception of
intentionality is that the intention is no longer conditional but conditioned
by our beliefs.

This substantive concept of intention has undeniable advantages over the
syncategorematic concept which renders the intention a part of the action. It
also recognizes those cases in which an intention is formed long before an
action is carried out, where the latter is not performed, as well as complex
acts and perhaps collective actions which require the contribution of several
agents. In Davidson's view however, the most important advantage is that it
appears to be better suited to his causal theory. Davidson (1980: XIII) maintains, in fact, that the intention is not a part of the action but a state or event
separated from the intended action. It is a state or event because, as mentioned earlier, an action is an event described with the help of intentional
concepts and the intention is the event that is its cause. To be its cause however,
the intention must be separated from the action as all causal relations are.
We are thus faced with two types of relations between the intention and the
action: a *logical relation* in the light of which we explain the emergence of
the intention to act and a *singular causal relation* between two events described
with the help of intentional concepts. To these two types of relations correspond two types of explanations which establish the connection between
reasons and the action. The problem we will now examine is concerned with
how to reconcile these two kinds of explanations.

THE EXPLANATION OF BEHAVIOR AND THE THEORY OF IDENTITY

The explanation of behavior is generally considered as an answer to the
question: "why?". There are as many answers to this question as there are
ways of characterizing behavior. For the manner in which a behavior is
described determines the kind of explanation required. If behavior is described

in terms of intentional actions, then the explanation will consist in retracing the agent's course of reasoning between the premises and the conclusion. It goes without saying that a single action can be described in several different ways depending on the desires, beliefs, and other attitudes that we attribute to the agent as we seek to interpret and explain his action. But the same behavior can just as well be described in purely physical, biological, and neurophysiological terms when it forms the object of the explanation of any one of these sciences. As pointed out earlier, the main problem brought up by Davidson in 'Actions, Reasons, and Causes' concerns the role of causality in the explanation of action. Indeed, Davidson's main thesis in this article is that explanations by rationalizations are types of causal explanations. Davidson reestablishes the concept of causality, against the objections raised by the Neowittgensteinians, with the purpose of dispensing with the concept of causality precisely where there are no laws linking reasons, intentions, and action. It is in this manner that Davidson exposes the narrow connection between the concept of description and that of causality and succeeds at once in reconciling two apparently contradictory points of view: Causality links two *singular* events whatever their description may be whereas the contexts of explanations where we have recourse to laws are opaque in the sense whereby the manner in which the terms of the relation are described is determinative for the explanation. An explanation that has recourse to the concept of law when events are described in a physical vocabulary has to do with types of events and it consists in subsuming a particular event having this or that property under a single type. It is precisely the ambiguity of the way in which the concepts of law and causality are used that is at the source of numerous misinterpretations of his theory of identity and the thesis of the irreducibility of psychological explanations to the kinds of explanations found in the sciences.

The thesis of the irreducibility of psychological explanations in terms of intentional attitudes to scientific explanations refers to what Davidson (1980: 220) called in 'Mental Events' his "anomalous monism". "Monism" refers to the ontological reduction and the identity of physical events to psychological events. This also means that the intentional idiom, just like the physical, is not an ontological category but a conceptual one – two descriptions of a single *descriptum*. This monism is qualified as "anomalous" in order to mark the irreducibility of the concepts used to identify psychological events to those which identify physical events. For, an event that can be described in a vocabulary that makes use of intentional concepts necessary to their identification does not fall under strict laws.

In 'Mental Events', Davidson articulates his anomalous monism on the following three principles.

P1   Causal interaction between the physical and the mental.

P2   The nomological character of causality.

P3   The anomalism of the mental.

The first premiss of his theory of identity, reemphasizes Davidson (1980: 215), "deals with events in extension and is therefore blind to the mental-physical dichotomy". It simply stipulates that certain mental states are the causes or effects of psychological events. The second principle stipulates that when two events are causally related, they can, when described in a physical vocabulary, instantiate laws. The third stipulates that when these events are described with the help of intentional concepts, they cannot instantiate laws. There are also no strict laws linking the events described in a psychological vocabulary to events described in a scientific vocabulary. The first two principles are concerned with the concepts of causality and of law, whereas the third is concerned more specifically with the relationship between the psychological and physical *concepts* or properties which Davidson claims to have taken into account in his concept of "supervenience". The anomalism of the mental represents, in fact, the attempt to conciliate the ontological identity and the conceptual difference. It constitutes Davidson's contribution to the epistemological debate opposing explanation in the sciences of the mind to explanation in the natural sciences.

Indeed, this debate formed itself around the question of the authentically causal character of explanations by rationalizations and of their explanatory power. Stoutland (1976), for example, pointed out against Davidson's causal theory that in speaking of intentional behavior, it is not sufficient to say that desires and beliefs cause behavior. These reasons must, in addition, cause behavior as reasons. Thus, according to him the oblique theory of causality implies that the reasons cause behavior as physical events and not as reasons. For the causal efficiency of an event would be indebted, not to its semantic properties, but to its physical properties. Consequently, by making reasons the causes of behavior, Davidson would have rendered the concept of intentionality useless if it is true that to act intentionally, an agent must be in the position of knowing if he acts for a reason.

If the reasons, as reasons, are not effective on behavior, then they won't have any explanatory power. This is, among other things, what Føllesdal maintains (1985). He criticizes Davidson for neglecting the difference between causal contexts, which are *transparent*, and contexts of explanation, which are *opaque*. Causal contexts are transparent to the extent to which one's way of referring to the antecedent and to the consequence of a causal relation is of little importance. On the other hand, when we explain behavior in terms of reasons, the way in which we refer to the terms of this relation is determinative if what is stated must have an explanatory value. For, if an explanation is to answer satisfactorily to the question "why?", it must have recourse to reasons. Thus, the reasons attributed to an agent in order to explain and predict his behavior presuppose, as mentioned earlier, that his behavior be conceived as an action. On the other hand, we also demand of an explanation that it informs us about the cognitive and emotional field of the agent. In addition, when we think that we know the reasons, which we express by virtue of an explanation, there is a gain which appears to be attributable to a relation that we establish between reason and action, which, in turn, appears to be essentially different from a causal relation. In other words, the assertion, for example, that reasons are in fact physiological causes of behavior, doesn't have the explanatory power that is demanded of it. If it is so, that is, if the characterization of behavior not only precedes but determines explanation, then the question concerning the latter must also be raised even before causality is, in principle, established. The question that must now be answered is whether or not these problems represent true objections to Davidson's theory.

The epiphenomenal objection is the result of a confusion not only between the nomological and dispositional concepts of causality, but also between the latter and the logical connection which we presuppose between reasons and the action. The nomological concept of causality is present in the second principle and is associated with the concept of "strict" law, in the sense whereby it doesn't use any "ceteris paribus clauses" or concepts as "dispositions", "tendencies", potentialities", etc. Thus, against those who maintain that we can dispense of causality in exchange for a strictly nomological explanation, Davidson (1987) replies that the dispositional concept of causality imposes itself precisely in the absence of scientific structures of explanation and strict laws: "appeal to causal concepts is appropriate to the explanation of action in part precisely because strict laws are not available." (1987: 42)[6] Indeed, the concept of law is a linguistic concept which takes the form of the conditional and links particulars in so far as they fall under the same type. It is the opposite of the concept of "singular causal relation" which is associated with explanations by rationalizations and which intervenes between events whatever their descriptions may be. Let us recall that an explanation by rationalization essentially consists in reconstituting the course of practical reasoning which goes from the desire/belief couple, the ends and the means, and the intention through which the agent commits himself to act, to what the interpreter or observer describes as an action. Now let us return to our classic example. We can explain why someone exercises by attributing to him the desire to be healthy, the belief that exercise will make him healthy and the judgment that there is something desirable about exercise. I would like to suggest here that the thesis of the irreducibility of explanations by rationalizations to explanations in terms of law[7] imposes itself only if we take into account the changes which we have alluded to and which concern his conception of practical reasoning conceived as an unconditional inductive judgement as much as it concerns his conception of intentionality, of pure intention, where he insists to a greater degree on the agent's relation to the desirability of a single action which he commits himself to carrying out. The argument is rigorously the same as the one which he applies to his first theory and as we will see, refers to the normative concept of psychological concepts.

The remaining problems are those concerned with the authentic causal character of explanations by rationalizations and epiphenomenalism. To say it quickly, Davidson points out against the epiphenomenal objection according to which the causal efficiency of reasons on behavior would be meaningless only if the latter caused it as reasons, the extensional character of causal relations.

For me, it is events that have causes and effects. Given this extensionalist view of causal relations, it makes no literal sense, as I remarked above, to speak of an event causing something as mental, or by virtue of its mental properties, or described in one way or another. (1993: 13)

This answer appears to contradict the thesis according to which explanations by rationalizations are types of causal explanations since, as mentioned earlier, the contexts of explanation are "opaque" in that they depend on the description of the phenomenon to be explained. But Davidson retorts by citing the case of deviant causal chains which require the concept of causal relations because it is possible that the beliefs and desires which we presuppose on the agent's part do not explain his behavior. We will soon see that the concept of causality can equally serve to identify the relation of the agent's thoughts to the events in the world.

In 'Thinking Causes', Davidson points out that if the doctrine of supervenience is correct, then the mental must make a difference. The concept of "supervenience", through which he claims to have captured the relation between the psychological properties which we attribute to certain objects and their physical properties, is the thesis according to which the ontological reduction does not necessarily lead to the nomological reduction. In 'Mental Events', Davidson defines it in the following manner:

That there cannot be two events alike in all physical respects but differing in some mental respect, or that an object cannot alter in some mental respect without altering in some physical respect. (1980: 214)[8]

At first glance, this definition implies that a change in the psychological properties of an individual is always accompanied by a change in his physical properties. This is directly derived from the ontological reduction. But what is the difference between psychological properties and physical properties? Can one and the same psychological property supervene on different physical properties? Are there several different ways of correlating the psychological with the physical? Let us examine a different formulation of supervenience which Davidson proposed in his recent 'Thinking Causes':

A predicate p is supervenient on a set of predicates S if and only if p does not distinguish any entities that cannot be distinguished by S. (1993: 4)

We can consider these predicates as concepts belonging to two systems of classification.[9] But a more appropriate way of approaching the doctrine of supervenience is the thought experience "The swampman". Relying heavily on H. Putnam's and T. Burge's causal theory of reference and recovering as his the denunciation of an internalistic, individualist, or solipsistic conception of thought and signification, Davidson recognizes that thought contents are partly determined by certain factors (social and physical) which are external to the individual and partly determined by their causal relations to objects. Against Putnam and Burge, Davidson maintains that they are nevertheless "internal" because they are identical to the brain's[10] physical states. "The swampman" is supposed to show that two agents "can be in all relevant physical respects identical while differing psychologically." (1986: 453)

Here is the story:

Suppose lightning strikes a dead tree in a swamp; I am standing nearby. My body is reduced to its elements, which entirely by coincidence (and out of different molecules), the tree is turned into my physical replica. My replica, The swampman, moves exactly as I did; according to its nature it departs the swamp, encounters and seems to recognize my friends, and appears to return their greetings in English. It moves into my house and seems to write articles on radical interpretation. No one can tell the difference. (1986: 443)

No one can tell the difference because the data delivered by the observation of his behavior are, in principle, insufficient to determine if it is the author of 'Radical Interpretation' or simply an artefact. We can suppose that when Swampman emits a given sound or behaves in a certain way, he is in a physical state that is identical to that of Davidson's. There is nonetheless a difference between the latter and his physical copy. This psychological difference is concerned with the manner in which the relation is established between the thought contents and the world of objects with which they interact. More precisely, the psychological difference, assuming that we can attribute thoughts to The swampman, is due to the role that the social, physical, and historical context plays, not only with regard to the way in which we acquire thoughts but especially in the determination of thought contents. We see it here, the story exposes how the relation is established between thoughts and the world. Davidson attempts to show that the psychological difference must be sought in the individual's causal interaction with certain aspects of the environing world.

A creature or object cannot have a thought about stars or squid or sawdust unless that thought somehow traces back causally to appropriate samples. There cannot be a memory of an event or person unless there has been causal commerce with the event or person. (1990: 15)

As in the case of the difference in signification of the words "water" and "twater", Davidson points out that the manner in which a thought is acquired or the usage of a word is learned, is itself constitutive of thought and signification. Like Putnam, we can say that the correct interpretation of what a person thinks is not uniquely determined by what is in her head but also determined by the history of causal intercourse that takes place with the objects and the persons of the world. That is why these thoughts are normally identified by the social and historical context in which they are acquired. But the ultimate difference between Davidson and his artefact, between the physical and psychological properties of an event, and the reasons that amount to thought making a difference, refer to the idea of norms which we shall now examine.

ON THE NORMATIVE CHARACTER OF RATIONALITY
AND THE ANOMALISM OF THE MENTAL

I now want to examine one of the arguments which Davidson uses to support his defense of anomalous monism, namely, the normative character of rationality and the concepts used in the explanation of behavior which, according to him, represents an inevitable obstacle to the reduction through rationalizations to other types of explanations using non-normative concepts and more particularly, to strict laws.

In 'Mental Events', Davidson (1980: 222) suggests that the indeterminacy of translation could serve as an argument in support of the anomolousness of the mental. He reemphasized that the heteronomous character of the concepts used to measure the mental refer to the radical translation through which Quine (1960) introduced his doctrine of the indeterminacy of translation. That which is equivalent in Davidson's thought is the radical interpretation concerning the interpretation of phrases, the attribution of propositional attitudes, as well as relations with the world. His theme is verbal and nonverbal behavior since the observation of behavior forms the basis of all attributions of intentional attitudes. His task consists in isolating beliefs from the meaning of phrases. Such is the major epistemological problem which must be dealt with. For, as pointed out earlier, the interpreter has access to his informant's beliefs only through the observation of his behavior. The evidence given by the observation of behavior underdetermine our interpretation of it. Here is where the indeterminacy, which Davidson applies, over and above meaning, to thought steps in. Davidson (1984: 139) sees it as a "trade-off" between beliefs and other attitudes that we attribute to an agent and the interpretation of his behavior. This trade-off can be conceived in terms of relations between interpretation and attribution. In this manner, several different interpretations can equally account for the observed behavior without ever being compatible with one another. And for Davidson as for Quine, there is no "fact of the matter" with regard to the choice of an interpretation over another.

However, in 'Three Varieties of Knowledge', Davidson reemphasizes that it is not the indeterminacy in and of itself which represents an argument in support of the anomolism of the mental.[11] Rather, it is the presupposition of rationality which it unveils in the explanation of action and the normative character of rationality and intentional concepts. It is perhaps in this sense that we must understand the following passage which I borrowed from 'Radical Interpretation':

This irreducibility is not due, however, to the indeterminacy of meaning or translation, for if I am right, indeterminacy is important only for calling attention to how the interpretation of speech must go hand in hand with the interpretation of action generally, and so with the attribution of desires and beliefs. (1984: 154)

Conceived in terms of relations between interpretation and attribution, the indeterminacy thus brings out the degree of rationality and stability that the interpretation presupposes in the agent's cognitive field. For, as Davidson points out in this passage, the beliefs attributed to an agent on the basis of verbal behavior must be consistent with those attributed to him in the interpretation of action. This presupposition allows us to significantly limit the breadth of the indeterminacy. This and other[12] constraints are presuppositions which are alien to a strictly nomological explanation of behavior.

As Quine noted in his *Word and Object* (1960: 59, 77–79), the "principle of charity" imposes itself on our linguist or interpreter in situations of radical translations, that is, when he is placed before his informant's sentences or behaviors which appear to him to be devoid of sense or basically irrational. Cases of irrationality include incontinent behavior, discussed earlier, whereby there is a visible deviation in relation to norms or standards of rationality. These norms, which we presuppose in the interpretation of action, are *constitutive* of the thoughts, beliefs, and intentions which we attribute to an agent in such a way that to attribute thoughts to an agent, we must presuppose that he adheres to the same norms of rationality to which the radical interpreter adheres.

To elucidate the normative character of rationality, let us examine two principles which Davidson (1991: 158) links to the "principle of charity". The first, the principle of consistency or coherence, is concerned with the relation among thoughts, and the second, the principle of correspondence, is concerned with the relation between these thoughts and the objects and events in the world. The first is a logical relation of beliefs among themselves in the cognitive field of the agent and between beliefs, desires, and intentions. The second is concerned with the relation between beliefs and the objects of the world to which beliefs refer. As we have seen, the latter relations are of the causal type.

The cases of irrationality expose the kind of relations considered as standards. Davidson's preferred examples refer to asymmetry and to transitivity but we can equally cite cases of incontinence whereby the agent does not behave as a function of the action which he judges to be the most desirable and thus goes against his own conception of what he judges to be reasonable. These cases of akrasia are particularly interesting here because they expose the logical and causal relations which are supposedly constitutive of intentional attitudes and the agent's rationality. As we have already pointed out, these cases are problematic for a causal theory of action because beliefs and desires are normally supposed to cause the action. In this case however, the agent has a reason that is not the cause of his behavior. In other words, we are dealing here with "a mental cause of an attitude, but where the cause is not a reason for the attitude it explains." (1985: 347). The case of "deviant causal chains" rigorously presents the same problems. In this case, the logical relation, presupposed in a continent behavior between the reasons and the action, is absent whereas only the causal relation remains. Cases of irrationality equally expose the principle of coherence since akrasia, self-deception, and other similar cases are essentially matters of internal inconsistency. For a belief or wish cannot be considered in isolation as irrational. They can be deemed irrational only in relation to other beliefs and desires.

These two principles of charity, to which all rational animals are supposed to subscribe, clarify our understanding of the normative character of the concepts used in the explanation of intentional behavior. "Normative" first in the Kantian sense: In normal circumstances, an agent considered as rational according to our own standards will act in a way which he deems to be the most desirable. He can, of course, deviate from these norms without being non-rational. But if he does, he will go against what he deems to be reasonable. This is another way of saying that we presuppose that most of his beliefs are true. Normative also in the sense whereby these kinds of relations are constitutive of beliefs, desires, and intentions through which we rationalise behavior – constitutive of in the sense whereby thoughts are defined in terms of intentional relations with objects: A belief is a belief because it is about something that is not the belief itself. To have a belief, however, we must have many. Furthermore, we presuppose certain relations which tell us something about the agent's cognitive field. Thoughts are usually identified by these kinds of relations. With regard to the relation of thoughts to objects, causal relations can serve to identify them: carrying an umbrella because of the belief that it will rain whereby the belief pushes the agent to act in the way that he does. This belief implies other attitudes on the agent's part, more particularly, the wish to remain dry, beliefs which are relative to the umbrella's function, to his ability to use it, to the weather, etc. A thought is such only within a network of relations.

But what are these norms? Davidson (1985: 352) admits to not being able to draw up a list. Nevertheless, the most acceptable model to date for the study of rationality and action is, for Davidson, the theory of decision. However, Davidson is far from thinking that the agent is rational as the theory of decision understands it. As we have already noted, we must, for strictly methodological reasons, maximize rationality. But constraints must equally be imposed on it by considering the agent's limitations. Thus, it is not the axioms of the theory of decision that are relevant for the explanation of rational action, but rather, it is the way in which its views with regard to the concepts of choice, decision, preferences, etc. can contribute to an elucidation of the problem related to the attribution of beliefs, intentions and desires.[13] Generally speaking, Davidson insists on two aspects of the theory of decision: The first is concerned with the agent's preferences relative to the different available alternatives. The agent evaluates these alternatives as a function of the probability of the results and the value attached to these results. The second aspect is concerned with the choice and decision through which the relation is established between the preferences and the action. We *presuppose* that, among the available alternatives, a rational agent will choose the one that has the "highest expected value". These two aspects appear to correspond to two of the elements which figure in the schema of the explanation of rational action presented earlier. Thus, an agent values a certain state of affairs; he evaluates different alternatives as a function of the probability of the results and the value assigned to these results; among the alternatives, he chooses the one that has the "highest expected value"; he therefore does "a" *because* it has the highest expected value. The order, from the premisses to the conclusion, formally corresponds to the schema of intentional action that we attributed to Davidson. However, there are differences, these are related to the different limitations introduced by the concept of the intention to act or the unconditional evaluative judgment. The first limitation concerns the agent's beliefs relative to the possible alternatives and to the value assigned to their results. For, according to Davidson, if the unconditional judgment must play a role in the determination of behavior, then an analysis of intentional behavior should take into account the fact that a given agent is limited by his practical skills, by the number of beliefs which significantly limits his alternatives or by the means he can imagine, and by the circumstances. It is in this sense that we have said that an agent's evaluative judgment is not conditional but conditioned by his beliefs and physical capacities, etc. The second limitation is concerned with the relation between reasons and action. The answer "*because* 'a' has the 'highest expected value'" to the question "Why does the agent decide on a certain action given the circumstances?" is not sufficient for the same reasons which forced Davidson to abandon the deductive conception of reasoning. For, to fulfill the desiderata of the explanation, the fact that there is something desirable with regard to the action must be included in the premisses. Thus, normativity intervenes precisely in the judgment on the desirability of the action. For, as the cases of irrationality show, it is only because we presuppose that the agent is rational in the light of our own norms that we can attribute thoughts to him and explain his behavior as the result of the course followed by his reasoning.

CONCLUDING REMARKS

Even in recognizing that the normative character be included in the unconditional evaluative judgment on the desirability of an action, the question remains: Why does the normative character of intentional concepts and rationality represent an argument in support of the anomalism of the mental and against the incorporation of these concepts to a system of strict laws? The first answer is that strict laws cannot combine normative concepts, that of the intention to act, for example, with nonnormative concepts (1991: 162). This is what the experience of thought "The swampman" shows, through which Davidson demonstrates that the determination of thought contents depends, in part, on the causal interactions with the social and physical environment. Davidson's argument rests essentially on the normative character of rationality and intentional concepts which enter in the description and characterization of behavior. "Normative" first in the sense in which the kinds of relations presupposed by the principles of coherence and of correspondence are constitutive of intentional concepts. But "normative" also to the extent to which these concepts and rationality are not simply the fact of a single interpreter, but are shared by a community of agents. In this respect, it is important to reemphasize the distinction Davidson makes among the three types of knowledge: knowledge of the self, that of others, and knowledge of the world. Descartes privileged the first, believing that the other two could be derived from it. For reasons discussed earlier, Davidson claims to have ruptured with both Cartesianism and positivism by affirming the irreducibility, but also the complementarity of these three varieties of knowledge. They form a "three-way relation" which he describes in terms of triangulation. Thus, my knowledge of the other is possible only if I presuppose that we share the same world:[14]

A community of minds is the basis of knowledge; it provides the measure of all things. It makes no sense to question the adequacy of this measure, or to seek a more ultimate standard. (1991: 164)

Indeed for Davidson, the intersubjective nature of the intentional and of the rational ultimately constitutes the source or the origin of the difference between the understanding of thought and the explanation of the physical world. The remainder has to do with truth.

Université du Québec à Montréal

NOTES

[1] Letter to O. Kraus cited by R. Chisholm 'Brentano on Descriptive Psychology and the Intentional', E. N. Lee *et al.* (eds.), *Phenomenology and Existentialism*, Baltimore, The John Hopkins Press, 1967, p. 13.

[2] As a reply to M. Dummett (1989: 18), Davidson proposes the following solution to Brentano's problem: "There is no need to suppose that if there are no such inner objects, only outer objects remain to help us to identify the various states of mind. The simple fact that we have the resources needed to identify states of mind, even if those states of mind are, as we like to say, directed to non-existent objects, for we can do this without supposing there are any objects whatever «before the mind»."

[3] See A. R. Mele's 'Recent Work on Intentional Action', *American Philosophical Quarterly* **29** (1992), 3.

[4] Indeed, Anscombe (1957: 1) distinguishes three uses of the term "intention": adverbial, "acting intentionally"; substantive, "having the intention to act"; syncategorematic, "acting with an intention". In his preface to *Essays on Action and Events* (p. XIII), Davidson reemphasizes that he had initially privileged the syncategorematic use to which the other two, he claimed, were reduced. His more recent writings privilege the substantive concept of intention whereby it is not a part of the action but a "state or event separated form the intended action" (1980: XIII). Thus, contrary to the position he had adopted in his 1963 article whereby acting with an intention was a syncategorematic expression which did not refer to any state, disposition or event, intention is a state separated from reasons. As he pointed it out elsewhere (1987: 39), the advantage of this concept of intention over the other two is that it recognizes those cases whereby an intention is formed before the intended action is carried out, whereby the intended action is not carried out, or when we want to explain how complex actions are controlled or "monitored".

[5] Davidson (1980: 97), (1984: 44), (1985a: 201).

[6] As Davidson explains (1992: 162) "Strict laws do not employ causal concepts, while most, if not all, mental concepts are irreducibly causal."

[7] Using as a target the kinds of explanations advocated by positivists, Davidson (1980: 274)

in 'Hempel on Explaining Action', points out that the concept of law takes the form of the conditional and answers the following kind of question: Why is it that a given substance (sugar) dissolves in a given liquid (coffee); the answer would no doubt sound like this: If something is soluble, it will dissolve in a certain kind of liquid; now, coffee is such a liquid and sugar is soluble; consequently . . . . According to him, a reason explanation "tells us nothing about all sugar-kind, though it tells us a lot about this cube. I think reason explanations are of the second kind."

[8] See Davidson (1985a: 242) for a different formulation of supervenience. We will also use Kim's (1984) and Laurier's (1989) articles as well as J. Kim's, E. Sosa's, and McLaughlin's comments on Davidson (1993) in the collection edited by A. Mele *et al.* (1993).

[9] The first system uses the concept of animal and the other of goat. Let us assume with Davidson (1990: 18) and (1985a: 242) that the objects to be identified form a series in which the first three elements are sheep and the forth is a goat. Before the last element, we can change the system and identify the elements of the series with the concept animal: animal 1, 2, 3, 4, . . . The idea here is that this system can indeed distinguish every item of the series by subsuming it under its concept, but it can't distinguish goat from sheep. This means that normative intentional properties or concepts which we use in a psychological vocabulary can elude the *descriptive* vocabulary of physics. In this manner, if it is true that an event cannot be considered as intentional simply because it has properties which cannot be described in physical terms, then intentionality cannot be reduced to physical properties. It follows then that the psychological difference should be sought elsewhere, other than in physical properties on which it nevertheless depends. It is also because physics does not make use of the necessary resources to "sort out" the psychological properties of an event that the mental cannot be captured by physics' schemas of explanation.

[10] In 'The Myth of the Subjective' (p. 164), Davidson reinterprets the moral of Putnam's fiction: ". . . two speakers may be alike in all relevant physical respects, and yet they may mean quite different things by the same words because of differences in the external situations in which the words were learned. Insofar, then, as the subjective or mental is thought of as supervenient on the physical characteristics of a person, and nothing more, meaning cannot be purely subjective or mental." Likewise, in 'What is Present to the Mind' (pp. 12–3), Davidson points out the consequences of his analyses on his doctrine of supervenience: "The point of this exercise may surprise us. It is that subjective states are not supervenient on the state of the brain or nervous system: two people may be in the same physical state and yet be in different psychological states. This does not mean, of course, that mental states are not supervenient on physical states for there must be a physical difference *somewhere* if psychological states are different. The physical difference may not be in the person; like the difference between water and twater, it may be (we are supposing) elsewhere."

[11] Davidson (1991) explains that it is because "indeterminacy turns up in both domains".

[12] I enumerated a certain number in Fisette (1992).

[13] See D. Davidson (1980: 273 and following) and (1985a: 199).

[14] Regarding the question of asymmetry between knowledge of the self (direct) and knowledge of the other (inferential), see Davidson (1986, 1989a).

REFERENCES

Anscombe, E., 1957, *Intentions*, B. Blackwell, Oxford.

Brentano, F., 1973, *Psychology from an Empirical Standpoint*, Humanities Press, New York.

Davidson, D., 1989a, 'The Myth of the Subjective', in M. Krausz (ed.), *Relativism*, University of Notre Dame Press, Notre Dame.

Davidson, D., 1986, 'Knowing One's Own Mind', *Proceedings and Addresses of the American Philosophical Association*, pp. 441–458.

Davidson, D., 1989b, 'What Is Present to the Mind', *Grazer philosophische Studien* **36**, 3–18.

Davidson, D., 1980, 'Toward a Unified Theory of Meaning and Action', *Grazer philosophische Studien* **2**, 1–12.

Davidson, D., 1963, 'Action, Reason and Causes', in Davidson (1980).

Davidson, D., 1980, *Essays on Action and Events*, Clarendon Press, Oxford.

Davidson, D., 1984, *Inquiries into Truth and Interpretation*, Clarendon Press, Oxford.

Davidson, D., 1987, 'Problems in the Explanation of Action', in P. Pettit *et al.* (eds.), *Metaphysics and Morality, Essays in Honour of J. J. C. Smart*, B. Blackwell, London.

Davidson, D., 1993, 'Thinking Causes', in A. Mele *et al.* (eds.) (1993).

Davidson, D., 1985a, 'Replies', in B. Vermazen and M. B. Hintkka (eds.), *Essay on Davidson: Actions and Events*, Oxford University Press, Oxford.

Davidson, D., 1985b, 'A New Basis for Decision Theory', *Theory and Decision* **18**, 87–98.

Davidson, D., 1990, 'Representation and Interpretation', in K. A. Mohyeldin Said *et al.* (eds.), *Modelling the Mind*, Clarendon Press, Oxford, pp. 13–26.

Davidson, D., 1991, 'Three Varieties of Knowledge', in A. P. Griffiths (ed.), *A. J. Ayer: Memorial Essays*, Cambridge University Press, Cambridge.

Davidson, D., 1982, 'Paradoxes of Irrationality', in R. Wollheim (ed.), *Philosophical Essays on Freud*, Cambridge University Press, Cambridge, pp. 289–305.

Fisette, D., 1992, 'Indétermination de la traduction et intentionalité', *Philosophie* **38**, 58–90.

Føllesdal, D., 1985, 'Causation and Explanation: A Problem in Davidson's View on Action and Mind', in E. LePore and B. McLaughlin (eds.), *Actions and Events: Perspectives on the Philosophy of Donald Davidson*, B. Blackwell, Oxford.

Hempel, C., 1962, 'Rational Action', *Proceedings and Addresses of the American Philosophical Association*, The Antioch Press, New York, pp. 5–24.

Kim, J., 1984, 'Concepts of Supervenience', *Philosophy and Phenomenological Research* **65**, 153–176.

Laurier, D., 1989, 'L'anomalisme du mental et la dépendance psychophysique' (Manuscript).

McLaughlin, B., 1993, 'On Davidson's Response to the Charge of Epiphenomenalism', in A. Mele *et al.* (eds.) (1993).

Mele, A., 1992, 'Recent Work on Intentional Action', *American Philosophical Quarterly* **29**(3), 199–218.

Mele, A. *et al.* (eds.), 1993, *Mental Causation*, Clarendon Press, Oxford.

Quine, W. G. O., 1960, *Word and Object*, M.I.T. Press, Cambridge.

Stoutland, F., 1976, 'The Causation of Behavior', *Acta Philosophica Fennica* **28**, 286–325.

ROBERT NADEAU

ECONOMICS AND INTENTIONALITY

1. THE PERVASIVENESS OF INTENTIONALITY IN ECONOMICS

A good way of characterizing what is usually called the 17th-century "revolution of modern science" is to focus on Galileo Galilei's theory of explanation. As is well known, he set aside three of the four Aristotelian causes (material, formal and final causes) in order to couch all sound scientific explanations in terms of efficient causes. In the second half of the 19th century a new scientific revolution occurred with Darwin's theory of evolution. As has been stated repeatedly, Darwinism also has something to do with the abandoning of teleology in science, as speciation is explained without any appeal to final causes. But in the last quarter of the 19th century a third scientific revolution occurred, this time in the social sciences. Many philosophers of science fail to notice or understand this intellectual event. This third scientific revolution is usually called the "marginalist revolution." The transformation of political economy into pure economics, and progressively, into mathematical economics had at least two distinctive features. First, this revolution broke out simultaneously but independently in three different European countries: with Carl Menger (1840–1921) in Austria, with William Stanley Jevons (1835–1882) in England, and with Léon Walras (1834–1910), who, in 1870, was the first to hold the Chair of Political Economy at the University of Lausanne in Switzerland.[1]

What is striking and should not be overlooked here is that this third scientific revolution did not reject teleology or the legitimacy of giving scientific explanation by making reference to final causes. In the realm of economic action (microeconomics), it even seemed to be theoretically sound to think that any individual economic agent, be it a personal or an institutional one, was guided by his "preferences" and was trying to maximize what is now called a "utility function." In a sense, this new conceptual framework was emerging because in the field of economics, theoreticians were working with a new, plainly subjective, theory of "value." John von Neumann and Oskar Morgenstern further emphasized this point in their *Theory of Games and Economic Behavior* (1944), a theory which is held to be applicable to situations of risk, and which takes as its most central explanatory principle the rule of the maximization of expected subjective utility. To put it in a nutshell, one could say that this very basic principle simply asserts that any rational economic agent behaves as if he would sum up all individual utilities, each utility being weighted or pondered by its probability of occurrence. As a result of this crucial turn of neoclassical economics, intentional language came

159

M. Marion and R. S. Cohen (eds.), Québec Studies in the Philosophy of Science II, 159–176.
© 1996 *Kluwer Academic Publishers.*

to form an essential part of the main conceptual framework of this scientific domain.

What exactly do we mean when we say that after the marginalist revolution, economists began to speak the language of intentionality? It may seem that a fundamental link to the idea of "intentionality" is to be found in "consumer choice theory," where it is related in the first place to the concept of *preference*. In this theoretical context, a choice is always assumed, as a matter of axiomatic definition or postulation, to express a preference, and all preferences held by one economic agent are transitive. It may be more accurate, however, to say that intentionality is the hallmark of the new theory of value propounded by the marginalist approach. As the marginalist revolution erects a brand new theory of economic value that is explicitly subjectivist (value being the result of a mental act, and utility being not an objective property of things in themselves but a currency for measuring preference relations), one can say that neoclassical economics is a theoretical framework thoroughly based on intentional concepts. Now, there are at least two radically different ways of looking at this situation. One can start with some preconception concerning what a "legitimate empirical science" must look like and try to see whether neoclassical theory passes the test, or one can choose to deal somewhat more directly with the epistemological specificity of economic science as captured by neoclassical theory.

Today Alexander Rosenberg is an eloquent and persuasive advocate of the first option. Rosenberg maintains that economics' epistemological backwardness can be largely explained by its unjustifiable complicity with common sense. According to this argument, the main epistemological defect of economics is that it does not cut nature at its joints or, in other words, that it fails to identify variables that are "natural kinds."[2] For Rosenberg and many other philosophers, it would seem that the "mind" has no place in social science. For many empiricist philosophers and positivist social scientists, the concept of mind that must inevitably be used when doing the kind of economics neoclassical theorists have developed is apparently one which only a kind of "unscientific" psychology still accepts. Such thinkers thus attack microeconomics' dependency on "folk psychology," arguing that it stems from a pseudoscientific or merely ideological point of view. According to Rosenberg's analysis, for instance, not only is the neoclassical theory of consumer choice the cornerstone, even the "paradigm" of the dominant trend in economics, but all subsequent refinements in microeconomics, even including game theory, is in fact nothing but "formalized folk psychology."

At the risk of failing to do justice to Rosenberg's dense and daring argumentation, I would like first to examine its central tenet. According to Rosenberg, fundamental economic theory posits that human action is entirely a function of the "desires" and "beliefs" of agents, and it does so to such an extent that, *ceteris paribus*, desires and beliefs causally explain all economic action. The only "law" cited in this theory could be formulated in the following way:

[L] = "Given any person x, if x wants d and x believes that a is a means to obtain d, then, under the circumstances, x does a".[3]

For Rosenberg, the problem here is that the microeconomic theory that provides the foundations for the whole scientific enterprise refers to "intentional" variables such as desires and beliefs, which are for all intents and purposes impossible to measure independently of the theory advanced by hypothesis. In other words, these variables are impossible to measure without first assuming that [L] is true. However, [L] which roughly corresponds to what is usually called the "rationality principle" does not seem to be testable, since in social science, and especially in economics, any attempt to do so inevitably takes the principle for granted. Rosenberg's analysis echoes in many ways the arguments of all those who have questioned the apparent emptiness of empirical content, and the proto- or pseudo-nomological status, of this principle, which seems to be indispensable in explaining economic action.[4] But his argument takes a very special and singularly important turn because it aims at challenging the conceptual framework which has helped constitute economics as it is known today.

What does it mean to argue that the terms designating the functional variables of [L] belong to the intentional vocabulary? In essence, this claim implies that there is no way to establish scientifically – in other words, on the basis of controlled experimental observations – that the beliefs and desires of an agent are in fact what make that agent act exactly as she does in a given situation. Since beliefs and desires cannot be described in terms which would make empirical control possible, these concepts have little informative value and the theory which puts them in a functional relation with the action to be explained or predicted can never have any serious explanatory power and can never succeed in increasing its predictive power.[5] Likewise, this theory proves uncontrollable, untestable, even entirely irrefutable. Thus it would be best to discard it and replace it with another founding conceptual framework that would not simply be a useful expedient. The replacement could be provided by experimental psychology but, according to Rosenberg, surely not by the behaviorist approach since it does not seem to be entirely free of intentional language. We might suppose that it could come from neurophysiology. I will discuss this topic in Section 3 below. For the time being, I would like to take a closer look at the fact of the matter, i.e. at what, more precisely, we have in mind when we speak of the intentional language of neoclassical economics, and of all theoretical economics in general that adopts the subjective theory of value originating from the marginalist revolution.

It seems to me that there is no need to dispute Rosenberg's severe and generally negative observations: his analysis is in a sense correct and the factual part of his conclusion is difficult to challenge. Certainly we should be grateful to Rosenberg for not having limited his remarks to a criticism of what he considers to be the rather "quaint" concepts of contemporary microeconomics, even if, in the end, his exploration of other possible avenues of research seems to provide only a very dubious solution to the theoretical and conceptual dead end in which he finds economics. My question is directed at a gap between the two main parts of Rosenberg's argument, namely, his observation of microeconomics' apparent failure, and his proposals for new avenues of research. Everything hinges on this part of the argument, that is, the moment when the phenomenon of intentionality and its role in economics is apparently taken into consideration and examined. For Rosenberg, the mental is simply unwelcome in economics; but this thesis requires more careful consideration.

I believe, unlike Rosenberg, that appeal to intentionality is here to stay in economics. I would even go so far as to assert that this reality of intentional phenomena is *constitutive* of the domain of economic theory and that its discovery marked the entrance of economics into the twentieth century. In order to establish this thesis, I would like to assert that Rosenberg's argument does not do justice to this phenomenon. Rosenberg rejects "intentionalist" terminology on the pretext that it lacks sophistication. One might say that to him it constitutes a kind of picturesque mythology without any real explanatory value. In fact, it might be asked whether, in condemning intentional language as he does, Rosenberg is not throwing the baby out with the bath water. What would be the point of giving up concepts of need, preference, and utility if, in so doing, one ended up giving up what these concepts were used, with varying degrees of success, to reveal, namely the irreducibility of social phenomena to pure physical phenomena? We might also ask whether Rosenberg is right to hold that the intentional terminology he denounces on the ground that it is epistemologically unsound corresponds to the psychologizing language of common sense, and that the concepts used in microeconomics, far from being technical concepts the definition of which is not obvious, designate mental entities the existence of which is dubious.

Let us begin with the last point. Rosenberg recognizes the formal value of microeconomic conceptualization. It is rather the *empirical scope* of intentional concepts which he questions. However, if he is ready to recognize that the axiomatic structure of microeconomics is logically valid and mathematically faultless, then should he not also recognize that the first virtue of an axiomatic approach stems from the fact that it allows one to neutralize all the usual connotations of the terms used in the various axioms? If this axiomatization truly constitutes a system of implicit definitions, should one not consider that, far from having their traditional or "popular" meaning, the concepts used in economics, though they are expressed using terms apparently belonging to ordinary language, have virtually nothing to do with what they usually mean from the point of view of common sense?[6] It is, in my view, clearly false to say that contemporary economists use those terms in exactly the same way as we do when we speak the language of common sense.

In spite of this fact, it is also important to see that all economic concepts are thoroughly intentional, whether we are talking, for example, about "marginal value," "order of preferences," "individual agent," "transaction cost," or "expected utility." Certainly, in any system of postulates, some terms are undefined, and it must seem entirely acceptable that those terms, insofar as they are given a basic or primitive status, remain indefinitely open to theoretical interpretation through other axiomatic or axiomatizable systems. In the history of economics, it is the marginalist revolution and its subjectivist approach to economic value that seems to have made it necessary to recognize the intentionality of economic phenomena. The adoption of such a subjectivist point of view, a point of view which seems to me inevitable and of which we have certainly not yet finished learning the inescapable consequences, complicates matters in economic methodology a great deal. Allow me to explain. One way to understand the meaning of the subjectivist perspective in economics is by seeing precisely how it contrasts with the "objectivism" of classical thinkers, an epistemologically naïve and methodologically unsophisticated perspective which is still widespread in some circles. To be a subjectivist in economics is above all to opt for an analysis in terms of market processes. However, the market, far from being describable as a "mechanism" in the physical sense, that is as a machine that would operate automatically to allocate rare and uncertain resources, should be seen, as Hayek asserts, as the most efficient way for each person to discover how he should employ his resources. Thus, to adopt subjectivism is equivalent to recognizing that it is impossible for anyone to make long-term, infallible predictions about the future state of her own needs, and also about the future state of real supply and demand, or the future state of the quantity of truly available goods and services, just as it is impossible to know now which technology will be available tomorrow and what scientific knowledge will determine its fine tuning.

This is because the cornerstone of subjectivist economics is the theory of subjective value. Here, value appears form the start as an undeniable mental fact since, far from boiling down to a pure and simple question of average production costs, the value of a given good is a function of the marginal utility which the person who acquires it ascribes to it. However, it must be noted that with such a concept of value, the notion of "cost" is also radically transformed.[7] For a given economic agent, any good has essentially an "opportunity cost," which theoretically corresponds to the value of the alternative consumption one forgoes when one makes a particular choice. Since this is determined subjectively and privately, it seems useless, or even impossible, to attempt to quantify this value in any precise way. Thus we conceive of it from the outset as something inaccessible as such to external observers. Of course, there is an undeniable and crucial methodological problem when it comes to understanding how an economic agent makes a price correspond to this subjectively measured value. Nonetheless, given such a point of view, all economic phenomena appear from the start as "manifestations of individual minds." There are individuals who try to maximize their well-being, who

analyze their respective situations as well as they can, who process the information which comes to them, who make decisions, and who, finally, act in
accordance with what they consider important, what they prefer, and the
situation in which they find themselves. These basic facts about human activity
in turn support a set of crucial assumptions. For example, these facts support
the belief that individuals have an economic life analyzable in terms of relations between ends pursued and the means used to reach them. These ends
are further analyzable in terms of mathematizable relations between costs
and expected benefits. And the same basic facts about human activity support
our assumption that there are such phenomena as the supply and demand
of goods and services, exchanges between individuals, money, appreciation,
salaries, annuities, and interest, as well as such macroeconomic phenomena
as unemployment levels, inflation rates, and trade balances between states.
Social existence appears to be fundamentally based on individual and subjective perceptions, which are themselves functions of the global situation in
which each person finds herself, but to which no other person has direct
access.

   If this analysis is sound, then a crucial epistemological fact that needs
recognition is that *most if not all theoretical terms used in economics are
intentional*. So, not only are the most obvious ones openly intentional, such
as "preference," "cost," and "risk aversion," but also those that we would
not recognize spontaneously as intentional, such as, to name a few, "credit,"
"stock," "debt," "supply," "demand," "inflation," "price," "returns," "transaction," "capital," "interest," and, as I will try to show in my next section, even
"money." If this is true, the problem is not simply that of saying how economists provide explanations in using [L] or the Rationality principle based
on beliefs and desires. The problem is that of assessing the fact that the
"objects" or the "things" that economic agents and theoreticians alike refer
to in economics are not, at least are not all, physical objects or material
things but mental realities or intentional entities. This issue is at once an
ontological and an epistemological one and can hardly be limited to a purely
methodological question.

   Now let me assume, for the sake of argument, that Rosenberg is right in
stating that the predictive power of the theory developed by neoclassical
economists cannot be improved. I will even assume, furthermore, that the
reason for this methodological mess can, as Rosenberg argues, be traced
back to what really causes this theoretical framework to be a blatant "empirical failure," namely the fact that all of its explanations are hopelessly based
on "the intentional stance." Instead of directly challenging the main philosophical arguments that are aimed against any "intentionalist brand" of
economic methodology, I will now try to give a *reductio* argument against
the very possibility of having both an economic theory and an exclusively non-
intentional conceptual framework for pursuing scientific research in the field
of socio-economic life. By a '*reductio*', I mean the kind of argument that would
lead anybody interested in pursuing research in the economic field to conclude

that it would be foolish or absurd for her to stop talking about certain phenomena if it were granted that in order to be able to speak about those phenomena she must use an intentional language. So, if it could be established that certain very peculiar phenomena are "in themselves" or *per se* intentional, then we would be confronted with the crucial choice of either totally ignoring them and giving up on trying to explain them, or endorsing the language required to do exactly that.

Let me give an example of what I take to be a relevant sort of *reductio*. If we grant, for example, that teleology is a hard fact of human action in society, in political or economic action for instance, then a *reductio* directed against any philosophy of social science that would condemn the use of intentional language would run as follows. If one wants to be able to make any sense of social life, one must adopt the only language that is adequate for talking about teleological action, i.e. a language of reasons, and not only of physical causes, a language that includes finality and not only causality (or a language that counts reasons among possible causes). This is exactly what resorting to an intentional language is meant to permit in economics.

This argument rests on just one basic principle: we must first have an idea of what it is we want to speak about before we fix the semantics and vocabulary of our descriptive and explanatory language. The opposite approach, i.e. one in which we weigh rigid constraints on the language, and then go from there to the objects, is unscientific: in science, reality comes first, and methodological constraints on language come second. What I have been arguing for in this first section is that intentionality should be acknowledged as an aspect of all important economic concepts developed since the marginalist revolution. Indeed, intentionality is pervasive in economics and as such completely inescapable. In the next section I argue that a central economic phenomenon, money, is a thoroughly intentional phenomenon and that, as such, it makes no sense to try to reduce it to something merely physical. If we cannot do economics without talking about money and if we do not want economics to be an unscientific activity, we must agree to rely on this kind of non-physical phenomenon and we must do it in a non-physicalist language like the language of intentionality.

2. MONEY AS AN INTENTIONAL REALITY

We all think we know what money is. One could perhaps say, *mutatis mutandis*, about money what Saint Augustine says in his *Confessions* about time: when I don't think of it, I know what it is; but when I come to think about it, I do not know any longer. As a very first and also a very crude approximation, I suppose that for each and every one of us, money is something that we can have in our hands, something that we can touch, smell, bite, and inspect. Money is something real, something that we can put into our pockets or wallets. One can toss a coin or play with it, just as one can count dollar bills and make a pile of them. To be sure, one can count coins and bills as easily as one counts many other physical objects. Money, then, looks and behaves like a certain sort of "thing." Money is something we can accumulate, put under a mattress or in a bank account. It is something we accept in exchange for our work, something we can use to pay back what we owe, something we can borrow, lose in a poker game, invest, and loan; it is even something that some of us can cherish above everything else. Ultimately, money is something that we can exchange for goods and services, something no one ever says he has enough of.

So, in a sense, money is like many other things in the world we live in, i.e. a type of physical thing with physical properties. A coin or a bill has, as such, a certain weight, a certain color, and even two sides on which human figures, landscapes and such are depicted. And, of course, one cannot fail to see that these physical properties are of crucial importance if we ask whether the money we now have in our possession is "real," as opposed to toy money, or "legitimate," as opposed to counterfeit money. This being so, it is then out of the question to challenge the fact that money presents itself as something that we can speak about in a physical language, or what Carnap, seeking to characterize the kind of language we could use for doing science in general, called a "thing-language."[8] It is obvious, then, that the first thing to say about money is that, from a certain perspective, it is part of common sense reality. Everybody seems to know what money is, what can be done with it, where it can be put, and so on. In a sense, people experience money as being a mere physical thing. Far from being a strange reality for them, money is a rather familiar one, and for that matter people can be very friendly with it – especially if it's not theirs – and give it homey nicknames (for example in French: "fric," "oseille," "pognon," "blé," and in English: "cash," "dough," "lolly," "moola").

So, if money is like anything else, then we should be able to speak of it the same way we speak of any other part of our everyday physical environment. But, when we think more wisely about it, we rapidly see that a "thing-language" of the kind Carnap tried to construct in the hope of clarifying "the methodological character of theoretical concepts" in natural science is not adequate to the economists' tasks. The reason for this is that for them, precisely, money is not just a "thing". It is not a kind of physical entity. Money is not a natural kind and economists would surely think it odd to try to transform it into one by means of some ingenious conceptual reduction. But how can that be? Would economists be completely mistaken in this regard? How can something that we experience as being "something" not be, in the final analysis, a material reality? How can money be something real and not be either one of the components of matter or a materialized compound of those bits? How can something be part of the furniture of our world and not ultimately be something that can be used as a building block of our physical environment? This, of course, has to be accounted for. Maybe money is just another fiction; but if money is a fiction, it is a fiction that works! Those who really think that money is not something real enough to be worthy of scientific explanation are invited to skip to the next essay in this anthology, for in what follows I shall assume the contrary.

When economists speak of money, they can hardly be said not to refer to exactly the same thing as ordinary people do. But if they use the same referring term, this does not mean that they have the same concept of money as ordinary people. Theirs is a much sophisticated, elaborate, and refined concept, for it is a technical one. One could say that their concept of money is in fact a physical concept since economists speak of "liquid money" and of a "mass" that expands or shrinks. In fact, they distinguish between three very different monetary aggregates and take into account different kinds of assets (coins in circulation, savings deposits, etc.). To be sure, when economists speak of money in a narrow sense, they see it as a pure medium of exchange, and when they speak of it in a broader sense, they see it as an overall store of value.

So, ultimately, when economists talk of money, they presuppose what I alluded to in my first section as the "marginalist theory of subjective value." Every consideration of the value of money must presuppose a state of society in which exchange takes place, and must take as its starting point individuals acting as independent economic agents within such a society, that is to say, individuals engaged in valuing things, i.e. transforming them into economic goods. This "economic valuation" is a sort of mental activity. What people accept as money in fact, and the value they attach to it, are essentially matters of individual valuation; as such these phenomena depend on the entire complex of economic activity in which individual valuations are made. This is the case even if we accept the somewhat mechanistic approach of Irving Fisher in his famous book entitled *The Purchasing Power of Money* (1911) in which he revived the old 'quantity theory' of money, arguing that the purchasing power of the money supplied could be predicted according to a simple formula in which the principal variable was the amount put in circulation. This approach was, as is well known, further developed by Milton Friedman, but I need not go into the details to make my central point.

I shall limit myself to asking the following question: what is the purpose of money? An answer to this question must inevitably say what kind of thing money ultimately is. As noted above, these questions are not of merely methodological but also of ontological and epistemological import. In this regard, money is a very unusual entity, since it is not a good that is produced or consumed, but a commodity that is acquired by people to facilitate the exchange of goods. Money does not produce things; indeed, the entrepreneur must give it away in exchange for production goods before any output can be generated. Nor can money be consumed: the sole function of money is to be once again given away in exchange for consumable goods and services, exceptions being perhaps the miser who wants to accumulate money for its own sake and the coin collector. But, in general, money is cherished for the purchasing power its represents and for nothing else. The sole purpose of money, oddly enough, is, then, to be exchanged. Money is purely a medium of exchange. But it can serve as such only because people can value it for what it stands for.

This is why the emergence of money in the economic system is easy enough to understand. The production and consumption goods which people want to acquire have different degrees of marketability: for some it is easy to find customers, while others have a narrower appeal that makes it difficult to find customers willing to offer (or vendors willing to accept) a mutually agreeable compensation at the right time and in the right place. So what is the hungry barber to do? It is a waste of time to search around for bakers in need of haircuts, but perhaps there are plenty of bakers who would be prepared to accept something else. The more marketable that 'something else' is, the more likely are bakers (and for that matter, butchers and professors) to accept it as payment even when they do not want the barber's special services. So even the bald baker will accept a commodity that he knows can be readily exchanged for the meat and lectures and other goods that he might want in the future. In general, the person who cannot instantly acquire what he or she needs because of this difficulty in finding a ready market can nevertheless improve matters by exchanging a less marketable good he or she wishes to trade against a more marketable one. The more marketable good can then be traded, more easily, for whatever is needed. Similarly, a person who wants to dispose of some commodity quickly (because it is perishable or expensive to store, or because a fall in its market value is anticipated) acts wisely to trade it against a more marketable good, even if this good does not satisfy his or her own needs directly.[9]

This system of indirect exchange becomes much more important as the division of labor grows. In our advanced and well-developed economy which is reaping the gains of a very high degree of specialization, individuals might well find themselves engaged in the manufacture of goods that are aimed at a very small and specific market and which have no interest or value for anyone other than those few buyers. It is not surprising, therefore, that the greater the division of labor, the more important is it to seek out intermediary goods that are more readily acceptable in the widest range of markets and so able to act as a medium of exchange that facilitates the very widest range of transactions. A medium of exchange is a good which people acquire neither for their own consumption nor for use in their own productive activities, but *with the intention* of exchanging it at a later date against those goods they want to use either for consumption or production. Eventually the competitive use of different intermediary goods will sort out one or a group of commodities that prove most acceptable as a medium of exchange. When an intermediary good becomes generally accepted, so that it operates as a common medium of exchange, we normally call it 'money.'

Exactly what will be used as money will depend on how generally acceptable it is for all potential agents on the market. It will then depend upon the individual valuations of those in the marketplace at that time. Today, we often imagine that only the government chooses what will count as money; but while the state has the power to decide what will be the legal medium of payment, only the fact of what people are prepared to accept will ultimately determine the common medium of exchange. If the government's own bank notes were to become worthless, people would certainly decide to deal illicitly using cigarettes, brandy or some other commodity as money. In both planned and free market economies governments cannot make something the common medium of exchange if people do not accept it as such. The essential characteristic of money, then, is to be a good acquired for the purpose of exchange. Money is not a factor of production, acquiring its value like other goods of production from the value of the goods it produces; it is not a good of consumption because the value to its holder consists in its being exchanged, not in being consumed; it is not even capital because it produces no benefits until it leaves the holder's hands. Its central function is as a medium of exchange, and all the other functions which are popularly ascribed to it are just particular aspects of this central function.

This characteristic of money stands in stark contrast to the commonsense view. Even if we were emphasizing the need for media of exchange to be durable and divisible, these physical properties would have to be seen only as incidental features of money, desirable as such, but secondary. They might, of course, help particular commodities such as gold and silver to become the common medium of exchange. When we are considering the economic issues surrounding money, however, we should not be misled by these incidental traits. The key characteristic of money is that it exists to facilitate exchange, nothing else. Although it is unusual, money is nevertheless an economic good. It is scarce, and there is a demand for it. People hoard it because they would like to be able to exchange it for the goods they will need in the future. And like all other commodities, money can be traded against other goods at a certain exchange ratio – a 'price.' In the case of money, the price is normally expressed a little differently, however, not in terms of the volume of goods or services that will exchange for a unit of money (e.g. how many apples to the dollar), but in terms of the number of units of money which exchange against another good (e.g. how many cents for an apple). In other words, a rising price of money means that what we call its "purchasing power" has decreased. Despite the terminological differences here, the principle is the same for money as it is for other goods. And to complete the equation, the demand people have for money, whether they choose to hold greater or smaller amounts of a medium of exchange, will affect its purchasing power, just as changes in the demand for any good will have a bearing on its price.

For the commonsense view, money is and probably always will be a certain commodity: something solid that can be acquired and hoarded. From that perspective, the value of the commodity money stems firstly from its physical properties, for instance for its industrial and commercial uses. Thus, the value of gold and silver seems to derive exclusively from their use as personal and household ornaments or in industry. This, of course, is plainly not the perspective of economic theory. The exchange value of a commodity money derives, in a sense, both from the industrial *and* the monetary uses to which people put it; as a factor of production or an elegant ornament *and* as a medium of exchange. Plainly, the fact that people want a good because it can be exchanged widely will certainly have a bearing on its market price, and this effect will be supplemental to the effect of its demand for consumers' and industrial uses. But economic theory is interested only in the commodity's character as a medium of exchange, and the value of commodity money is of importance for monetary theory only in so far as it depends on the peculiar economic position of money, and on its function as a common medium of exchange.

Money has no meaning whatsoever for economics unless units of it ultimately represents a certain amount of purchasing power that will be accepted or that economics agents *believe* will be accepted by somebody else. To underscore the fact that money is an intentional entity (an entity related to the fact that we humans have minds and for that reason can form intentions and share beliefs), we note that money works only because it stands virtually for something other than what it is materially speaking; money works because it has an *intrinsic relation to something else* and because it is nothing but this relation. And this relational property of "standing for" is precisely what economic theorists have to focus on when explaining what money is. They cannot succeed in doing this unless they speak of money in an intentional language. And this argument about money could be generalized as it surely applies to many if not all other economic phenomena. For this reason, intentionality seems unavoidable in economics.

3. THE NEUTRALITY OF ECONOMICS WITH REGARD TO THE ONTOLOGY OF MIND

Economics simply has no effective means of social engineering that would prove beyond reasonable doubt that this science is empirically sound. As a result, it is tempting to think that if economics could do away with mind and the language of mental life, it would accomplish the conceptual revolution that could finally free it from folk psychology and allow it to enter the pantheon of those sciences considered more prestigious because they are founded on experimental observation. Since, whether we like it or not, the concepts and theories of economics commit us to the existence of the mind, perhaps we should hope that with a change in the status of psychology itself, economics would also benefit from a fundamental turnaround. The question of economics' apparent dependence on psychology thus becomes acute. It would be easy to believe that, from what was said above, economics is, in its foundations, at the mercy of theories which are now debated in psychology and philosophy of mind.

Furthermore, it could be thought that the conceptual foundations of economics are even weaker since this discipline is inevitably affected by future developments in scientific psychology. In any case, to accept this argument comes down to claiming that economics is obliged, whether it likes it or not, to take the side of a particular research program in psychology, and that it is obliged, unwillingly perhaps, to enter a debate over which it has no authority but in which certain results could have consequences in its own domain. Should we take it for granted, then, that the economist should, as an economist, either be a behaviorist, or be a cognitivist, or perhaps even more radically, a disciple of neuroscience? It could be claimed indeed that because it is committed to the existence of the mental life of human beings, economics risks being inevitably affected by the present upheaval in philosophy of mind. Must we agree that the orientation, and even the future, of economics risks being seriously affected by the result of a debate which, at the present time, pits dualists of all sorts against hard-line monists? Must the economist pledge his or her alliegiance to epiphenomenalism, parallelism, or interactionism, the anomalism of the mental, eliminative materialism, physicalist materialism, or emergentist materialism? This is the final question I wish to take up.[10]

We must immediately acknowledge that if the economist takes an interest in what I have tried to show is a thoroughly intentional phenomenon, this is above all because she wants to understand something which presents itself, paradoxical though this may seem, as "unintentional." In effect, the economist is not interested in explaining human voluntary and conscious thought, nor is she interested in explaining the psychological processes by which individuals produce their deliberations and calculations. She is even less interested in revealing the mechanisms of the central nervous system that lead individuals to make the decisions that seem to suit them. Thus, economic theorizing should be considered independent of research in behavioral psychology, cognitive psychology and neurophysiology. I would like to argue that economics is not only independent of those issues, but also simply *neutral* with respect to at least some of the possible results of the scientific debates and philosophical controversies presently surrounding this domain of research.

In effect, the economist is, by definition, only interested in individual action because she is interested in the phenomena of social existence, that is, in human life in groups insofar as it gives rise to institutions and organizations. The economist must explain the collective behavior of individuals who bring about, through their reciprocal interactions, the existence of those institutions and organizations. Financial organizations, monetary systems, and markets are all paradigmatic examples of interactions and coordination processes between individuals. Each one forms a kind of more or less "spontaneous order," to use Friedrich von Hayek's phrase, and if it is true to say that none of these institutions and organizations would have seen the light of day without conscious and deliberate interventions by individual or collective agents, it is also true to say that none of them in fact represents the system of expected consequences and desired results of any one of those agents.[11] If individual human action is indispensable for order to appear in the social grouping of a multitude of individuals, the goals that each individual consciously pursues inevitably run up against other converging or diverging wills. Institutions are born out of the effect of coordination which occurs when wills collide. All of society, in each of its temporal states, is nothing but the consequence of individual undertakings. Certainly, while such an analysis is well-adapted to the goal of explaining economic phenomena, it is not very promising as far as practical spin-offs are concerned. Perhaps we should even go so far as to grant that, as an empirical science, economics has a precarious, or at least a very ambiguous, status. Even if economic analysis can be used, for example, in developing monetary or fiscal policies, and thus even if economics is clearly related to empirical states of affairs, it is difficult to consider the statistical examination of public finances to be as rigorous and precise a test for economic theories as are clinical and experimental studies in physical and biological sciences. But experimental control is not the sole criterion of a science's empirical status.

However, I must at this point address myself to a line of thought that might be tempting to some. One may still ask whether economics will be sooner or later faced with a *fait accompli*, which would be the case should "mentalism" disappear from scientific psychology. If psychology were more or less absorbed by biology, would economics, which, as I have said, is committed to the existence of the mind, collapse as a result of loosing its object and its legitimacy? I don't think so. Even if we were to consider the *mind* as something "material" or physical, it would still be mistaken to think of a mental phenomenon like money as being a mere physical and not a thoroughly intentional reality. However, this said, no economist has to endorse philosophical dualism since the mind does not constitute for him something like an irreducible substance. The kind of dualism that seems to be forced upon economics, if it is a genuine dualism, has to do with the logic of science: it is "in practice" that the economist, the monetary theorist for instance, finds herself forced into a kind of dualism, since the language in which she can speak of physico-chemical things, Carnap's "thing-language", is not at all adequate for speaking about the intentional reality of economic phenomena.

It is thus easy to claim that economics has taken up the fight for a mentalist psychology which thinks of the human mind as an autonomous entity irreducible to the system of functions and processes which neuroscience reveals. It is altogether possible to reproach economics for methodologically presupposing the existence of the mind without effectively proving its causal efficiency. It is fair enough to say, finally, that economics suffers the same deadly defects as all intentional psychology which only seems interested in cognition, which neglects emotion as a possible source of behavior and action, and which tends to compartmentalize the mind in order to structure it into independent abilities. This might be so. But suppose it were true to say that the will is seated in the brain, more precisely in the frontal cortex, as is suggested by the fact, known since the 1930's, that lobotomized patients lose the ability to plan and make decisions. Suppose, thus, as Mario Bunge suggests, that "will power can be excised with a scalpel" and that it can be concluded, as he does preemptively, that "intention is a process which takes place in the brain."[12] Does this change anything for economics? Maybe it would if economists were trying to explain mental events causally and to construct theories about the psychogenesis of beliefs and desires. However, I believe that this view is incorrect and stems from the illusion that economics *presupposes* a psychological theory, when in fact it does not. It would then be plainly erroneous to claim that, since economics is committed to the existence of the mind, it would lose its privileged viewpoint on reality the day it is shown that intentional phenomena can be completely explained in terms of neurons. The reason is that fundamental economic theory, though it supposes the existence of individual utility functions and subjectively held information, leaves open the question of how these intentional processes are realized in the human organism. The fact that economics uses "intentional" concepts neither presupposes that they are unanalyzable, nor requires that they be analyzable (i.e. that they can be explained away) in psychological or biological terms.

The pertinence of this argument can be better seen when it is emphasized that microeconomics does not only posit intentional entities: it also posits, of course, the existence of material objects and of so-called "objective" constraints, which are in fact subjectively evaluated, and which influence agents and their decisions. The technology available at a given time is an example of one kind of constraint. While, then, it can be argued that in their domain, economists refer more or less naïvely to the existence of a "world of physical things," it would be completely wrong to criticize them for not seeing that this macroscopic universe of substances is in fact illusory and that it would be better to adopt immediately the concepts of the new physics of elementary particles, as if that had any bearing in economics. I argue that in this case economists, insofar as they are economists, remain entirely neutral on the question of what constitutes the ultimate nature of physical matter. Indeed, neither is it an economist's goal, nor is it within his competence, to take a position on the structure of matter, as, for example, continuous or discontinuous, corpuscular or undulatory. Furthermore, he has no need to do this in his own work. The only thing which really counts is that his theories be and remain compatible with any physical theory the scientific community decides to endorse today and in the future. A parallel argument can be used to show that economists can just as well dispense with taking a position on the ultimate nature of mental life and the right way to explain human behavior. While economists are expected to respond to the question of the nature of money or explain how the market functions, which might entail an explanation of how agents make economic decisions, they are never required to take a stance in debates taking place elsewhere, and with respect to which, in spite of any interest manifested by economists, insofar as they are economists, they remain entirely neutral. Thus, whatever advances are made in biology and psychology, the economic phenomena which we are trying to understand and which we now explain with more and more complex mathematical models (I have in mind phenomena such as value, cost, money, price formation, market coordination, financial speculation, rational anticipations, etc.), will never become biopsychological phenomena or, even more radically, structures explicable in neurophysiological terms. Even if, due to revolutionary discoveries in neuropsychology, psychophysical dualism definitively gave way to psychoneural monism, the epistemological status of a science like economics would not be affected one iota: its scientific task would remain the same, and its ontological view would remain absolutely intact. It would still be moved by a conception of reality that presupposes the existence of human mental life. This is why, unless it abolishes itself as a scientific enterprise, economics need not stop speaking the language best suited to the objects it has chosen to scrutinize, in other words, that of intentionality.

4. CONCLUSION: INTENTIONALITY IS INEVITABLE IN ECONOMICS

Thus, what forces economics to adopt intentional language is the fact that economics is "subjectivist" from the very beginning: this discipline suppose the existence of the individual human being as a "mind," and this is why the intentional vocabulary of mental life has appeared, and still appears today, to be indispensable in economic theory. The very first objects of economic analysis, such as the perception of the strategies of other economic actors, the planning of an action oriented toward the attainment of a goal, deliberation, calculation, and decision – all these conceptual objects that the economist must try to hold together in a network of theories forming a system – require the fine tuning of a technical language which cannot be reduced to the language of any of the natural sciences. This is what explains why, compared with practitioners of the physical and biological sciences, the economist finds herself in a much more complex and difficult methodological situation. The objects of these theories are certain phenomena of the subjective mental life of individuals living in society, and these phenomena can be virtually considered properly publicly unobservable as such. It is thus not surprising that these phenomena seemed at first accessible only through direct knowledge or introspection. This theoretical stance has now been replaced by Paul Samuelson's theory of revealed preferences, which relies completely on the choices actually made by economic agents.

Here a comparison between the social and physical sciences reveals nonetheless an important epistemological asymmetry. As Hayek has shown, the physicist, like the economist, is obliged to resort to unobservable entities to be able to explain adequately the phenomena which concern him. This is how he is led to postulate the existence of electrons, quarks and other elementary particles, as well as that of electro-magnetic fields, gravitational fields, etc. Similarly the economist postulates, for the needs of her own analyses, preferences, individual values, rational deliberation processes, calculations, but also costs, markets and partial equilibria. However, while the physicist can have only indirect and impersonal knowledge of what he uses as explanatory principles, the economist is in a wholly different position; in the end, she can only know what it is to be a rational economic agent directly, by acquaintance, as Russell would say, but I would add "by acquaintance with herself," i.e. through personal experience. What the economist talks about, and what she attempts to objectify, to some extent at least, in the mathematical language of the models she constructs, is something which cannot be properly analysable in terms of physical events. For it is an intentional reality. Because this subjective mental reality constitutes the ultimate empirical basis for economics, intentional language was finally adopted by economists. It is also for this reason that it is impossible to dispense with it, even today, in this scientific discipline, and perhaps also in all other social sciences. Consequently, economics' complicity with "folk psychology" is not a nuisance and it is not just a methodological constraint. It is both an ontological necessity and an epistemological requirement.[13]

Université du Québec à Montréal

NOTES

[1] Carl Menger published in 1871 his seminal book *Grundsätze der Volkwirtschaftslehre* [*Principles of Economics*]. Léon Walras is known for his independent development of the marginal utility approach to the theory of value in 1873, and he also created what is now called the "General Equilibrium Theory" (*Éléments d'économie politique pure*, 1874–7). But Stanley Jevons has claim to priority for having first expounded the marginal utility approach to value in 1862.

[2] This argument has been set forth on many occasions by Rosenberg. See for instance Rosenberg, 1980a, 1980b, 1988 and 1992.

[3] Rosenberg, 1988, p. 25.

[4] For a discussion of what is at stake here, see Nadeau, 1993.

[5] Rosenberg's whole argument is now fully developed in his 1992 book. Previous steps in the same direction had already been taken in his 1980 and 1983 papers. With respect to the latter, Rosenberg was surely right to argue in his 1986 paper that Wade Hands had not properly understood his main point (see Hands, 1984).

[6] I should add that for Rosenberg, expectations and preferences "are just cognates for the beliefs and desires that figure in [L]" (Rosenberg, 1988, p. 27).

[7] See Vaughn, 1980.

[8] See Carnap, 1956.

[9] The point of view I am defending here is closely akin to André Orléan's approach to money (see Orléan, 1991 and 1992). For instance, describing the modern form of money, he writes: "For any agent, *i*, that agent's acceptance of a worthless sign, money, in exchange for a commodity, depends on that agent's expectations about the future acceptance of the same sign by another agent, *j*. The particular qualities of the sign hardly matter, for what is essential in determining agent *i*'s decision is *i*'s expectation about the behavior of *j*. Agent *i* will only take the money if he knows that *j* will accept it in turn one day. But, to the extent that *j*'s acceptance of the money also depends on *j*'s expectations about a new agent, *k*, *i*'s acceptance of money depends on *i*'s expectations about *k*'s acceptance of the monetary sign. It should be obvious that this reasoning hardly stops with agent *k*. Thus, the acceptance of money depends on an infinite chain of expectations about the expectations of other agents. 'Specularity' is the name we give to this kind of situation where the agent's behavior is based on their reciprocal expectations about each other's behavior" (Orléan, 1992, pp. 123–4).

[10] For an excellent survey of these points of view and the psychological stakes involved, see Bunge and Ardila, 1987.

[11]   See Hayek, 1967.
[12]   Bunge and Ardila, 1987, p. 215.
[13]   I would like to acknowledge financial support received from *Fonds FCAR* (Government of Quebec) and from the *Social Science and Humanities Research Council of Canada*. I would also like to express my gratitude to Paisley Livingston, Don Ross, Chantale LaCasse, Gérald Lafleur, Michel Rosier and Yves Gingras who kindly accepted to help me revise a first draft of this paper.

REFERENCES

Bunge, Mario and Ardila, Rubén, 1987, *Philosophy of Psychology*, Springer, New York.

Carnap, Rudolf, 1956, 'The Methodological Character of Theoretical Concepts', *Minnesota Studies in the Philosophy of Science* I, 38–76.

Hands, D. Wade, 1984, 'What Economics is Not: An Economist's Response to Rosenberg', *Philosophy of Science* 51, 495–503.

Hayek, Friedrich von, 1967, 'The Results of Human Action But Not of Human Design', in *Studies in Philosophy, Politics and Economics*, Routledge & Kegan Paul, London; University of Chicago Press, Chicago; University of Toronto Press, Toronto, chap. 6, pp. 96–105.

Nadeau, Robert, 1993, 'Confuting Popper on the Rationality Principle', *Philosophy of the Social Sciences* 23, 446–467.

Orléan, André, 1991, 'La monnaie et les paradoxes de l'individualisme' *Stanford French Review* 15.3, 271–295.

Orléan, André, 1992, 'The Origin of Money', in F. J. Varela and J.-P. Dupuy (eds.), *Understanding Origins. Contemporary Views on the Origin of Life, Mind and Society, Boston Studies in the Philosophy of Science*, Vol. 130, Kluwer Academic Publishers, Dordrecht/Boston/London, pp. 113–143.

Rosenberg, Alexander, 1980a, 'Obstacles to Nomological Connection of Reasons and Actions', *Philosophy of Social Sciences* 10, 79–91.

Rosenberg, Alexander, 1980b, *Sociobiology and the Preemption of Social Science*, The John Hopkins Press, Baltimore.

Rosenberg, Alexander, 1983, 'If Economics Isn't Science, What Is It?' *Philosophical Forum* 14, 296–314.

Rosenberg, Alexander, 1986, 'What Rosenberg's Philosophy of Science Is Not', *Philosophy of Science* 53, 127–132.

Rosenberg, Alexander, 1988, *Philosophy of Social Science*, Westview Press, Boulder, CO.

Rosenberg, Alexander, 1992, *Economics: Mathematical Politics or Science of Diminishing Returns?* University of Chicago Press, Chicago.

Vaughn, Karen, 1980, 'Does It Matter That Costs Are Subjective?', *Southern Economic Journal* 47, 702–715.

HOW COULD ANYONE BE IRRATIONAL?

The question "how could anyone be irrational?" probably sounds rather odd to those who consider examples of irrationality to be pervasive in human behavior. In any case, since truly irrational behavior would be totally unpredictable and not amenable to scientific analysis, the social sciences are based on the idea that there is at least some rationality in human behavior. This is the reason why what is usually called the principle of rationality plays a key role in the foundations of the social sciences, and especially of economics. In its most straightforward formulation, this fundamental principle affirms that human action is rational, generalizing the idea that people, after all, have reasons for doing what they do. But such a claim suggests that any action whatsoever is rational in the sense that it has a rationale. However, this seems to overshoot the mark, for if it were no longer possible for someone to act irrationally, what would it mean for an action to be *rational*? Even without raising this problem in these terms, many theoreticians of the social sciences have been bothered by this paradoxical situation and have proposed various ways to cope with it. In this paper, I would like discuss a few problems raised by the views of some among them on this issue.[1]

THE RATIONALITY OF GEORGE DANDIN'S MARRIAGE

Ludwig von Mises' radical approach to this question lets us clearly see why it is so difficult to solve the problem in a satisfactory manner. One can even say that this economist put the matter in such a way that he literally dissolves the meaning of the question by making a proper answer *a priori* impossible. Mises starts by defining action as "conscious behavior" or as behavior oriented towards an end which is nothing other than the "removing or alleviating" of a "state of dissatisfaction".[2] Since economists normally, when they refer to rationality, refer to what Max Weber called *Zweckrationalität*, namely the adaptation of means to any end (by opposition to a conception of rationality which implies an evaluation of this end), Mises does not hesitate to claim that "action" and "rational action" are strictly equivalent.[3] From his point of view, any action is rational since it is nothing but a conscious alleviation of a state of dissatisfaction by using means, whatever they are, to reach the end (a state which is desired because it is different from this state of dissatisfaction) that they provide by the very fact of taking the action. When I buy something, I am by this very fact reaching a new state which is to me more satisfactory than the earlier state; and reaching this end (a more satisfactory state) through this means (buying) is rational because I clearly have reasons to prefer being in the second state, since otherwise I would have not taken

177

M. Marion and R. S. Cohen (eds.), Québec Studies in the Philosophy of Science II, 177–192.
© 1996 *Kluwer Academic Publishers.*

this action. True, an economic agent can take actions the unexpected consequences of which turn out to be harmful. Let us consider, for example, George
Dandin, the hero of one of Molière's comedies, who decided to sacrifice his
entire fortune to marry a lady of the nobility only to discover that the fact
of being known henceforth by an appreciably longer name, "Monsieur de la
Dandinière", was not sufficient to provide him the consideration he was looking
for. But to conclude that Dandin's marriage was irrational simply because it
had not permitted him to attain his goal (being a highly respected person) is
confusing an action with a whole strategy. Indeed, becoming a respected per-
sonage is one of the vague and general goals of many agents. However, while
Dandin's very expensive marriage was indeed an appropriate means towards
his end of being associated with the nobility, it nevertheless failed to help
him attain his more general goal of being respected. The end of a specific
action has to be a member of the set of available ends such as "having his
name changed for a longer one associated with nobility" (a situation which
is preferred over the previous one, "being a rich money holder"). Thus Mises
would surely say that Dandin's action was rational (and had a rationale) since
it was clearly motivated by the end that was reached through it. Indeed,
according to Mises, "as far as there is scarcity of means, man behaves rationally, i.e., he acts. So far there is no room left for "irrationality."[4]

Most methodologists of economics would not agree with such an apriorist
view of rationality, but this view was the basis of the standard argument –
popularized by Robbins' influential book[5] – used to undermine the standard
objection to marginalist economics. To those who found the conception of
rationality on which this economic theory seemed to be based inapplicable
to human agents – because it treated rational human beings as mere *homini
oeconomici* – Robbins' answer was that the notion of rationality required by
marginalist analysis implies nothing about the nature of the ends as such. Thus,
if a philanthropist gives his (or her) whole fortune to poor people instead of
profitably investing it, because he (or she) feels – at least, when the donation
takes place – more happy as a generous donor than as a mean investor
(otherwise the donation would not occur), this philanthropist would be a
perfectly rational utility maximizer whose actions could be theoretically
analyzed with the tools of economics. Given such a conception, Dandin's
"transaction" could also be characterized as perfectly rational since it is
oriented to the satisfaction of bearing a name associated with nobility, which
was – at least on his wedding day – greater for him than the satisfaction of
being a rich money holder. But given such a conception, how could anyone
be irrational at all?

Any action is explained by its end, and the most apparently irrational
action is itself oriented to an end, even if, in the extreme case, this end could
be nothing but the exquisite thrill of having performed an action apparently
totally free of any end.[6] However, the problem for economics or other social
sciences based on such an approach is that, being valid *a priori*, they are
excluded from the *empirical* sciences. Consequently, it was not surprising that,

with the rising prestige of empirical science in the forties and the fifties, many economists, like Hutchison and Lester, took as their target such a conception of rationality, of which Fritz Machlup was, by this time, the most articulate champion.[7] Both of them emphasized, from their respective points of view, the oddity of analyzing, with the help of a theory assuming rationality *a priori*, the behavior of real economic agents who, in so many cases, seem far from being rationally guided by profit (or income) maximization. For a consistent empiricist, the principle of rationality could *not* be accepted at face value. But a consequence of refusing to consider human action as *a priori* rational is that a line has to be drawn between cases of truly rational behavior and eventual cases of *irrational* behavior which, if they exist, would be potential falsifiers of an overly optimistic principle of rationality. In this context, Karl Popper's attempt to draw such a line by admitting empirically attested cases of irrationality while not rejecting the principle of rationality was rather heroic, coming from a philosopher whose name was associated both with the idea of falsificationism as the correct version of empiricism and with the idea of a situational logic (based on rationality) as the only valid basis of social sciences.

THE FLUSTERED DRIVER

Indeed, in a highly controversial but very short paper,[8] Popper attempted to clarify the status of the principle of rationality, a principle which plays the key role in the situational logic which, Popper claims, corresponds to the basic method of the social sciences. As an empiricist, Popper insisted on keeping his distance from a position such as that of Mises and strongly rejected the claim of those maintaining that the rationality principle "is *a priori* valid, or *a priori* true."[9] More precisely, Popper attempted to establish the empirical character of the rationality principle (which he sometimes prefers to call the "principle of acting adequately to the situation") of which he had formulated his own version. This was: "agents always act in a manner appropriate to the situation in which they find themselves".[10] Popper did not hesitate to affirm that such a principle is empirical, but he also affirmed that it is *manifestly false.*

However, as is well-known, this indefatigable defender of falsificationism recommends that this false principle *not* be treated as falsified. Many have found Popper's position to be blatantly contradictory or, at the very least, extremely ambiguous, and this has been the source of a long debate among philosophers of social sciences, but this question will not be discussed here.[11] What does interest me, however, is the nature of those cases which, according to Popper, falsify the principle of rationality, given that only a genuinely *irrational* action could falsify this principle. Thus, to provide an example of such an action, Popper had to draw a line between rational and irrational action and, as an empiricist, he had to do it without simply resorting to a value judgment about what merits to be called rational. Even if Popper maintained that the principle of rationality is manifestly false, the task of finding an example of irrational behavior was not so easy because, according to him, this principle was also a "good approximation to the truth" and was *only occasionally* contradicted. Now in what sort of cases is it contradicted? Popper maintains that it is contradicted, for example, in the case of a "flustered driver" who after an unsuccessful search for a parking place, desperately tries to park his car in a space which is clearly too small. This driver, according to Popper, manifestly *does not* act in a manner which is appropriate to the situation in which he finds himself. However, if Popper maintains that the rationality principle is still a "good approximation," it is because he considers that such cases are not really representative. To anyone who would object that cases of maladapted or even stupid actions are rather pervasive among human beings, Popper would answer that many seemingly inappropriate responses to a situation are nonetheless rational, as long as, from the perspective of the agents, they are appropriate responses to the situation *"as they see it."*[12] According to Popper, Freud showed that the neurotic responds in a way which is completely appropriate to his situation *as he himself sees it*. Popper suggests that in a similar way a pedestrian might throw himself into the path of an oncoming cyclist in order to avoid being hit by a car. There is no doubt that if the pedestrian had a better view, he could avoid both accidents; however, taking into consideration what he is in position to see, his response to the situation is completely appropriate. Since Popper's intention is clearly to establish a contrast between the flustered driver and the pedestrian who unknowingly throws himself in the cyclist's path, he surely meant that the driver behaves in a manner *which is inappropriate* to the situation *even as he himself sees it*, otherwise he could not claim that the driver is more irrational than the pedestrian. The only way to reconcile Popper's stand on each of the two examples is to admit that he takes the verb "to see" in its literal sense, or in any case, in its *strictly cognitive* sense and that, in his mind, the driver sees perfectly well that he cannot park his car in such a small space. In such an interpretation, one has only to suppose that the driver of the example is so aggravated that he makes desperate maneuverings in order to attempt it anyway, and, afterwards, he has to struggle to drive out of the cramped space into which he has needlessly squeezed the car. In Popper's view, this would be a totally inappropriate response to a situation, and consequently would be completely different from the response of the pedestrian who is not in a position to see anything other than the car that is about to hit him.

So, between the rational pedestrian and the irrational driver, Popper proposes to draw a rather fragile line. I say that this line is fragile because it seems to be quite possible to examine the flustered driver's perception of his situation in order to show that his behavior *was* appropriate to the situation as he sees it, if we extend slightly the content of what Popper calls "the situation" and if we include in it the agent's own psychological state. Perhaps, for example, he simply needed to blow off steam and could explain his reaction in the following way: "It does me good to show how absurd it is to try to park in a city which is so badly administered!". But Popper leaves no room for such a refinement of his psychological analysis of the flustered driver. Apparently, in his view, what the driver and the pedestrian "see" is nothing other than the *external* aspects of their situation and not such internal aspects as psychological needs. Popper readily concedes that the driver may not be aware of all of the facts about the situation, but since he *saw* that the parking space was too small and still let himself go for a few minutes, Popper estimates that, in contrast to the pedestrian hit by the bicycle, he acted in a way which was irrational because inappropriate to the partially known situation. He concludes at once that such examples falsify the universal validity we might be tempted to attribute to the principle of rationality, but the important point for the present discussion is that the very fragility of his distinction illustrates the difficulty of finding room for irrational behavior once room is made for an acceptable principle of rationality of human action.

THE IRRATIONALITY OF INERT OR IMPULSIVE BEHAVIOR

Popper's position has been rejected by most methodologists because the type of contrast he proposed between cases of rationality and irrationality is unconvincing and because the idea of maintaining a falsified principle together with a falsificationist philosophy seems contradictory. However, to my knowledge, nobody else has attempted to give an alternative empirical interpretation of the principle of rationality by singling out cases of irrationality as potential falsifiers for such a principle. For example, when he analyzed, in his 1962 paper,[13] the working of an economic model based on irrational behavior, Gary Becker did not try, like Popper, to reconcile observed irrational behavior with the principle of rationality. Rather, he considered irrationality as a possible alternative to rationality and boldly tried to show that irrational behavior fits as well as rational behavior when it comes to deriving the essential tenets of economics. Here again, my point will not be to discuss his audacious thesis[14] but to evaluate the candidates that he proposes as paradigms of irrational behavior. Becker adopted a conception of rationality characterized by the capacity to adapt to a situation, a conception that seems quite compatible with Popper's conception according to which rational agents always act in a manner appropriate to the situation in which they find themselves. Becker's way of attacking this question was to build an economic model featuring two extreme cases of irrationality (i.e. of misadaptation of an action to a situation): the case of "impulsive" behavior and the case of "inert" behavior, corresponding respectively to whether decisions are taken at random or always maintained as identical no matter how the parameters of the situation might change. If both types of behavior are called "irrational," it is clearly because they are totally unresponsive to changes in the situation. With the help of these models and of the usual tools of economic analysis, Becker manages to show that constraints imposed by budget lines force such impulsive or inert consumers to limit their behavior to what is compatible with opportunity sets made available to them after any change in price, and that this very fact implies that "regardless of the decision rule used" the average behavior of these consumers will be reflected in a negatively inclined demand curve. In other words, this typical economic result is derived whether we suppose that the agents follow the irrational rules associated respectively with inertness and impulsiveness or that they follow the rule dictated by rationality.

In a more controversial line of argument, Becker also maintains that other essential theorems of microeconomics can be derived from such models without resorting to any principle of rationality. But, leaving aside the question of the validity of such an argument, I would like to assess the meaning of such cases of irrationality. It is fairly clear that behavior patterns that are just mechanically repetitive when the situation changes, or behavior patterns which consist in mechanically reacting at random to any type of situation can hardly be characterized as rational behaviors. There is a *prima facie* reason to classify as irrational any behavior that, not being responsive to changes in a situation, does not seem to be adapted to the realization of any end. However, the important question here is to know whether this absence of rationality is due to the role, respectively, of tradition and of randomness in the decision-making process, or if it is due to the *mechanical* character of the behavior. Suppose that Becker's impulsive consumers, instead of resorting mechanically to random processes, explained their behavior this way: "We want to attain happiness with the help of our limited resources, but we think that, given the very limited state of our knowledge, making decisions at random is the most promising strategy for reaching such a goal for two good reasons. First, they avoid troublesome and generally useless computations, and second, on average at least, they tend to eliminate extreme situations and consequently allow us to avoid situations most inimical to happiness." Similarly, Becker's inert consumers could freely interpret an old Hayekian thesis and say that traditions are a far more reliable guide than any "rational" computation. These individuals, *purposefully* deciding not to respond to new information about their situation, could be easily characterized as rational by someone who considers that adopting a well-conceived strategy to reach a goal is being rational. At the very least, there is no longer a *prima facie* reason to reject their rationality.

Naturally, such a strategy might turn out to be very poor as an income maximizing tool, and on this ground it might make sense to conclude *post facto* that it was not really adapted to this goal and consequently, not a rational strategy. But what is meant by the irrationality of inert and impulsive behaviors, if they could be subjectively justified as goal oriented, and objectively characterized as not clearly misadapted to the attainment of this goal? The only sensible answer seems to be that, by contrast with the standard income maximizing behavior (the typical behavior of *homo oeconomicus*), these behaviors do not respond to changes in situations and available information. There is little doubt that the behavior described by a standard *homo economicus* model is characterized by constant adaptation, thanks to perfect information, to the parameters of a situation. However, this is nothing but a purely mechanical (or causal) model where behavior is thought to be maximally efficient, given perfect information, in reaching a preassigned end. It has little to do, however, with intentional human action and with the principle of rationality for which, as we have seen, it is a substitute. As Jon Elster put it, after briefly discussing two types of behavior akin to Becker's inert and impulsive behavior: "Both traditional and random behavior belong to the causal rather than the intentional image of man".[15] The principle of rationality concerns intentional human action the ends of which are not preassigned and in which the assignment of ends depends both on information available and on a hierarchy of superior goals. George Dandin's marriage and the pedestrian's movement were human actions because they were rationally oriented to ends (bearing a name associated with nobility, being out of the way of a car) that are dependent on available information and on superior goals (being respected, staying alive) which could have been different (having as much pleasure as possible, committing suicide) and could change without being ultimately submitted to a rationality rule. In brief, the analysis of human action and of the principle of rationality has to be located somewhere between mere mechanical efficiency and the moral assessment of superior goals. It is at this level that it is difficult to see how can anyone be irrational. At the mechanical level, some causal routes are more efficient than others for reaching an end, and consequently the alternative less efficient route could be called irrational, given the assignment of this end. However, in a more strict way of speaking, the rationality should be attributed only to the human action of the engineer or the programmer who determined this particular route. But if these persons have all the required information and nevertheless choose the so-called irrational route, is it not because they have another end in mind (saving either money or programming time, being unconventional, obtaining some byproduct, etc.) and act rationally to reach it?

If rationality is almost a category mistake at the mechanical level and if irrationality is difficult to characterize at the level of human action, what about the moral assessment of ultimate goals? According to Elster, it does not make sense to reject the possibility of assessing ultimate goals when discussing human action, a task which could be done, in principle, through what he calls a "broad theory" of rationality.[16] It is no doubt quite sensible to claim that not all ultimate goals are equally rational and that the question of assessing the rationality of those goals is an important one for philosophers. But it is also clear that the principle of rationality on which the social sciences are based does not refer to this ultimate and normative kind of rationality and has to be defined in terms of the adaptation of *means* to an end. This does not mean that what I call the ultimate goal cannot be the end considered when assessing the rationality of an action; the crucial point is that the type of rationality required by the social sciences is the one which concerns the appropriateness of some means to a given end, whether this end is ultimate or not. In any case, Jon Elster also contributed to such a "thinner" discussion of ratio-nality, while it was in a somewhat larger framework that he related means and ends and attempted to characterize irrational behavior in a more satis-factory way.

In contrast to Becker and like Popper, Elster defends a conception of ratio-nality and irrationality that strictly concerns intentional behavior; but in contrast to Popper and like Becker, Elster presents the germs of irrationality as existing wherever a behavior is not a maximizing one. For Elster as well as for Becker, maximizing behavior is the paradigm of rational behavior, but, since such behavior is an intentional behavior according to Elster, it is not, in his view, associated with the mechanical behavior of an *homo oeconomicus* as it is for Becker. Rather, it is typified as the behavior of angels, which are allegedly perfect but *intentional* decision-makers. Human beings, by contrast, are "neither angels (i.e. fully rational) nor animals (i.e. essentially myopic); they are imper-fectly rational creatures able to deal strategically with their own myopia.[17]

Since an imperfectly rational behavior is one that is partially irrational, elements of irrationality are pervasive in human behavior – as testified, for example, by the undeniable cases of weakness of the will (the *akrasia* of Aristotle) one meets so often. In Popper's analysis of irrationality, such a factor does not play an important role, because in assessing the rationality of an action, Popper, like Mises, does not consider long term goals of individuals. Popper and Mises discuss single actions and raise the following question: "can an action be irrational in the sense of being ill adapted to its own end, given the perception of the situation by the agent?" To this question, Mises answered "no" and Popper answered "possibly, look at the flustered driver, but very exceptionally". We have seen that both of these answers are not fully satis-factory. In any case, Becker and Elster raised the question differently and did not limit their discussion to the case of one single action. For them, being rational is adopting a maximizing strategy as ideal economic agents do, according to economists. This is the reason why in their respective theories, where rationality is characterized as *perfect* rationality, cases of irrationality look pervasive. When one puts the matter this way, the idea of degrees of perfection (or of imperfection) in rationality associated with degrees of adap-tation of an action is a quite natural one. I have concluded that Becker's approach, by replacing any form of intentionality by a mechanical force, empties the notions of rationality and irrationality of their very meaning by leaving no agents' intentions to be assessed on the basis of these notions. On the contrary, intentionality is at the center of Elster's analysis. Here, agents have a general goal to be reached in an optimal way through strategic behavior and express their intentions to reach such a goal. Consequently, we may contrast the perfectly rational agent (the angel) who, having a goal in view, acts in an optimal way, with the imperfectly rational human beings who tend not to persist in their determination to reach their goal and who tend consequently to adopt behavior that is irrational in the sense of being not optimal. In this context, irrationality is a matter of *limits* of information and of intelligence, and still more a matter of weakness of the will. Now, cases of weakness of the will are easy to find, but this approach nonetheless raises some slightly paradoxical problems for someone who wants to characterize irrationality along these lines.

In his analysis of weakness of the will, Elster is led to discuss the stratagems that can be used to overcome such a limitation to rational behavior. As a paradigm for such stratagems, he chooses the case of Ulysses binding himself to his mast to be able to hear the sound of the Sirens without being seduced by them. Ulysses is clearly not an angel and he knows that his will is too weak to resist the seductive Sirens' song. Being insufficiently "rational" to overcome directly what he perceives as an obstacle to the attainment of his goal (going back home), he uses the stratagem of being bound and decides to reduce temporarily his options in order to avoid a fatal threat to his ultimate goal. But why would it be less rational to use such a stratagem than directly to face the danger and triumphantly overcome it? This last course of action is possibly more glorious and more honorable, but is it more rational? Rationality is not sheer virtue; it is the ability to take the proper means to reach a given end. After all, the rational entrepreneur is not necessarily a paragon of either virtue or honor. He is rather a paragon of skill in maximally reaching his goal. And if an entrepreneur is successful thanks to the use of stratagems, it would surely be counterintuitive to characterize him or her as less rational than another person who acts in a more straightforward way but is happy enough to overcome the dangers associated with such a way and to be equally successful. The entrepreneur who proceeds a bit blindly in a straightforward way will tend to be characterized, rather, as a naive fellow unaware of danger and successful only by chance. Ulysses had such good self-knowledge that he knew that he was totally unable to resist the Sirens' song. By itself, self-knowledge is a plus. Thus, to reach his goal, two ways were open to him: to become temporarily deaf by filling his ears with wax (a strategy he reserved to the members of his crew) or to be bound to his mast with the help of his crew. If Ulysses chose the second way, it was because it permitted him to *maximize* his satisfaction, since a second goal (experiencing the sound of the Sirens' song) could be reached this way without sacrificing the first (going back home). How could anyone discount the rationality of such behavior?[18]

In any case, self-binding is just one type of stratagem among other *indirect* stratagems. Let us consider the case of Horatius who succeeded alone (after the death of his two brothers) in killing the three Curiatius brothers by resorting to a stratagem which, while in no way implying self-binding, was still less glorious but surely as rational as Ulysse's. Horatius' knowledge was remarkably precise. He knew that he was able to beat each of the Curiatius brothers separately (each of them having been wounded in the first assault) but unable to beat them together. He knew enough about the gravity of each of the brothers' wounds to be able to conclude correctly that they would run after him at respectively different speeds. Finally, he knew enough about their respective psychologies to be able to conclude correctly that the less seriously wounded ones would not adopt the slower speed of the more heavily wounded ones. Horatius, at the (calculated) risk of looking like a coward, wagered a temporary reduction of his capital of honor against the gain both for Rome and for his future honor and decided to simulate flight, in order to be in position to kill each of the three Curiatius brothers separately. It would be difficult to find in literature or in mythology a better example of rationally calculated behavior. It is true that in this case the indirect stratagem was devised to compensate for physical weakness rather than for weakness of the will, but it shows that a high level of rationality can correspond to a capacity to use indirect means as a substitute for direct ones. In any case, a strong will is just as much an endowment as strong arms, and since nobody would argue that one who cleverly invents an ingenious system of pulleys to raise a weight is less rational than the brute who easily raises it by the sheer strength of his arms, why should Horatius be less *rational* than the powerful giant who would have killed the three Curiatius brothers together? Why should Ulysses be less *rational* than the virtuous and angelic captain whose will would have been strong enough to be almost unaffected by the Sirens' song?

For Elster, the point seems to be that if rationality is the capacity to use the proper means to reach a goal, *full* rationality is the capacity to reach this goal *directly* by more immediate means (like remaining imperturbable while hearing the Sirens' song). However, if generalized, such a view of rationality would imply a negation of any constraint, since a limitation of the will could be considered as a constraint, as is a limitation of physical strength or a limitation of physical resources or of wealth. The typical rational entrepreneur who accumulated his fortune painfully in spite of various constraints would thus be considered less rational than the one who was so rich at the beginning that he could adopt without significant budgetary constraints any course of action he found appropriate.

THE ALCOHOLIC AND THE MOVIE ADDICT

Even if we forget this last objection concerning the role of constraints, another difficulty with Elster's view of irrationality concerns the determination of *what counts as an ultimate goal*. Is this ultimate goal the one that the agent presents as his or her ultimate goal? That solution would make the criterion of rationality too dependent on an unreliable self-appreciation. Will it be a long-term goal in opposition to short-term temptations? But how long should the term be? This type of solution raises well known problems which, by the way, were not problems for Mises and Popper because they considered only the end of a single action and not the ultimate goal of a whole set of actions. A standard example, evoked by Elster,[19] seems, however, quite favorable to his thesis, namely the example of the alcoholic who wants to stop drinking but who, by weakness of the will, takes another drink. It seems clear that the alcoholic is less rational than he would have been had he resisted the temptation, but the matter is not this simple, as a different example can illustrate. Suppose I want to become one of the most prolific philosophers of my generation and that I am convinced that constant work is a necessary condition to reach such a goal. Suppose further that I am also a zealous movie fan. Since seeing a movie is always a very pleasant and easy experience and working on a philosophical paper is a very painful one, one might think that deciding to see an extra movie instead of working is just like the alcoholic's having another drink. Given my goal, it clearly seems that it would be much more rational not to attend an extra movie and to continue working on a paper. But, at the very moment I decide to see this extra movie, I think that, after all, I have no serious chance of reaching my excessively ambitious goal and that it is more sensible to see this movie, which not only will be a source of pleasure and beneficial rest, but which will also improve my general background knowledge. If, after some time, I realize that my chance of reaching my goal has dwindled considerably, largely because I have seen too many movies, I may be accused of weakness of the will, but to conclude that another course of action would have been more rational, one has to maintain that this goal can still be considered as my *ultimate* goal. But suppose that, in the wisdom of a more advanced age, I reconsider the whole thing and I conclude that life is so short that, given my own limits, my ambitious goal was ridiculous and that I was right when I decided to improve my style of living and my general culture by attending more and more movies. On what grounds would one characterize as irrational my past decisions to see these movies? It is true that an objection to this line of argument could be that, at the very moment that I decided to go to the movie, "to become a prolific philosopher" was then my *main* goal (whether or not this goal turned out to be my ultimate goal). But on what grounds is it possible to concede such priority to this one among all my goals at the precise moment when this priority seemed to be (temporarily) rejected? Why would it be unacceptable to say rather that it was the goal (doomed to be perceived as an ultimate goal in my late wisdom) that was taking shape in these moments and that it was *not irrational* to act in a way commanded by this emergent ultimate goal? Short of resorting to a value judgment establishing that being a prolific philosopher is more meritorious than just being a person who got a lot out of movies, and short of declaring that rationality consists in adapting oneself to the fulfillment of the most meritorious goals, it seems difficult to oppose rationality and irrationality on this ground.[20]

THE ANT AND THE GRASSHOPPER

Elster also analyzes the case of irrationality related to *time preferences*, which, while also related to weakness of the will, might seem less dependent on value judgments. In this case, however, rationality is identified with consistency: "a rational actor on the standard definition is simply one who has consistent and complete preferences at any given point in time".[21] Once this definition of rationality is admitted, it seems easy to point out where irrationality is to be found: "It is widely, but far from universally, agreed that for an individual the very fact of having time preferences, over and above what is justified by the fact that we are mortal, is irrational and perhaps immoral as well."[22] Even if it is difficult to discount the exact effect of one's mortality on one's time preferences, it seems reasonable to maintain that, assuming this factor discounted, a rational fellow would tend to distribute his consumption equally over his lifespan. I would like to illustrate the essential idea with the help of a famous fable which suggests that the grasshopper is less rational than the ant. One could even insist that the ant would be still more rational if it wisely managed to consume its entire resources by exactly equal amounts each day of its life. But the trouble with such a view is that it still supposes a *value judgement* to uphold that the adoption of a particular goal (the ant's goal) is more rational than the adoption of another one (the grasshopper's goal). Applied to this example, Elster's thesis would be that the way the grasshopper-type is discounting the future is in itself irrational. Someone who quickly spent a large portion of a fortune and complained thereafter of being left with too few resources for the rest of his life would look rather irrational indeed. But suppose that for someone the very fact of saving is enough to spoil any pleasure in life; why would it be so irrational for this person to spend heavily now for full enjoyment in life and to add later to this unspoiled pleasure the extra unspoiled satisfaction that the meagre resources left available can provide? Furthermore, since the ant-like behavior could result from pure habit and the grasshopper-like behavior could, in such a case, result from a computation of the optimal way to distribute over time the diminishing spendings that provide the optimal satisfaction never spoiled by saving, there is no point in founding such a view of rationality and irrationality on the fact that a wise computation is associated only with the ant-type of behavior. Here again, it seems that only a value judgment could permit one to draw a line between rational and irrational actions. Defining rationality by a maximizing process implies the predetermination of a *maximandum* and the determination of this *maximandum* can hardly be done without an authoritatively pronounced value judgment.

The present inquiry on the nature of irrationality could have been pursued much further and many other arguments aiming to characterize irrational behavior could be discussed in the same fashion.[23] However, the objective of the present paper is not to propose a survey of theoretical discussions of irrationality, which naturally would have to include many more than the few authors sketchily discussed here. Nor is its objective to deny the difference between rational and irrational behavior, and still less do I offer an apology for irrationality.

The present paper instead challenges the possibility of a strictly positive characterization of irrationality, i.e. a characterization in which no value judgments are made. It is clear that one can have good reasons to consider that some behavior is irrational for being genuinely destructive from the point of view of a value taken as a norm, without having significant compensatory positive effects from the point of view of any other equally acceptable value. One can safely claim that it is more rational to contribute to increasing what is unquestionably recognized as a good (happiness, health, wealth, etc,) than to take steps that could have negative effects on the possibility of enjoying such a good, but this type of claim implies value judgments about what is good (about what value should be taken as a norm) and indirectly about what is rational.

However, methodologists of economics – a science that assumes some type of rationality – typically aim to avoid explicit references to value judgments when characterizing rationality because such judgments tend to undermine the scientific character of economics. As long as they unquestionably *assume* that wealth is the only value to be considered by economists, they are in a position to define rationality in an apparently technical manner by equating rational behavior with wealth maximizing behavior (through revenue maximization or profit maximization). But since the narrow association of economics with wealth maximization has been regularly challenged, it has become current practice to apply the tools of economics to the maximization of any value (happiness, health, satisfaction, etc.), making it clear that any of these values can be considered as a norm to characterize rationality in last resort. A way out of the problem is to maintain that economists' tools could be applied to the (rational) search for any values insofar as those values are supposed to be given once and for all before the economist's investigation. In that perspective, consistency (in time) becomes the hallmark of rationality. But this methodological decision to denounce any case of nontransitivity as evidence of irrationality could not be taken without some overvaluation of sheer consistency. How else could one consider that it is rational *consistently* to maximize impoverishment or pain ("De Gustibus Non Est Disputandum"), when maintaining that it is irrational successively to maximize variable types of satisfaction according to changing moods? On what grounds should the rigidity associated with consistency be declared more rational than adaptability to changing moods? Clearly, one can strongly object to the laxity implied by such an approach and claim that it is unacceptable to see rationality in any inconsistent behavior and that consistency, at least, counts among the commendable values that should be associated with rationality. But such a legitimate objection is an open value judgment; my maximal thesis is that such value judgments are unavoidable when distinguishing between rationality and irrationality.

The point is that once one discards the idea of defining rationality by the mechanical maximization of a predetermined variable (an operation typically attributed to *homo oeconomicus* but done more efficiently by a computer), it becomes almost impossible clearly to distinguish rationality and irrationality without resorting to value judgments. It is clear that this problem of precisely identifying what is excluded by the very idea of rationality has not been seriously considered by most economists and philosophers. If the present paper has shown convincingly that the ingenious proposals of the few theorists who have attempted to solve this problem without resorting to sheer value judgements were not really successful, the remaining options seem to be the following: either one defines rationality through strict maximization of a predetermined variable, but one has to admit that a science based on such a view of rationality can hardly describe what is going on in a world where (rational) decisions are usually not taken in such a mechanical way; or one defines rationality through the subjective adaptation of action and, while leaving moralists to evaluate the rationality of actions in a broader sense of the word, hardly avoids the conclusion that any action is rational in the limited (or thin) sense of being subjectively adapted to its immediate end. However, in the latter case one has to renounce invoking an *empirical* principle of rationality. In the first case, the principle of rationality would (almost) always be false, in the second, it would always be true. In both cases, the science based on such a principle of rationality cannot be an empirical science, in the first case for being (almost) always falsified and in the second for being unfalsifiable. Certainly, one can be irrational and many people act irrationally, but it is far from certain that irrationality understood in a sense that is significant for a science pretending to analyze actual economic behavior can be characterized otherwise than by resorting to sheer normative judgments.

Université de Montréal

NOTES

[1] A first version of this paper has been presented to the 19th Annual Meeting of History of Economics Society held at George Mason University at Fairfax, Virginia in June 1992. The author thanks Mary Baker, Jon Elster, Erik Litwack, Paisley Livingston, Robert Nadeau, Malcolm Rutherford and Bruce Toombs for their helpful comments as well as the SSHRC (Ottawa) and the Fonds FCAR (Quebec) for financial assistance.

[2] Mises, Ludwig von, 1976, *Epistemological Problems of Economics*, New York, New York University Press, pp. 23–24.

[3] *Ibid*, p. 23.

[4] Mises, Ludwig von, 1944, 'The Treatment of "Irrationality" in the Social Sciences', *Philosophy and Phenomenological Research* 4, 3(June), p. 544.

[5] Robbins, Lionel, 1952, *An Essay on the Nature and Significance of Economic Science*, 2d edition, London, Macmillan & Co (1st ed. 1932), see pp. 94–99; Robbins does not couch his argument in terms of rationality, but this notion is implied by the way he uses the notion of "economic men".

[6] This kind of "acte gratuit" was carefully described by André Gide, for example, in his novel *Les caves du Vatican*.

[7] Machlup, Fritz, 1955, 'The Problem of Verification in Economics', *The Southern Economic Journal* 22, 1 (July), pp. 1–21. Machlup was engaged in two memorable debates about this

issue, one with Hutchison (see *The Southern Economic Journal* **22**, 4, 1956, pp. 478–493), the other with Lester (see *American Economic Review*, 1946–47).

[8] Popper, Karl, 1985, 'The Rationality Principle', in D. Miller, ed. *Popper Selections*, Princeton University Press; previously published in French translation under the title "La rationalité et le statut du principe de rationalité", in Classen E. M. (ed.), *Les fondements philosophiques des systèmes économiques*, textes de Jacques Rueff et essais rédigés en son honneur, Paris, 1967.

[9] *Ibid.*, p. 360.

[10] *Ibid.*, p. 361.

[11] However in a paper entitled 'Popper and the Rationality Principle' and published in *Philosophy of the Social Sciences* (vol. 23, dec. 1993, pp. 468–480), I attempted an interpretation (partly reused in the present paper) of Popper's ambiguous positions; for a different view on the same question, see also the paper by Robert Nadeau, 'Confuting Popper on the Rationality Principle', in the same issue, pp. 446–467.

[12] Popper, Karl, quoted article, p. 363.

[13] Becker Gary S., 1962, 'Irrational Behavior and Economic Theory', *Journal of Political Economy* **LXX**, Feb., pp. 1–13; reprinted in Becker, Gary S., 1976, *The Economic Approach to Human Behavior*, Chicago, The University of Chicago Press.

[14] This thesis gave rise to a brief controversy with Israel Kirzner. See Kirzner, Israel M., 1962, 'Rational Action and Economic Theory', *Journal of Political Economy* **LXX**, Aug., pp. 380–385 and the replies from each author: Becker, Gary S., 1963, 'A Reply to I. Kirzner', *Journal of Political Economy* **LXXI**, Feb., pp. 82–83; Kirzner, Israel M., 1963, 'Rejoinder', *Journal of Political Economy* **LXXI**, Aug., pp. 84–85. I have analyzed the implications of this debate in Lagueux, Maurice, 1992, *Kirzner versus Becker: Rationality and Mechanisms in Economics* in: *Perspectives on the History of Economic Throught. Selected Papers from the History of Economics Society Conference 1991*, R. Hebert, ed., Aldershot, Hants., Edward Elgar Publishing, 1993, pp. 23–37. An argument from this paper is reproduced in the present text.

[15] Elster, Jon, 1984, *Ulysses and the Sirens, Studies in Rationality and Irrationality*, New York, Cambridge University Press, Paris, la Maison des sciences de l'homme (1st edition, 1976), p. 140.

[16] Elster, Jon, 1983, *Sour Grapes, Studies in Subversion of Rationality*, New York, Cambridge University Press, Paris, la Maison des sciences de l'homme, section I.

[17] Elster, Jon, 1984, *Ulysses and the Sirens, Studies in Rationality and Irrationality*. New York, Cambridge University Press, Paris, la Maison des sciences de l'homme (1st edition, 1976), p. 86.

[18] In a personal communication, in which he reacted to an earlier version of this part of my paper, Jon Elster observes quite correctly (and illustrates this very clearly with many examples drawn from another book from him, *Nuts and Bolts for the Social Sciences*, New York, Cambridge University Press, 1989) that "there are always costs involved in binding oneself" and he concludes from this "therefore, it would be better if one could just realize the aim directly". Of course, it would be *better*, but would it necessarily be more rational? If someone is impeded by some constraints from realizing "the aim directly" and, after taking account of all the costs involved, manages to realize it in a very clever but indirect way, in what sense is it sensible to say that his response is less rational than the direct response of a person who does not have to face these constraints? This is the question I try to reformulate (and the point I try to substantiate) in the following paragraphs.

[19] Elster, Jon, 1986, 'Introduction', in Elster, Jon ed. (1986), *Rational Choice*, New York, New York University Press, p. 15.

[20] The question of weakness of the will (or of akrasia) has been a frequently discussed topic in the recent literature. For example, in a remarkable book entitled *Irrationality, An essay on Akrasia, Self-Deception, and Self-Control* (New York, Oxford University Press, 1987), Alfred Mele argue that akratic action is in itself irrational action. As made clear by the arguments of the present text, I maintain against Mele that one must dissociate these two characterizations of action. However, a serious discussion of the ingenious arguments proposed by Mele, and by

other analysts of akrasia, would require another paper which would also have to consider Donald Davidson's views on this question.

[21] Elster, Jon, 1984, *Ulysses and the Sirens, Studies in Rationality and Irrationality*, New York, Cambridge University Press, Paris, la Maison des sciences de l'homme (1st edition, 1976), p. 65.

[22] *Ibid.*, p. 66.

[23] In particular, the cases of inconsistent behavior described by Tversky and Kahneman (see, for example, Tverksy, Amos & Daniel Kahneman, 'The Framing of Decisions and the Psychology of Choice', in Elster, Jon, ed. (1986), *Rational Choice*, New York, New York University Press, pp. 123–141) should be carefully considered but it seems that the question of the predetermined character and of the identity of the *maximandum* implied by the notion of rationality challenged by these researchers would still be a crucial consideration.

ALAIN VOIZARD

"IF COWS HAD WINGS, WE'D CARRY BIG UMBRELLAS." AN ALMOST NUMBER-FREE NOTE ON NEWCOMB'S PROBLEM

With David Lewis, I see Newcomb's problem as a prisoner's dilemma for space cadets: a secular, sci-fi successor to the problems of predestination that exercised such thinkers as Jonathan Edwards (1703–58). It is a problem to which I (unlike Lewis) would apply the thought of Esther Marcovitz (1866–1944): "If cows had wings, we'd carry big umbrellas."[1]
R. C. Jeffrey

1. INTRODUCTION

Ever since it was first made public in 1969 by R. Nozick,[2] William Newcomb's 1963 problem has haunted decision theorists. It has given rise to "Newcombmania",[3] causing an innumerable number of attempts at devising a solution to this problem, which shook all of decision theory. Some authors[4] have proposed to change standard and purely evidential Bayesian decision theory in order to accommodate Newcomblike situations. They contend that only a *causal* decision theory may face up to the challenge posed by Newcomblike situations, but the change implies a major overhaul of the standard theory (or, rather, of the *family* of standard evidential theories). Many believe though, and I for one, that we should not get carried away. After all, the standard theory has a lot to offer: its mathematics is good, the theory itself is elegant, intuitive, simple, and useful: all of which are important epistemological advantages. But it also has the further advantage of not relying, as its *causal* counterpart does, on the attribution of probabilities to counterfactuals, as they are needed to acknowledge the causal dependencies which the theory is designed to make perspicuous. Now *that* is a tremendous advantage, especially if we consider the problems which it behooves us to solve in order to deal with the semantics of counterfactuals: the displaced fog would only grow thicker around the bend. Simply put, the question is: why change a good theory for one which appears to get things in line with our intuition in a bizarre problem based on an improbable situation sure never to obtain?[5] Such a move towards a new theory is still in dire need of a stiff defence. This is manifestly so if we take account of the fact that, if one sets his sights straight, noncausal decision theory yields the right decision (though an astonishing one) for an agent in Newcomb's predicament after all – that is: *take one box only.*

My aim here is to clarify a few *strictly conceptual* muddles. I hope that, after some of the mess is cleared up, the reader will come to accept that standard and purely evidential Bayesian decision theory stands alone in yielding

193

M. Marion and R. S. Cohen (eds.), Québec Studies in the Philosophy of Science II, 193–213
© *1996 Kluwer Academic Publishers.*

an advice based on readily *available* data, and, hence, the only sound and dependable advice. If I do not succeed in doing this, then I hope I will not add to the mess. Finally, my argumentation rests rather heavily on the trivial presupposition that one cannot expect decision theory to help assign initial probabilities to an agent's beliefs; decision theory is not a full-fledged theory of rationality, though it certainly is part of one. In other words, I take it that rationality has a purely instrumental value and that decision theorists should be Humian through and through, i.e. none should be so bold as to believe it to be in his field's grip to determine, piecemeal, the rationality of a given agent's beliefs. It is, though, a theory which may play a *rôle* in indicating to an agent, *given* his assignment of a (conditional) probability to each member of a set of mutually exclusive and collectively exhaustive outcome statements (one constraint being that the probabilities assigned to them add up to one in such a way that book can't be made against him), and *given* his desirabilities, or his utility assignments (another structural constraint being the ordering and transitivity of preferences), what to do given these assignments (the last constraint being composability of expected utilities). The *rationality* of his action being thus relativized to the body of beliefs and desires which are his is thus purely instrumental and may very well fall horrendously short of *ideal* rationality – whatever that may be in view of the difficulties which accompany the 'ideally rational agent' fiction. So, if, to some, the query found in Newcomb's problem does not seem satisfactorily answered, it is, in my opinion, because in cases such as this one decision theory is expected to offer answers to questions it cannot answer. Decision theory should not be expected to list the criteria for beliefs to be rational. It will guide one, given his beliefs, desires, and a set of possible courses of action, as to how one should act.[6]

In Section 2, I will start by presenting a standard version of Newcomb's problem. Then, in Section 3, I will opt for what may be called a no-position position by dismissing the idea that any overarching conclusion may be drawn regarding an agent's rationality, or regarding the value of evidential theories of decision in general, from the choice of one or the other horn of the dilemma encountered in Newcomb problems. I will, in subsections of that section, offer two different interpretations of the standard problem presented in Section 2 in order to keep apart two different stories which are not to be confounded.

Section 4 will examine the Principle of Maximizing Conditional Expected Utility (PMCEU) and show how, depending on which interpretation one opts for, it can be read as a normative principle or as a descriptive principle. It will turn out, unsurprisingly, to be a normative principle for the agent to abide by, but of course relativized to his belief set; it will be a guiding principle in the theorists' hands when explaining (rationalizing) an agent's behaviour. I claim that instability bred by an oscillation between adopting one or the other of the two interpretations presented in Section 3 is what leads to non-ratifiable decisions.

In Section 5, I will sort out some of the difficulties which stem from the relations between the Principle of Dominance (PDOM), the concept of stochastic independence and that of probabilistic causality. I will show how causal theories of decision presuppose the agent's capacity to identify causally relevant correlations, and I will claim that this presupposition divests Newcomb's problem of what philosophical interest it may have. It will turn out that this purely conceptual analysis of what is relevant in the examination of a decision situation such as that of Newcomb's, indicates that correct understanding of subjective probabilities should lead to the idea that we ought to adopt a stance which is, through and through, Humian: (i) no belief is in itself rational or irrational; (ii) no end pursued by a given agent is, again, in itself rational or irrational and (iii) taking full measure of what subjective probabilities are, we should do away with specifically *causal* beliefs, reframing them in terms of the lacks of independence between statements the agent is willing to treat as relevant data.[7] If the foregoing is correct, it will appear that we should be able do without *causal* decision theory.

2. THE PROBLEM

The setup: There are two boxes (B1 and B2) on a table. B1 is made of clear plastic and visibly contains a thousand dollars (henceforth $T). B2 is opaque and contains either a million dollars ($M) or nothing ($0).

Options: The agent is asked to choose between taking either both boxes, or B2 alone.

The problem: A soothsayer has foreseen the agent's choice and has, accordingly, placed either a $M or $0 in the opaque box. He has put a $M in B2 if he predicted the agent will take B2 only; he put $0 in B2 if he predicted the agent will take both boxes. Also, if the soothsayer predicted the agent will decide to randomise his choice by, say, flipping a coin, then he put $0 in B2. *Ex hypothesi* (i) the agent enjoys money, his one interest in life lies in making a fast buck (let us be Humian about this); *and* (ii) the soothsayer has an excellent prediction record. He has been known to make correct predictions in all cases[8] that have hitherto been presented to him. The agent knows all this and he knows that the soothsayer knows it too, etc. The only thing the agent is left in the dark about is the prediction itself. Knowing all this, the agent comes to believe that the choice he decides to act upon will have been predicted by the soothsayer. This means that, whichever decision is made, making that decision will be news in disguise that *this* was the content of the soothsayer's prediction. Explicitly: to make the decision to take both boxes will be news in disguise that there is $0 in B2; conversely, to make the decision to take B2 only is, likewise, tantamount to *learning* that there are a $M in B2. Now, if you were the agent and were asked to cast a stone in either one of the two buckets, what would *you* do?

3. INTERPRETATIONS

3.1. *Exploiting Collateral Knowledge: The No-Position Position*

Now, if that were the whole story, there would then not be much of a problem. The decision being incumbent upon the agent, it seems it should only depend on whether he thinks the soothsayer has correctly predicted his behaviour or not. Now since his decision should only depend on that, he can decide either to take B2 only or both boxes, and in either case he would be acting rationally.

He would be acting rationally in choosing only one box in so far as he believes the soothsayer was right in predicting he would take only B2, and believes also that were he to act otherwise the soothsayer would have predicted this move and he would not become a proud millionaire. But he would also be acting rationally were he to take both boxes since he might believe that the soothsayer has not correctly predicted his choice and so, by taking both boxes, and in the best of cases, he could become a proud millionaire + *one thousand dollars*; he knows that if the soothsayer has correctly predicted his action, then there will not be a $M in B2, and so again, to take both boxes is preferable to taking only B2 since a $T is preferable to $0.

There would be no paradox here because, as mentioned, all would depend on whether the *agent* thinks the soothsayer has or has not correctly predicted his behaviour, and this we can readily determine by observing the agent's behaviour. So, if we consider decision theory as a descriptive theory, then the paradox does not arise. There is no conflict between either the agent's subjective probability function and/or his subjective desirability function on the one hand, and his action (whichever course of action is chosen) on the other. Any action can be described and correctly qualified as rational simply by making sufficiently charitable hypotheses relative to the agent's weighted calculation of probabilities and desirabilities. On this account, both taking one or two boxes may come out compatible with standard decision theory and, on that basis, both choices could be declared rational.

This of course is possible only under the assumption that we ask of decision theory that it be no more than a descriptive theory of an agent's rational action under certain conditions, given his desires and beliefs. The seemingly irrational one-boxer may be dubbed 'rational' since that which is needed to explain his action presupposes the application of a Davidsonian charity principle[9] for third person belief/desire ascriptions, and it is this principle which supplies us with enough leeway to reinstate the one-boxer into the community of rational agents. I concede, though, that if we show ourselves too charitable by, let's say, multiplying the number of relevant parameters in the analysis of the agent's behaviour, the danger of our interpretation's losing much predictive power lurks. Nevertheless, seen in this light, everything is fine: whether you take one or two boxes, *you're o.k.* As I have said, I will argue for this interpretation, ultimately basing my claim primarily on the trivial fact that it is more conservative a measure to refrain from attributing a causal influence of any perceived cause C on any perceived effect E than to assert it. More precisely, my claim is that the right answer is to take one box only, at least if one is ready to stick to the problem's story (what I wish to call the *strong* interpretation). Taking both boxes is still a live option but only if one understands the problem under its *moderate* interpretation, one which takes it for granted that there is causal independence between the agent's action and the content of B2. This is not a very interesting route for it is dangerously close to a total disrespect of the problem's specifications.

If the discussion between one-boxers and two-boxers appears to be irremediably deadlocked, it is in part because of the following two different and incompatible interpretations of the problem. I would like to defend the idea that the most philosophically interesting interpretation is the upcoming strong one, and that under this interpretation one can only be a one-boxer. One can only be a one-boxer because I do not think we can take for granted that, in such a situation, the agent can tell statistical correlations apart from true relations of causation. There are good reasons for this. In this Newcomb problem, the only interpretation one can give of the agent's probability assignments is a subjective interpretation (i.e. they are expressions of his beliefs given the specifications of the problem and the information that is made available to him). Now the information being what it is, the agent cannot tell if the dependence which seems to obtain between the acts (take one or both boxes) and the states (there is $0 or a $M in B2) is due to a series of purely fortuitous coincidences, or to causal propensities. Things being so, it should lead us to think that, given the probability settings in Newcomb's problem, it is perfectly rational to choose one box only (in fact it is irrational to choose both). Indeed, subjective probabilities being what they are, the agent has no grounds for thinking that the acts and states are independent of each other. When the two-boxer proclaims this independence, he is in fact changing the terms of the problem.

3.2. *The Strong Interpretation*

Under the strong interpretation, Newcomb's problem is a *Gedankenexperiment* and, as such, it can ask of us that we accept, *ex hypothesi*, almost anything (that locomotives fly or what not . . .) *and* that we stick to the story of the problem. That is, that we do not dismiss any of the constraints stipulated in its formulation on grounds of implausibility or on some other non-philosophical grounds. Under this interpretation, it is a problem which asks us to reflect on what a rational decision should be in a situation where an agent's choices have been correctly predicted.

If we are not to treat this problem lightly, then we cannot dismiss the stipulation according to which a prediction has been made and that it has consequences on there being or not a $M in the opaque box. Thus, we cannot say that the prediction does not influence the structure of this story. We *must*

take into account the fact that the prediction does count. Under the strong interpretation one can only be a one-boxer.

What the strong interpretation presupposes is not that the prediction could
cause the agent to act in a determine manner, nor that the agent's action *causes*
(backwards in time) the $M to be or not in B2.[10] What it does *strongly* take
into account is the fact that, as stipulated, *the prediction will affect the content
of* B2. This interpretation has been criticized because it was thought irrational to believe that a prediction could bear on an agent's decision to choose
one or both boxes. But this is not exactly what this interpretation underscores since it does not suppose that there is a causal link of sorts between
the prediction and the agent's actions. What it does nevertheless presuppose
is that there *is* a link between yesterday's prediction and the opaque box's
content today. There is no magic here: the soothsayer himself put the money
(or $0) in the opaque box, and it is stipulated that he relied on his predic-
tion for that. It is also said that his predictions are *very* reliable. I call this
the strong interpretation because I will suppose that we accept the terms of
the problem (whatever intellectual gymnastics this implies). So let me reformulate the problem in order for the soothsayer's infallibility to be given
some plausibility, albeit *gedankenexperimentell*-plausibility.[11]

Let us suppose that our soothsayer has no paranormal foreknowledge of
events, but simply happened to get his hands on a computer-like and ever so
slightly unreliable futuroscope which allows him to "see" future. All that the
soothsayer needs to do is type the agent's name on the keyboard and up pops
the answer on a screen: either 'two-boxer' or 'one-boxer'. The soothsayer thus
has access to the future of the actual world; he can *see* it, not only a possible
state of the future, but the future of the actual world: the future as it *will* be.
What should one do in those circumstances, knowing that the futuroscope is
quasi-infallible? One has to be a one-boxer. Why? Because if the agent takes
only one box, there will be a $M in it. But if he takes both, the soothsayer
will not have put a $M in the opaque box. But, we should feel like replying,
why does the agent not choose to be tricky, i.e. to take the opaque box, find
the $M (or $0) and then take the thousand dollars in B1 that's there for him
to grab? Because if he did that, there would not have been a $M under the
opaque box for *him*. Why should he leave a thousand dollars behind and not
touch it? After all, the soothsayer cannot change the fact that the $M has
already been placed in the opaque box, so why not *take the money and run*?
Because it appears that leaving the $T behind is the condition for there being
a $M in B2! The soothsayer's futuroscope would have shown that the agent
was the kind of person who would try to pull that trick,[12] and therefore the
soothsayer would not have put a $M in the opaque box in the first place.
This argument, I believe, is cogent. Any two-boxer can be said, under this
interpretation of the problem, to be either irrational or to misunderstand the
problem.[13] Now, if this amounts to acting upon a choice liable to change the
evidence in such a way as to bring about the "good" news that there are a
$M in B2, so be it.[14]

3.3. *The Moderate Interpretation*

This, of course, can't be the whole story. What stands at the core of Newcomb's original problem is a time factor: the soothsayer has *already*, before the time at which the agent is to make his decision, placed either a $M or $0 in B2. Insistence upon this is what makes a paradox out of this problem. He has placed money, or nothing, in B2 *yesterday* as it were, and he cannot change that. So the question, it seems, is: why not be a two-boxer? Why not go for B2, find a $M (or $0), then grab the contents of B1 and, again, *take the money and run*? After all, it has already been decided whether or not there will be a $M in B2, so by taking both boxes the agent can be sure that, even if the worst comes to the worst, he'll still come out a thousand dollars ahead. It seems that *it* is the only rational thing to do, since by acting upon such a choice the agent is sure to fare better.

Under this moderate interpretation, Newcomb's problem is but a teaser. I wish to call this interpretation 'moderate' since it will not suppose that the agent is bound to the terms of the problem. But still, the soothsayer places either one $M in B2 or nothing at all according to what he predicts or according to what his defective futuroscope foresees in his place.

How are we to tackle this problem? At choosing time, it makes no difference whether we believe in the soothsayer or not. It doesn't even matter whether he has consulted his futuroscope or not – the whole thing could even be, as far as we are now concerned, a hoax. He will still have put either a $M or nothing in the opaque box yesterday as it were, and he cannot change that. What is the agent to do? He might want to take only B2, thinking perhaps that his doing so will influence there being a $M in B2. Would that be rational? No. In order for this to be rational, he must believe in the soothsayer's accuracy in predicting his choice. But why should that matter? After all, if the agent was to choose only B2, and the soothsayer had predicted he would and had accordingly put a $M in B2, why should the agent pass up the opportunity to grab the $T in B1, granted the soothsayer cannot make it that the money will not be there any more, as that was settled yesterday? For someone to take only B2, under this interpretation, I maintain, would be irrational.

Here are some wrong turns and absurd thoughts to be found in the two-boxer's reflections. Two-boxers contend that in order for someone to be a one-boxer (under the foregoing interpretation), one must think either that the soothsayer's prediction causes the agent to choose B2, or that choosing B2 is not only a sign but a cause of there being a $M in the opaque box.[15] Two-boxers exploit these absurd thoughts in their many criticisms levelled at one-boxers. They, in fact, maintain that a one-boxer must entertain these thoughts in order to be, precisely, a one-boxer.

Of course, if the correct interpretation is the moderate one, one-boxers must have very strange beliefs indeed; so I fully agree that to act upon such a choice, under such an interpretation, would indisputably need to be irrational. The important question is then: what, exactly, could account for such overtly irrational behaviour? I think the answer to this question lies not in the domain of decision theory but is linked rather to the agent's beliefs regarding issues concerning freewill and its conflict with determinism. The fundamental point which then needs to be addressed is: when the soothsayer predicts at time t_0 that at time t_1 the agent will act upon decision d_i, is his prediction true at any moment prior to t_1? It is well known that if we give 'yes' for an answer, we expose ourselves to a battery of quasi-insoluble difficulties. If we say of a prediction that it was true before the instant at which it is fulfilled, then we must accept that had a true prediction been made yesterday that you would now be reading this paper on Newcomb's problem, then it is impossible that you should not be doing so now. But this certainly cannot be our problem, for the moderate interpretation which leads to the absurd thoughts and lands Newcomb in the predestination debate, does not suppose the agent is bound to the terms of the problem. And the consequences of this are great. Indeed, if the agent disregards the probabilistic dependence which, *ex hypothesi*, obtains between the acts and the state of B2's content, then *there is no problem*. Any one-boxer with a taste for money is, in this case, irrational.

If we treat Newcomb's problem as one of predestination, one may draw any of the following conclusions: (i) there does not – rather, there cannot – exist such a soothsayer as he cannot err; (ii) equivalently: true predicted choices are a contradiction *in adjecto*;[16] (iii) there will never be a $M in B2 for $\mathcal{U}$-rational[17] agents. Hence, this treatment lends support to the view according to which, one way or another, irrationality is richly rewarded.[18] But not to the view according to which one should choose to take both boxes.

4. PMCEU, NORMATIVITY, THE AGENT AND HIS INTERPRETER

Unpacking the one-boxer's argument yields the following. The desirability matrix should look something like this:

	P_1 Predicts one box	P_2 Predicts two boxes
T_1: Take one box	$ 1 000 000	$ 0
T_2: Take two boxes	$ 1 001 000	$ 1 000

The problem states that the soothsayer's record is *very* good. Let us, accordingly, set the probability that he has correctly predicted the agents choice at close to one, say 0.99. So the conditional probability $Prob(P_1/T_1) = 0.99$ and, of course $Prob(P_2/T_1) = 0.01$. Also, $Prob(P_2/T_2) = 0.99$ and $Prob(P_1/T_2) = 0.01$. The probability matrix should thus look something like this:

	P_1 Predicts one box	P_2 Predicts two boxes
T_1: Take one box	0.99	0.01
T_2: Take two boxes	0.01	0.99

This way – desirability units being each worth a buck –, the subjective expected utility (SEUA$_1$ = Σ_iProbT_1DesT_i) of taking one box is:

$$0.99 \times d(\$1\ 000\ 000) + 0.01 \times d(\$0) = \$990\ 000.$$

And that of taking both boxes is:

$$0.01 \times d(\$1\ 000\ 000 + \$1\ 000) + 0.99 \times d(\$1\ 000) = \$11\ 000.$$

So is, by applying the Principle of Maximizing Conditionalized Expected Utility (PMCEU), the case made for one-boxers.

The foregoing argument, coupled up with the overwhelming intuition that it is irrational to leave B1 behind, is the argument found all over the two-boxer literature which purports to show that the evidential theory is flawed. This, though, is a somewhat weak argument for it presupposes that if the agent uses an evidentiary theory, he will actually assign a probability of 0.99 to $Prob(P_1/T_1)$ and to $Prob(P_2/T_2)$ (or anything similar which respects $Prob(P_1/T_1) \gg Prob(P_2/T_1)$). But is there a reason for this?

If this probability assignment of 0.99 is attributed to $Prob(P_1/T_1)$ and to $Prob(P_2/T_2)$ *ex ante*, by the agent, then we must think of him as understanding the problem under its strong interpretation. If so, we cannot say that he is wrong in exploiting PMCEU and applying it normatively, less we be ready to admit that by doing so we consider irrational his belief according to which $Prob(P_1/T_1)$ = 0.99. Now, neither our judging his belief irrational, nor his attributing 0.99 to $Prob(P_1/T_1)$, has to do with the decision theory adopted (nor with PMCEU), and hence the theory cannot be held at fault here: the decision to take B2 only should not be seen as evidence against the sound-ness of evidentiary theories; and the agent's attributing 0.99 to $Prob(P_1/T_1)$ should not be seen as evidence of his irrationality *since he understands the problem under the strong interpretation*. In other words, the agent's choosing one box only is nothing over and above testimony that he is treating the problem as a *Gedankenexperiment*.

Now, if a third party is, *ex post*, judging the agent's action, after he has chosen B1 only – let us call this third party the *interpreter* –, then, using PMCEU descriptively, in order to rationalize the agent's action (to *explain* it), the interpreter may charitably surmise that the agent has assigned the 0.99 figure *because he understood the problem under the strong interpreta-tion*. But notice that it is only then, i.e. when all the power of hindsight analysis is made available, that the agent may be said, from the interpreter's point of

view, to be irrational. He may be said irrational for believing something tantamount to backward causality between act and state – state dependent itself
upon the prediction. Notice that by substituting the interpreter's point of view
to the agent's, we have also substituted the moderate interpretation for the
strong one. Notice also that it is only the *ex post* analysis, jointly with the
moderate interpretation, that allows the probability assignment to be regarded
as irrational. So, as noted in Section 3, to adopt the interpreter's stance, to
judge the probability assignment irrational, and to adopt the moderate interpretation, all represent as many departures from Newcomb's problem.

Let us now suppose that the two-boxer is interpreting the problem under the
strong interpretation, and that he is examining it *ex ante* – i.e. he treats the
problem as a genuine *Gedankenexperiment*. What he must believe in order
for the evidential theory to be discredited, yielding as it does a one box
option, is that it is necessary for the agent to assign 0.99 to $Prob(P_1/T_1)$. The
problem with this is its implying the actual existence of such a soothsayer
as one who cannot err! Indeed, for the two-boxer to efficaciously counter
the evidential theory, he must think that $Prob(P_1/T_1)$ *must* be set at 0.99 by
anyone relying on an evidential theory. But to do so, he must give this probability a frequentist interpretation: setting the probability at 0.99 is a result
of the soothsayer's past record.[19] Or, perhaps, he gives it a subjective interpretation, but, if so, for the probability assignment to be 0.99 he must believe
that convergence theorems hold and that a 0.99 probability is obtained through
conditionalization on initial assignments which may have been different from
the current 0.99. Now this, I fear, is a wild departure from the terms of the
problem for it implies that the two-boxers attribute, to the one-boxers, the belief
in the *existence* of such a soothsayer – more than the problem asks of anyone.
If one does not follow the foregoing train of thoughts, then one *need not* assign
0.99 to $Prob(P_1/T_1)$ (nor to $Prob(P_2/T_2)$) and hence Newcomb's problem *need
not* be a problem for evidential theories. In other words, two-boxers are
wrong in thinking that if one adopts an evidential theory, one must make his
subjective probability assignments match the ones found in the problem's probability matrix as given above. This hardly qualifies as a genial argument, but
recall that subjective probabilities are degrees of belief. Applied to conditionals,
they are *a propensity to believe in . . . .* $Prob(P_1/T_1) = 0.99$ is to be interpreted as a measure of belief that T_1's holding makes P_1's holding 0.99-true.
Or, in other words, if T_1 is the case, then I should be willing to buy from
(or sell to) you a bet with 99:1 (1:99 *resp.*) odds. Now, the probabilities are
said to be subjective *precisely* because we deem it unnecessary to probe into
the agent's thinking in order to evaluate the reasons which were used in
determining them. The agent's initial assignments are regarded as raw data,
it is *not* prior knowledge which upon deliberation yields some other probability assignments to be acted upon, and then judged rational or not.

The two-boxer assumes, it appears, that the piece of knowledge according
to which the predictor has never failed (i.e. that he is scoring 99% on prediction tests) must induce the belief that we should assign a 0.99 probability

to the correctness of his prediction. How is this assignment induced? Is it *caused* by the telling of a story according to which there is a soothsayer who . . . ? The point here is that there is no mechanical procedure which yields a 0.99 probability figure from the data offered in Newcomb's problem, nor can there be one if one assumes we are dealing with subjective probabilities. If we stick to subjective probabilities, and acknowledge that we should not, offhand, prejudge the rationality (or irrationality) of any particular assignment, wild as it may be, then there is no problem regarding the rationality of the one-boxer's behaviour. Best of all, we can, this way, as equally well explain the two-boxer's behaviour as we can the one-boxer's.

There must thus be settled a preliminary philosophical dispute pertaining to the interpretation of the probabilities to be used. For the two-boxer's criticism to have some power, it must, unfortunately, include some confusion relative to the probability function at work. To sum it up: either (i) the two-boxer interprets the probabilities as relative frequency; or (ii) he interprets them as subjective but he must then either forego the illusion that *necessarily Prob(P₁/T₁)* = 0.99, or accept that his position presupposes the existence of a soothsayer and that convergence theorems apply. In both cases, the two-boxer is caught between a rock and a very hard place.

The reader will have guessed that the probabilities I wish to use are subjective probabilities and that they meet the following constraints (all of which are not, of course, independent):

1. $0 \le Prob(A) \le 1$
    $0 \le Prob(A/B) \le 1$

2. *Prob(A)* = 1 *if, and only if,* A *is* **T** (i.e. logically true)[20]

3. *Prob(~A)* = 1 − *Prob(A)*

4. *Prob(A)* = *Prob(A & B)* + *Prob(A & ~B)*

5. *Prob(A ∨ B)* = *Prob(A)* + *Prob(B)* − *Prob(A & B)*

6. A *and* B *are independent of each other if, and only if*
       Prob(A & B) = *Prob(A)* × *Prob(B)*[21]

7. *Prob(A/B)* = *Prob(A & B)/Prob(B), if* B *is logically false, then Prob(A/B) is undefined.*

8. *Prob(A₁/B)* = [*Prob(A₁)* × *Prob(B/A₁)*]/[*Prob(A₁)* × *Prob(B/A₁)*] + . . . + *Prob(Aₙ)* × *Prob(B/Aₙ)*] (Bayes' theorem) *where Prob(Aᵢ)* > 0 *and Prob(B)* > 0.

9. *Prob(A₁ ∨ A₂ ∨ A₃ ∨ . . . ∨ Aₙ)* = ∑ⱼ*Prob(Aᵢ)* = 1 *for any complete partition of collectively exhaustive and mutually exclusive A's.*[22]

With previous holding, and the holding of a preference ordering between the members of a complete partition of collectively exhaustive and mutually exclusive outcomes (with transitivity such that, ≻ expressing a preference relation, ((A₁ ≻ A₂) & (A₂ ≻ A₃)) ⊃ (A₁ ≻ A₃)), we have all that is needed to rationalize any agent's *action* in a decision situation. However, we have nothing which could yet allow us to assess the rationality of the given agent's *beliefs*. That is, we are in possession of what is needed to occupy the normative position from which the rationality of an agent's action (given *and* relative to his beliefs) can be judged; but as regards his individual beliefs, we can only take note of them.

What, in my opinion, this standard Newcomb problem brings to the fore is that one must distinguish between the normative *rôle* decision theory (unqualified) vies for, the descriptive *rôle* which lies within its reach, and both of these together from its utility from an agent's point of view. Newcomb situations clearly illustrate that it is never anything but the agent's own beliefs which drive the decision machinery, and this imposes stringent limitations upon decision theory's contentions regarding normativity. It shows also that there can be a radical asymmetry between an agent's and a third party's analysis of a decision situation (the third party we called the agent's *interpreter*).

This, of course, is no ground-breaking remark, but we must hold steadily in mind that though as agents we may look at decision theory as a guide for action – to be reasonable, on the occasion of, let's say, one *tough* decision – the course of action we decide to act upon will be a function of *our* beliefs. We, in ideal conditions, so to speak, act so as to maximize the utility of our action (unqualified again, be it conditional expected utility, subjective expected utility, $\mathcal{U}$-utility, $\mathcal{V}$-utility etc.), but whatever our decision function maximizes, and however it maximizes *it*, the function does so on our set of beliefs rather than on any *ideal* set which, as philosophical folklore has it, would be an ideally-rational-agent's-set.

If there is another sense in which decision theory is not normative, it is in the following trivial sense: once an agent has computed desires, probabilities, beliefs and utilities, and the calculations have yielded an option, the agent is not bound to act upon that choice! Smokers know a thing or two about this. Even though PMCEU yields a one-box solution, this does not mean that the agent must then take one box only.[23]

Decision theory is but a tool. If what was said in this section is right, then (i) a solution to the dilemma of espousing either an evidential or a causal theory goes by way of defining constraints to be yielded to when ascribing beliefs; and the understanding that passing a judgement on the rationality of an agent is always a measure of our own (instrumental) rationality. And, (ii) a matter we have not yet touched upon, belief in the existence of a relation of causation is a belief which needs to be buttressed by other beliefs; most notably perhaps by the belief that it is in our power to distinguish relations of causation from purely statistical correlations.

5. PDOM, INDEPENDENCE AND CAUSALITY

Two-boxers who were willing to assign a probability of 0.99 to the soothsayer's prediction, instead of relying on a principle of expected utility, claimed we should rely on a principle of dominance (PDOM): if an outcome O_2 (of act A_2 above) is always better than an outcome O_1 (of act A_1 above), then it is never rational to choose O_1. This appears to be the case here since whichever of P_1 or P_2 obtains, A_2 is in both cases preferable to A_1. So, it seems, is the case made for two-boxers. But this principle can only be appealed to when acts and states are probabilistically independent of each other. Here, the acts are A_1 and A_2, but the states are not P_1 and P_2 but rather M_1 and M_2, i.e. there being or not money in B2, respectively. Reformulating Nozick's[24] probabilistic independence principle, we notice there is no state S in Newcomb's problem that is independent either of A_1 or of A_2 (A and B hereafter):

$$\sim(\exists S)(\exists A)(\exists B)[prob(S \text{ obtains}/A \text{ is done}) \neq prob(S \text{ obtains}/B \text{ is done})].^{[25]}$$

PDOM is not applicable here since the states are not probabilistically independent of the acts; which, by the way, is precisely what Newcomb's problem *states*, so, in a way, things are just as they should be. It is because of this dependence that the standard evidential theory yields what is seen as the wrong answer. There is, though, what we might wish to call *causal independence* between acts and states, and this fact is what a sound theory should, according to the causal theorists, make perspicuous. PMCEU, it seems, is not sensitive to causal beliefs that an agent might have. Hence the substitution of so-called $\mathcal{U}$-utility (calculated from probabilities of counterfactuals)

$$\mathcal{U}(A) = \sum_j prob(A \;\square\!\!\rightarrow\; O_j)\mathcal{D}O_j^{[26]}$$

to $\mathcal{V}$-utility (calculate from conditional probabilities)

$$\mathcal{V}(A) = \sum_j prob(O_j/A)\mathcal{D}O_j.$$

With ($A \;\square\!\!\rightarrow\; O$) standing for "if I were to do A, then O would happen", subjunctive conditionals find their way into the theory. This is not surprising since subjunctive conditionals are fine-tuned for expressing relations of causation. All causal theories based on subjunctive conditionals presuppose, though, that the agent has the capacity to identify relations of causation, and distinguish them from mere statistical correlations. Indeed, any probability function allowing the assignment of a probability to ($A \;\square\!\!\rightarrow\; O$) *is* the expression of this causal knowledge (the assignment itself is a measure of the propensity of A's causing O). This procedure violates the specifications of Newcomb's problem in which the agent knows only that the soothsayer has an excellent record. So, however *rational* the causal strategy may seem, it is but a way of avoiding the problem which is specifically designed to bring the agent to believe that acts and states are not independent of each other.[27] However ludicrous this dependence may be, it is perfectly in line with the story and it is rational to believe in it *so long as* you do not question the rationality of believing in the possibility of there being such a situation as Newcomb's. Indeed, for the problem to arise, it must be formulated in such a way as to bring the agent to believe either that his action is somehow *caused* by the soothsayer's prediction (that somehow the two are hardlinked), or to believe his acting in a certain way will cause the soothsayer to place either a $M or $0 in B2 (perhaps through the magic of backward causality). This way the agent may come to believe that whatever he does, the soothsayer will have predicted it and will thus believe that there is no walking away, as it were, from the soothsayer's prediction: the soothsayer's prediction was true, and if true at time of prediction, then nothing can now be done to alter its truth value. And, therefore, nothing the agent does, can change that because as Ryle would say *it was to be*.[28] So, it seems that to assume causal independence between the soothsayer's prediction and the agent's action is equivalent to denying the soothsayer's capacity to predict an agent's act. This, I agree, is the rational way of looking at things, but it does not comply with the problem's formulation.

Of course, if the agent does not believe in the soothsayer, then *there is no problem*. We can be sure that the agent will assign a probability equal to, or less than 0.5005 – i.e. less than the cutoff probability at which both acts have the same expected utility ($500,500). As Lewis has it, any "know-it-all", that is, any agent who knows exactly which *dependency hypothesis* holds cannot have a Newcomb problem drawn-up for him. If one knows which "maximally specific proposition about how the things he cares about do and do not depend causally on his present actions"[29] holds, there *cannot* be a problem. But notice that this strategy is, again, tantamount to a dismissal of the problem since it consists, roughly, in contending that any agent ready to accept that it is logically possible that there be such a soothsayer has got to be so irrational that we should not bother trying to explain his action. This dismissive attitude can be given some substance by asking the following questions: Why should we believe the soothsayer's record to be good out of his *capacity* to predict and not out of sheer luck? How are we to tell the difference between the two?

That causal theories based on the use of subjunctive conditionals presuppose that the agent knows which dependency hypothesis holds appears in Gibbard and Harper's theory. For *M* standing for "there are a million dollars in B2", A_1 for the act which consists in taking one box and A_2 for the one consisting in taking both, they treat the problem thus:

$\mathcal{U}(A_1)$ and $\mathcal{U}(A_2)$ depend on the probability of *M*, which in turn depends on the probabilities of A_1 and A_2. For any probability of *M*, though, we have $\mathcal{U}(A_2) > \mathcal{U}(A_1)$. For let *M* be μ; then since *M* is causally act-independent, $prob(A_1 \square\!\!\rightarrow M) = \mu$ and $prob(A_2 \square\!\!\rightarrow M) = \mu$. Therefore

$$\mathcal{U}(A_1) = prob(A_1 \square\!\!\rightarrow M)\mathcal{D}\$1,000,000 + prob(A_1 \square\!\!\rightarrow \sim M)\mathcal{D}\$0$$
$$= 100\mu + 0(1 - \mu) = 100\mu$$
$$\mathcal{U}(A_2) = prob(A_2 \square\!\!\rightarrow M)\mathcal{D}\$1,001,000 + prob(A_2 \square\!\!\rightarrow \sim M)\mathcal{D}\$1,000$$
$$= 101\mu + 10(1 - \mu) = 91\mu + 10$$

Thus $\mathcal{U}(A_2) - \mathcal{U}(A_1) = 10 - 9\mu$, and since $\mu \leq 1$, this is always positive. Therefore whatever probability *M* may have, $\mathcal{U}(A_2) > \mathcal{U}(A_1)$, and $\mathcal{U}$-maximization prescribes taking both boxes.

Two things must be underscored here, both of them appear at the beginning of the quote. First, it is clear that knowledge of which dependency hypothesis holds is presupposed: Gibbard and Harper state that $prob(A_1 \;\square\!\!\rightarrow M) = prob(A_2 \;\square\!\!\rightarrow M) = \mu$! Second, but I will not examine this further, it is at best curious that they should have the probability of there being \$M in B2 *depend* on the probabilities of the possible acts. There being or not a \$M in B2 depends on the prediction, and unless one believes that the act depends on the prediction (something at least a *causal theorist* should not want), one cannot cogently believe that the probability of M depends on the probabilities of A_1 and A_2.

Though it is clear that all causal theories which are based on probabilities assigned to subjunctive conditionals presuppose that the agent masters relevant causal dependencies (be it only because "analysing causation by counterfactuals [. . .] presupposes that we understand causation"),[30] not all causal theories are based on counterfactuals. Skyrms' theory, for instance, is not. His theory, though, not only presupposes that the agent is able to tell relations of causation from mere statistical correlations, but it can also be reformulated in terms of subjunctive conditionals. He offers:

$$\text{Value } (A) = \sum_{j} \text{DB}(K_j) \sum_{i} \text{DB}(C_i \text{ given } A \;\&\; K_j) \text{ Value } (C_i \;\&\; A \;\&\; K_j).^{31}$$

Causal knowledge is here embedded in K_j, i.e. in the proposition that causal "propensities are fixed in the jth possible way".[32] Now, in order for the causal propensity (CP: a probability function representing a causal propensity) for $(C_i$ given action $A)$, to be represented in terms of degrees of belief (DB above: a probability function representing the agent's degree of belief), K_j propositions must correspond "to the possible causal propensity distributions within the field of our degree-of-belief probability function".[33] So, it seems to me that Skyrms' agent is thought of as able to identify causally relevant background factors in any given situation.

Causal theories of decision are thus faulty in two ways: (i) they don't comply with the problem's specification, i.e. they wilfully neglect the data according to which there is a very strong correlation between act and state, which data furthermore represents almost all the agent has to go on, *and* (ii) they presuppose that the agent is able to draw a distinction between statistical correlations and causal ones, and if this is taken for granted, then there is no Newcomb problem because $Prob(P_1/T_1)$ will certainly be set at less than 0.5005 (i.e. the cutoff probability), *even by someone who favours an evidential theory*. At this point, I would like to pursue this matter and indicate why we should never presuppose that a given agent is capable of evaluating causal dependencies in such a way as to distinguish them from statistical correlations. Now, if the foregoing and what follows is correct, then it appears we should adopt a theory which is evidential and not causal.

It should be obvious to all that one cannot explain the concept of causation in terms of such classical notions as temporal precedence, necessity, and the like. This translates into the fact that causality ought only be treated in terms of probabilities, and that it be thought of as a cognitive concept and not as a relation holding between real events. As Wittgenstein quite bluntly puts it in *Tractatus* §5,1361 "Superstition is the belief in the causal nexus". It should likewise be obvious that, if one takes full measure of reasonable Humian doubts, then (i) singular causal facts are true in virtue of generic causal facts, and (ii) generic causal facts are reducible to regularities.[34] This means that in any one singular case of presumed causation, we cannot *know* that C (standing for 'cause') will cause E (standing for 'effect'). Even in a case where $C \mapsto E$ ('$\mapsto$' standing for the relation of causation) is a proven relation of causation (i.e. all previous tokens of C's have been followed by tokens of E's while background factors have been sifted through), one ought not to assign a probability of 1 to $(C \mapsto E)$ in any given singular case yet to come. There are many sceptical reasons for this, some Humian, some classical. One of them is very nicely put by Jeffrey:

"Dogmatism" [. . .] is an ancient term for the view that judgement is and ought to be a matter of assertion and denial, of belief [. . .] and disbelief. Probabilism is a relatively recent view [. . .] that judgement isn't or shouldn't be a matter of believing but of – what to call it? – probabilizing. Will it rain today? I don't know; I'm not in a position to assert or deny it. But still I may have an action-guiding probabilistic judgement, e.g. my odds on rain might be 7:3, my probability for rain might be 70%.

100% probability corresponds to infinite odds, 100:0.That's a reason for thinking in terms of odds, i.e., to remember how momentous it may be to assign a hypothesis probability 1. It means you'd stake your all on it's truth, if it's the sort of hypothesis you can stake things on. To assign probability 1 to rain is to think it advantageous to bet your life on it in exchange for any petty benefit. We forget that because of the ambient dogmatism in which we ordinarily speak and think. We imagine that we assign probability 1 to whatever we'd simply state as true, but of course it's not so. If you ask me the time, I'll glance at my watch and tell you, even though I don't utterly rule out the possibility that I've misread it, or that the things has gone haywire. When I say "Ten past two", I'm dogmatizing, not probabilizing.[35]

Now, if we cannot assign a probability of one to any singular relation of causation, whence the distinction between statistical correlation and causal dependencies? This does not imply that I do not believe that relations of causation exist, I believe they have an epistemic existence. We can even ascertain the holding of a relation of causation if we control background factors. What we cannot assert is that $Prob(C \mapsto E) = 1$, and we cannot, conversely, assert that $Prob(C \mapsto E) = 0$. So, for the reflexive individual who expresses his beliefs in terms of subjective probabilities, relations of causation, unless there be a set of past experiences on which to base the belief in them, should be treated on a par with statistical correlations. If there are past experiences, and there has been some control of background factors, then probabilities assigned to proven causal statements should be given a frequentist interpretation. Notice though, that such is not the case in Newcomb's situation. If we recall the distinctions made in the two interpretations presented in Sections 3.2 and 3.3, then we may say that the one-boxer's *ex ante* probability assignments to $Prob(P_j/T_i)$ are to be interpreted as true subjective probabilities expressing the agent's opinion on a matter he does not, for lack of information, fully grasp. This, I believe, is as it should be, as one cannot know *a priori* which correlations are expressions of relations of causation and which are not. Given the data in Newcomb's problem, it is only reasonable, lacking all information to the contrary, that the agent should interpret the soothsayer's extraordinary record as the expression of some relation it would be wise to think will hold in his case.

Taking Gibbard and Harper's theory as example, we may show that when causal theorists deny there is a relation of causation between the act and the state, e.g. $Prob(A_1 \ \Box\!\!\rightarrow M) = Prob(A_2 \ \Box\!\!\rightarrow M) = \mu$, they, in fact, attribute to $(A_1 \mapsto M)$ a probability of 0. Indeed, since $Prob(A_1 \vee A_2 \vee \ldots \vee A_n) = 1$, then, if $Prob(A_1 \ \Box\!\!\rightarrow M) = Prob(A_2 \ \Box\!\!\rightarrow M) = \mu$, then $\mu = Prob(M)$, so causal independence and probabilistic independence are here indiscernible. This entails that they attribute a probability of 1 to $\sim(A_1 \mapsto M)$ a move dangerous in itself, and also one which violates Newcomb's specification: it changes the problem's probability structure around (what does the probability matrix look like *now*?). Indeed, to assert this is certainly to assert much more than what one ought to in view of the unworldly situation which our agent faces in Newcomb's problem.

Our theory should be evidential for it appears that a causal theory should only be applicable where and when the agent has explicit knowledge of which causal dependencies obtain. But knowledge of these is made available only when all causally irrelevant background factors have been appropriately filtered out so as to ascertain that $(C \mapsto E)$. But then $Prob(C \mapsto E)$ is based on past experience so it should, accordingly, be given a frequentist interpretation. If this is the case then causal theories of decision should suffer, it appears, from the same problems responsible for the downfall of frequentist interpretations of probabilities: it is difficult, ironically, to imagine how they could be of use in situations the causal architecture of which has not been ascertained. It seems then that $Prob(A \ \Box\!\!\rightarrow S)$ can only be given a frequentist interpretation, which interpretation conflicts with the only one possible in Newcomb's situation: the subjective interpretation. Therefore, contrary to what causal theorists claim, we should not believe that causal theories can handle a greater number of cases than their evidential counterpart. It is not true that causal theories can treat both the deviant Newcomblike situations *plus* the standard unproblematical ones.

Because his probabilities cannot be given a subjective interpretation, but only a frequentist one, it appears that an agent relying on a causal theory must feel Newcomblike situations harrowingly distressful: he has no useful probability function at hand. So, if the evidential theory's results are surprising in Newcomb's problem, it can only be because the situation is intractable. If cows had wings, what would *you* do?[36]

By clearing some of the conceptual mess surrounding Newcomb's problem, we notice it is in effect, as Jeffrey says, a "problem for space cadets".[37] Many more difficulties await if we wish to defend an evidential theory, for what has been said does not solve all prisoners' dilemmas. But the removal of these conceptual confusions, I believe, goes a long way in indicating the road to be followed in improving on past theories or *logics* of decision.[38]

Université du Québec in Montréal

NOTES AND REFERENCES

[1] Jeffrey, R. C. (1965/1983), *The Logic of Decision* (2nd Ed.), Chicago, University of Chicago Press, p. 25.
[2] Nozick, R. (1969), 'Newcomb's Problem and Two Principles of Choice', first published in Rescher, N., ed. (1970), *Essays in Honour of Carl Hempel*, Dordrecht, D. Reidel, pp. 114–146. Also in the following: Moser, P. K., ed. (1990), *Rationality in Action*, Cambridge, Cambridge University Press, pp. 207–234; Campbell, R. and L. Sowdon, eds. (1985), *Paradoxes of Rationality and Cooperations: Prisoner's Dilemma and Newcomb's Problem*, Vancouver, The University of British Columbia Press, pp. 107–133. For a follow-up see his (1974), 'Reflections on Newcomb's Problem', *Scientific American*, March, pp. 102–108; and (1993), *The Nature of Rationality*, Princeton, Princeton University Press, pp. 41–50.
[3] The term is Levi's. See Levi, I. (1982), 'A Note on Newcombmania', *The Journal of Philosophy* 79, pp. 337–342.
[4] Best known are:
Gibbard, A. and W. Harper (1978), 'Counterfactuals and Two Kinds of Expected Utility', in Hooker, C. A., J. J. Leach and E. F. McClennen, eds., *Foundations and Applications of Decision Theory*, Vol. 1, Reidel, Dordrecht. Also found in: Campbell, R. and L. Sowdon, eds. (1985), pp. 133–158; Gärdenfors, P. and N.-E. Sahlin, Eds. (1988), *Decision, Probability and Utility*, Cambridge, Cambridge University Press, pp. 341–376; and of course in Harper, W. L., R. Stalnaker and G. Pearce, eds. (1978), *Ifs*, Dordrecht, D. Reidel Publishing Co. All references will be to the *Ifs* version.
D. Lewis (1981), 'Causal Decision Theory', *Australasian Journal of Philosophy* 59, pp. 5–31. Also found in: Lewis, D. (1986), *Philosophical Papers*, vol. 2, New York, New York University Press; Moser, P. K., ed. (1990), pp. 235–263; Gärdenfors, P. and N.-E. Sahlin, eds. (1988), pp. 377–405. See also his (1979a), 'Counterfactual Dependence and Time's Arrow', *Noûs* 13, pp. 455–476; (1979b), 'Prisoner's Dilemma is a Newcomb Problem', *Philosophy and Public Affairs* 8, pp. 235–240; (1981a), 'Are We Free to Break the Laws?', *Theoria* 47, pp. 113–121; (1981b), 'Why Ain'cha Rich?', *Noûs* 15, pp. 337–380; (1983), 'Levi Against *U*-Maximization', *The Journal of Philosophy* 80, pp. 531–534.
Sobel, J. H. (1966), 'Dummett on Fatalism', *Philosophical Review* 75, pp. 78–90; (1970), 'Utilitarianism: Simple and General', *Inquiry* 13, pp. 194–449; (1975), 'Determinism: A Small Point', *Dialogue*, pp. 617–621; (1985a), 'Predicted Choices', *Dalhousie Review* 64, pp. 600–607; (1985b), 'Circumstances and Dominance in a Causal Decision Theory', *Synthese* 63, pp. 167–202; (1986), 'Notes on Decision Theory: Old Wine in New Bottles', *Australasian Journal of Philosophy* 64, pp. 407–437; (1988), 'Infallible Predictors', *Philosophical Review* 97, pp. 3–24; (1990), 'Newcomblike Problems', Uehling, T., ed., *Midwest Studies in Philosophy XV*, Notre Dame, University of Notre Dame Press; (1991), 'Non-dominance, Third Person and Non-action Newcomblike Problems and Metatikles', *Synthese* 86, pp. 143–172; many of which are to be found in (1994), *Taking Chances, Essays on Rational Choice*, Cambridge, Cambridge University Press; and 'Predicted Choices' (mimeo).
Skyrms, B. (1980), *Causal Necessity*, New Haven, Yale University Press; (1982), 'Causal Decision Theory', *The Journal of Philosophy* 79, pp. 695–711; (1984), *Pragmatics and Empiricism*, New Haven, Yale University Press; (1990), 'The Value of Knowledge', in Savage, C. W., ed. (1990), *Minnesota Studies in the Philosophy of Science, vol XIV: Scientific Theories*, Minneapolis, University of Minnesota Press, pp. 245–266.

[5] As B. de Finetti puts it: " [. . .] je juge plus utile un schéma simple comme repère dont j'admets de m'écarter pour des raisons accessoires plutôt qu'une construction complexe qui, afin de tout embrasser, ferait perdre la perspective de ce qui est essentiel". de Finetti, B. (1961), "Dans quel sens la théorie de la décision doit-elle être «normative»?", *La décision* (*Colloques internationaux du Centre National de la Recherche Scientifique*), mai 1958, Paris, Editons du C. N. R. S., pp. 164–165.

[6] To expect more reminds me of a student who took a class in logic expecting it would help him *become* rational!

[7] My agent is thus subject to be taken in by spurious causes, epiphenomenal causes etc., he is just like you and me: fallible.

[8] In his initial presentation of the problem, Nozick (1969) does not state that the 'being' – as he has it – is infallible, only that he "has often correctly predicted the choices". The important point is, however, to lead "you [the reader] to believe that almost certainly this being's prediction about your choice in the situation to be discussed will be correct". A few people have made a great deal of this (see for instance Sobel, J. H. (1988)). The distinction, *pace* Gibbard and Harper, is crucial. See note 13 below.

[9] Davidson, D. (1984), *Essays on Actions & Events*, Oxford, Clarendon Press.

[10] As J. H. Sobel puts it – but as he acknowledges, the metaphor is Seidenfeld's (1985), 'Comments on Causal Decision Theory', *PSA 1984: Volume Two*, East Lansing, Michigan Philosophy of Science Association – this would amount to ordering money from a menu. See his (1988), p. 3, note #1. Regarding backward causality, see, amongst others: (i) Forrest, P. (1985), 'Backward Causation in Defence of Free Will', *Mind* 94, pp. 210–217; (ii) Bar Hillel, M. and A. Margalit (1972), 'Newcomb's Paradox Revisited', *British Journal for the Philosophy of Science* 23, pp. 295–304; (iii) Lewis, D. K. (1979a); (iv) Dummett, M. A. E. (1954), 'Can an Effect Precede its Cause?', in *Truth and Other Enigmas*, London, Duckworth, 1978, pp. 319–331; and (v) Mackie, J. L. (1977), 'Newcomb's Paradox and the Direction of Causation', in *Canadian Journal of Philosophy* 8, pp. 213–225.

[11] This formulation was suggested to me in conversation by François Lepage.

[12] He would have *seen* it on the screen.

[13] Gibbard and Harper maintain that even in the case of the infallible predictor, it would be irrational not to take both boxes. I think their argument is flawed. It runs like this: If the agent knows for sure that he will take only one box, then he knows for sure that he will be a millionaire. If he knows he will be a millionaire, why should he pass up the opportunity to be a millionaire + $T? *That* is flawed. What this argument purports to show, if anything, is that their predictor is not, contrary to their own hypothesis, infallible. Had he been infallible, there would not have been a $M in the opaque box in the first place. Here, their point of view is abetted by the phrase "the agent knows for sure". The agent's knowledge has no bearing on the matter, it is only what he does that matters. See Gibbard, A. and W. L. Harper (1978), pp. 182–83.

[14] See Stalnaker, R. C. (1972), 'Letter to David Lewis. May 21, 1972', in Harper, W. L., R. Stalnaker and G. Pearce (1978), pp. 151–152.

[15] This last point made in Sobel, J. H. 'Predicted Choices' (mimeo).

[16] Both of these points were made by J. Howard Sobel in (1988) and in 'Predicted Choices' (mimeo), respectively.

[17] See Section 5 below.

[18] See Gibbard, A. and W. L. Harper (1978), p. 183; and Lewis, D. K. (1981b).

[19] Note that we need not worry about the one-boxer's interpretation of the probability assignment because, *ex hypothesi*, he is willing to put up with the *ex ante* recommendation yielded by PMCEU.

[20] I am well aware of the difficulties engendered by this but I would much rather not have it impossible that I turn into a butterfly before I die. T only has probability 1; bequeathing probability 0 to only $\mathbf{F}$ – i.e. the contradiction or logically false stands alone in having 1 - *Prob*T. This leaves the future open. Or, as Wittgenstein in a very matter-of-fact way puts it: "It is a hypothesis that the sun will rise tomorrow: and this means that we do not *know* whether it will

rise. [. . .] The only necessity that exists is *logical* necessity." Wittgenstein, L. (1921), *Tractatus Logico-philosophicus*, London, Outledge and Kegan Paul Ltd., §§ 6.36311 and 6.37.

[21] $Prob(A \& B) = Prob(A) \times Prob(B)$ is of course Kolmogorov's concept of independence. See Kolmogorov, A. N. (1933), *Foundations of the Theory of Probability*, Chelsea Publishing Company, New York, 1956. This crystal-clear concept I hold as basic, and I wish to favour it since other interesting candidates (i.e. $Prob(A/B) = Prob(A)$ sometimes called *stochastic* independence; and $Prob(A/B) = Prob(A/\sim B)$ sometimes called *probabilistic* independence) are all provably equivalent to it, except for their requiring that at least the two first, when not all four of the following constraints hold: (i) $Prob(A) > 0$; (ii) $Prob(B) > 0$; (iii) $Prob(\sim A) > 0$; and (iv) $Prob(\sim B) > 0$.

[22]   I assume we are always dealing with creatures who base their decisions on a finite number of beliefs. But if, for practical purposes and technical reasons, we are inclined to assume countable additivity, we will be confronted with well known objections. See de Finetti, B., *Theory of Probability*, 2 vols., New York, John Wiley & Sons, vol. 1, pp. 118–119.

[23]   Thus is Eells' tickle defense still a live option. But there is still a problem with *it*. Just as in many causal theories, it supposes that the agent will come to recognize that his act is causally independent of both the soothsayer's prediction and of the content of B2. His agent is too ticklish. See Eells, E. (1982), *Rational Decision and Causality*, Cambridge, Cambridge University Press; (1984), 'Metatickles and the Dynamics of Deliberations', *Theory and Decision* 17, pp. 71–95; (1984b), 'Newcomb's many solutions', *Theory and Decision* 16, pp. 59–105; and Eells, E. and E. Sober (1986), 'Common Causes and Decision Theory', *Philosophy of Science* 53, pp. 223–245.

[24]   Nozick, R. (1969), note 12.

[25]   Note that $\sim(\exists S)(\exists A)(\exists B)[prob(S$ obtains$/A$ is done$) \neq prob(S$ obtains$/B$ is done$)]$ can be simplified as $Prob(S/A) = Prob(S/B)$. Since, $\sim(Prob(S/A) = Prob(S/B))$ being logically equivalent to $Prob(S/A) \neq Prob(S/B)$, it is equivalent to

$$(\forall S)(\forall A)(\forall B) \sim\sim (Prob(S/A) = Prob(S/B))$$

which in turn may legitimately be abridged as

$$Prob(S/A) = Prob(S/B).$$

Now, acts *A* and *B* being, in Newcomb's problem, mutually exclusive and collectively exhaustive, then, unless the agent does not act upon a choice, $Prob(S/A) = Prob(S/B)$ may be expressed, without loss as $Prob(S/A) = Prob(S/\sim A)$, which is itself provably equivalent to the concept of independence I wish to favour (i.e. $Prob(S \& A) = Prob(S) \times Prob(A)$).

[26]   Contrast the concept of independence I favour with Gibbard and Harper's *causal independence* formulated as $(A \,\square\!\!\rightarrow S) \leftrightarrow S$, then for each S we have $Prob(S) = Prob(A \,\square\!\!\rightarrow S)$ where $(A \,\square\!\!\rightarrow S)$ is to read as "If I were to do A then S would happen"; contrast also with their *stochastic independence* where $Prob(S) = Proba(S/A)$ is formulated as $Prob(A \,\square\!\!\rightarrow \,\mapsto O/A) = Prob(A \,\square\!\!\rightarrow O_i)$; both of which presuppose a capacity to positively assess statements of the sort $(A \,\square\!\!\rightarrow X)$.

[27]   Recall Nozick's formulation. See note 8 above.

[28]   Ryle, G. (1954), 'It Was To Be', in *Dilemmas*, Cambridge, Cambridge University Press.

[29]   Lewis, D. (1981), in Gärdenfors, P. and N.-E. Sahlin (1988), pp. 384.

[30]   See Gärdenfors, P. (1988), *Knowledge in Flux*, Cambridge (Mass.), MIT Press, pp. 191.

[31]   Skyrms, B. (1982), pp. 697.

[32]   *Idem*.

[33]   *Idem*.

[34]   This passage is from Cartwright, N. (1988), 'Regular associations and singular causes', in Skyrms, B. and W. L. Harper, eds. (1988), *Causation, Chance and Credence*, Boston, Kluwer, vol. 1, pp. 79. The passage is used out of context here and does not represent Cartwright's opinion.

[35]   From Jeffrey, R. C. (1988), 'How to Probabilize a Newcomb Problem', in Fetzer, J. H., ed., *Probability and Causality*, Dordrecht, D. Reidel Publishing Co., pp. 241; the same argument

appears in (1992), 'Introduction: Radical Probabilism', in his *Probability and the Art of Judgement*, Cambridge, Cambridge University Press, pp. 1.

[36]   There is, I believe, a solution to the difficulties posed by our having to forego causal talk. It is to treat statistical correlations and causal relations as lack of probabilistic independence, and then isolate causal relations by exploiting the Popperian concept of *factual content* defined as the probability of the statement's negations. But there is much work to be done here.

[37]   See quoted passage at beginning of text.

[38]   Thanks are due to Daniel Laurier (Université de Montréal), François Lepage (Université de Montréal) and the participants at their seminar on conditionals, counterfactuals and decision theory for very helpful discussions of some of the matters touched upon here. Thanks are also due to Paul Dumouchel who has read an earlier version and commented on it. My greatest debt is to Hugues Leblanc (Université du Québec à Montréal) who has very adroitly questioned many points I put forward and has, by so doing, made me reframe many important issues. He has also read an earlier version of this paper. I am sure we still strongly disagree on some of them. I remain sole responsible for the errors – either in judgement or in reasoning – this paper may still contain. The *Fonds pour la Formation de Chercheurs et l'Aide à la Recherche* provided financial support for which I am grateful (F. C. A. R. Programme de soutient aux équipes de recherche # 94ER0637–07).

J. NICOLAS KAUFMANN

THE BELIEF-DESIRE MODEL OF DECISION
THEORY NEEDS A THIRD COMPONENT:
PROSPECTIVE INTENTIONS

The purpose of the paper is to make a contribution, from the point of view of the analytical philosophy of action, to a critical evaluation of the standard model of rational choice, i.e. the model of expected utility maximization. This model is actually confronted with serious difficulties. The one I will take up is a problem raised by the economist Strotz in his seminal paper on «Myopia and Inconsistency in Dynamic Utility Maximization» (1955/56), developed by Machina (1989) and more recently by McClennen (1990).

Strotz was probably the first economist to pay attention to a certain form of apparently irrational *economic* behaviour not accounted for by existing theory, the irrationality of which he did not locate in a particular decision but within the agent. Spendthriftiness, gambling or drug addiction and other forms of akratic behaviour of people being trapped in a consumption pattern they wish to abandon but cannot, the sufferings from a tragic conflict of Faustian decision-makers who may at the same instant both pay for cigarettes and pay for a smoking cure,[1] are Strotz's paradigm cases.

Strotz was not really interested in the traditional philosophical problem of weakness of the will and the available techniques of control over irresistible desires, but in a general problem of irrationality in situations where we are supposed to apply the standard decision rules to dynamic choice problems. Strotz's question was: What does it mean to maximize a utility function when the problem is one of *intertemporal* utility maximization? He put the problem of an economic decision-maker who must choose among several consumption plans (or investment plans, saving plans or plans of income expenditure), subjected to standard constraints and by relating them to time, the plan which represents a maximum, given the agent's preferences. Since such plans imply further choices and admit the possibility of reconsidering at some later decision point – note that according to Strotz a rational agent *must* reconsider if he is to maximize Φ_τ (where $\tau > 0$ is the time at which the further choice is made) –, the solution to the maximum problem at $\tau > 0$ may be completely different from the solution obtained at $\tau = 0$. Strotz therefore questions the criterion of utility maximization when applied to contexts of dynamic choice: «To continue to obey a fixed consumption plan just because it was optimal when viewed at an earlier date is not rational if that plan is not the optimal one at the present date» (170/71). But, if the solution were different for each time point, an agent may be driven to abandon at each instant a plan he just adopted. In this case, why lose time in planning and deliberating? On the other hand, rationality demands more than the myopic behaviour of the agents

215

M. Marion and R. S. Cohen (eds.), Québec Studies in the Philosophy of Science II, 215–227.
© 1996 *Kluwer Academic Publishers.*

imagined by Strotz, even though «there is nothing patently irrational about the individual who finds that he is in a intertemporal tussle with himself – except that rational behaviour requires he take the prospect of such a tussle into account» (171). The idea of rationality demands a remedy for the kind of myopia illustrated, either by «precommitment» (precluding future options in order to conform future «choices» to the plans actually chosen), or by «consistent planning» (which takes into account possible future disobedience, realizing that the possibility of disobedience imposes a further constraint on the set of attainable plans).

The objective of the following analysis is to show that the standard model of expected utility maximization may conflict, in dynamic choice situations, with a general rationality condition: the condition of dynamic consistency, a condition pertaining not only to decisions considered separately, but pertaining to plans containing a series of decisions distributed over time. It is possible to identify the source of the difficulties in the conceptual base of the standard model of expected utility maximization and to show that the conceptual resources of this model are insufficient to overcome the obstacles. Finally, the hypothesis will be advanced that the problem must find a solution by incorporating in the desire-belief model (with utilities as intensities of desires and probabilities as degrees of beliefs) a third component, that is, intentions, particularly «prospective intentions» as theorized by Brand (1984), Bratman (1987) and Mele (1992).

1. EXPECTED UTILITY MAXIMIZATION
AND DYNAMIC CONSISTENCY

It is possible that an agent, who applies the rationality criterion of the standard model, behaves in an indefensible manner, as the examples of Strotz[2] nicely illustrate. Here is the case of Mary[3] who considers a set of plans, complete with regard to all the relevant details, located on the following decision tree.

Table 1 represents a situation of dynamic choice, that is, a sequence of choices distributed over time and deployed on a decision tree (T) with chance points (O) and decision points (□). The branches terminate by the results corresponding to the paths (plans) located on T. The tree begins at n_0 and terminates with the relevant time horizon beyond which events do not make any difference for the adoption of a plan by Mary. The decision tree T contains all the information which Mary will use to evaluate the set of plans among which she has to choose. According to plan s, Mary will get $ 2500 at endpoint 1, or, if unlucky at chance point n_1 $(-E)$, she will get $ 0 at endpoint 4; or, if she is unfortunate at chance point n_4, she will get $ 0 at endpoint 2. Following plan r, Mary could get $ 2400 at endpoint 3, or, with bad luck at chance point n_1 $(-E)$, get 0 $ at endpoint 4. Following plan r^+, Mary could get $ 2401 at endpoint 5 with luck (E) at chance point n_2, or get $ 1 at endpoint 6 when $-E$ happens at chance point n_2.

TABLE 1

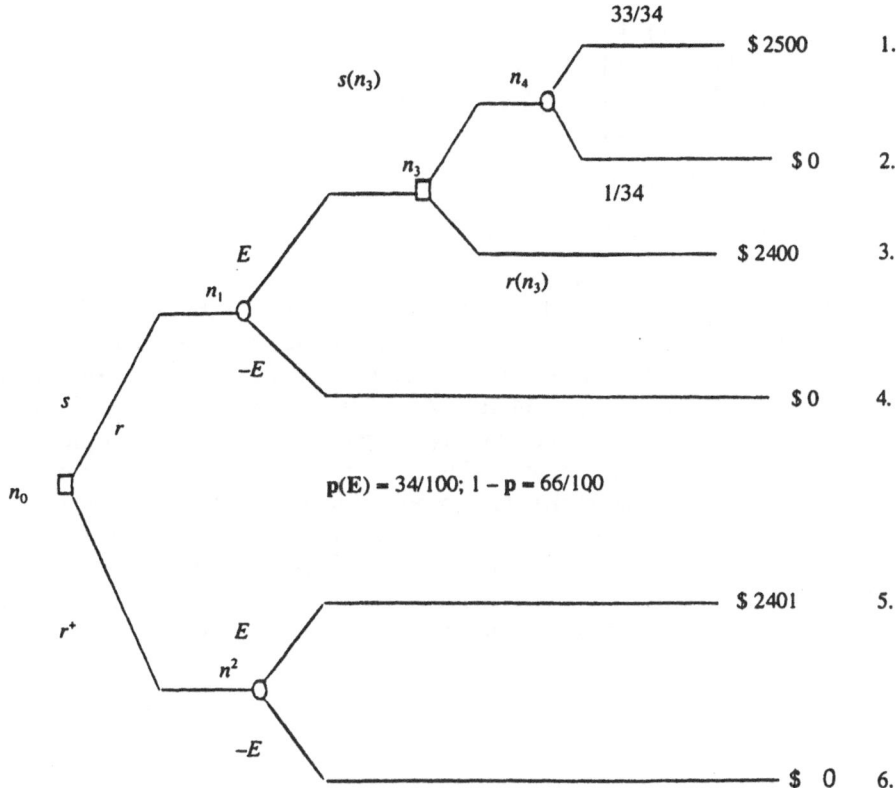

To evaluate the plans at decision points different from n_0, three perspectives may be adopted, following McClennen (1990, 104 ss.), to continue the plans which remain on the truncated tree at n_i: (a) the plans on the truncated tree at n_i could be seen as the continuation of the plans judged acceptable at the starting point n_0 [= the set $D(S)(n_i)$]; (b) given the preferences at n_i ($n_i \neq n_0$), one could ask what would be, from the point of view of n_i, the best choice among the options yet open at n_i [= the set $D(S(n_i))$]; (c) finally, without considering what brought her to n_i, one could ask which choice to make at n_i, abstracting completely from the partial execution of the plan previously decided upon, and choosing as if confronted with a choice problem *de novo* [= the set $D(S(n_i)^d)$]. In summary, Mary may adopt an *a*historical perspective (cases b and c), or she may adopt a historical perspective (case a) when choices to be made later on are situated on a tree with a past, and when the evaluation is made in the light of prior choices and the situations created (or closed) by them, in the light of the risk borne and in taking into account the collateral counterfactual possibilities which could have been realized if the agent had chosen otherwise.

In order to make his case, McClennen supposes that Mary has the following preferences for the following options:

o_1: [$ 2400, 1]
o_2: [$ 2500, 33/34; $ 0,1/34]
o_3: [$ 2400, 34/100; $ 0, 66/100]
o_4: [$ 2500, 33/100; $ 0, 67/100]

where $o_1 \succ o_2 \succ$ and $o_4 \succ o_3$. By adding a new option,

o_3^+: [$ 2401, 34/100; $ 1, 66/100],

Mary would now prefer $o_4 \succ o_3^+ \succ o_3$.

If Mary follows the rationality criterion max $\Sigma\, \mathbf{p}_i u\,[r_{ij}]$ of the standard model of expected utility maximization (utilities are assumed to be a linear function of dollars for the sake of illustration), at n_0 Mary must choose the plan s to reach n_1, a point with $\mathbf{p}(E) = 34/100$. At point n_3, she will opt for $s(n_3)$ to try to obtain $ 2500 which is her most preferred result. Arriving at n_3 following plan s with an opportunity to reconsider how to continue in the light of the *same* information she had at the starting point (the probabilities of the chance events and the preferences at each decision point were known at the outset), and considering what is still open to her at n_3, Mary must choose, in conformity with the given rationality criterion, the plan $r(n_3)$, and must abandon the plan s. At n_3, Mary may continue along the truncated tree:

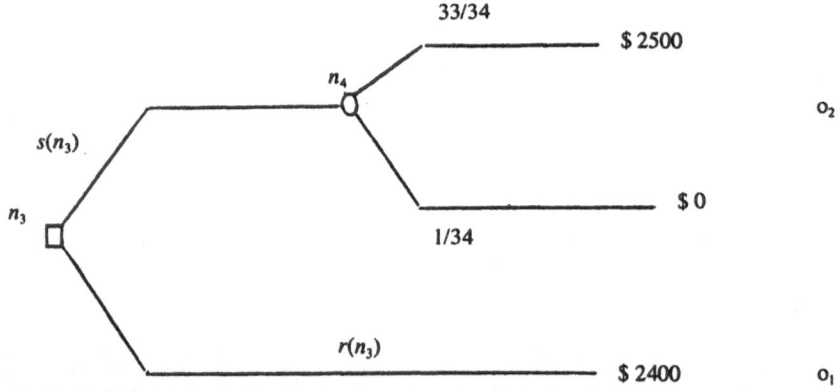

This implies a choice between the two options o_1 and o_2. In the light of her preferences, Mary prefers o_1 to o_2. The same rationality criterion which favours plan s, and the possibility of reconsidering at n_3, compel Mary to continue with plan r and thus to violate the condition of *dynamic consistency* which Strotz and his followers consider an essential condition of rationality.

(DC)   An agent is dynamically consistent if her choice at n_i, planned from the point of view of n_0, coincides with the choice the agent will make arriving at n_i: $D(S)$ at n_i coincides with $D(S(n_i))$.[4]

Consider a second expected utility maximizer Clara who,less myopic than Mary, could anticipate at the outset (n_0) that, given her preferences, she will choose on arriving at n_3 to continue on r abandoning the plan s judged best at the starting point. Clara anticipates in the *ex ante* perspective her choice at n_3 to continue on r abandoning the plan s judged best at the starting point. Clara anticipates in the *ex ante* perspective her choice at n_i, takes account of this «fact» as a constraint (from the perspective *ex post*) in adopting a plan at n_0. This means that the actual self must make a concession to the future self. In order to conform to the condition of dynamic consistency, Clara can make a sophisticated choice at n_0, either by «precommitment», as did Ulysses when he ordered the crew to attach him to the mast of the ship so that he will not be beguiled by the Siren's song into changing his course, or she agrees to be constrained by her future self, the concept of consistent planning of Strotz, and chooses plan r at the beginning, which is not optimal when judged from n_0. Thus, consistency would be guaranteed by the sophisticated agent who can «precommit» or plan consistently. But then, the choice of plan r corresponds to the option o_3 = [2400 \$, 34/100: 0 \$, 66/100] which is dominated by o_3^+ of plan r^+.

2. DYNAMIC CONSISTENCY AND THE SEPARABILITY CONDITION

In either the case of the myopic choice violating the criterion of intertemporal consistency of choices made by the same agent, or the sophisticated choice by which the agent tries to remedy this default, the role played by the rationality condition implied by the criterion of expected utility maximization, i.e. the *separability condition*,[5] is crucial. Machina distinguishes two forms: The *replacement separability* follows from the additive structure of the expected utility preference function, and the fact that the contribution of each outcome/probability pair [r_{ij}, p_i] to the sum is independent of the other outcome/probability pairs. The *mixture separability* follows from the fact that the contribution of each outcome/probability pair to expected utility can be interpreted as the utility of its outcome $u[r_{ij}]$ multiplied by its probability p_i. Both properties are implied by the independence axiom.[6]

The separability condition can be illustrated using the classical decision problem which takes the following form. If an agent prefers a_1 over a_2,

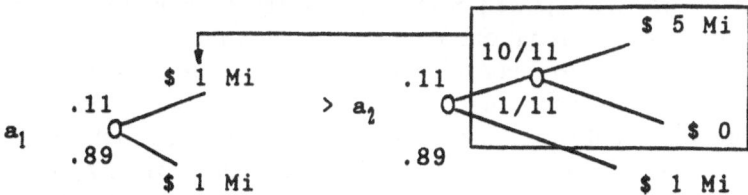

this agent would be prepared to substitute the (framed) sub-lottery of a_2, that is, [\$ 5 Mi, 10/11; \$ 0, 1/11] for [\$ 1 Mi, 0.11] in a_1. Since she also prefers a_3 to a_4

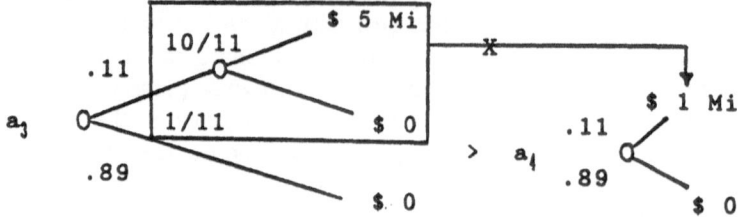

she would not be prepared to make the same substitution. The reason is that there are some aspects in the decision problem which are not separable.

Introducing a choice point between the two chance points generates the typical elementary dynamic choice structure with the following plans.

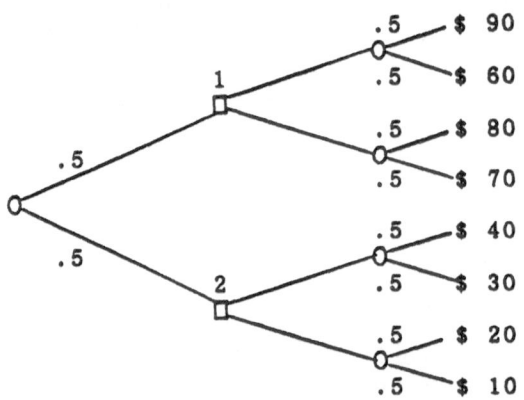

The strategies which specify how the agent would choose at each of the two choice nodes (1 and 2), where U or L denote a choice of the upper or lower branch, are listed below.

strategies	probability distribution	values
$[U, U]$	($90, 0.25; $60, 0.25; $40, 0.25; $30, 0.25)	u (distrib)
$[L, U]$	($80, 0.25; $70, 0.25; $40, 0.25; $30, 0.25)	u (distrib)
$[U, L]$	($90, 0.25; $60, 0.25; $20, 0.25; $10, 0.25)	u (distrib)
$[L, L]$	($80, 0.25; $70, 0.25; $20, 0.25; $10, 0.25)	u (distrib)

If $[U, U] \succ [L, U] \succ [U, L] \succ [L, L]$, then $[U, U]$ would appear to be the best plan. After Machina, one could ask what would happen, if our agent had the opportunity to reconsider her choice at decision point 1? After the occurrence of the chance event, there is, in a certain sense, «new» information, but the agent already made up her mind *conditionally* to the circumstance concerning the outcome (point 1 or 2) of the chance event. At this point, a reconsideration would be expected to yield the same result chosen at the starting point. As Machina points out, the question is how the agent *should* consider. It is evident that, at n_1 or n_2, the tree is no longer the same as at the beginning; it will be reduced. To be rational, should our agent reconsider her choice, «snip» the decision tree at (that is, just before) the current choice node, throw away the rest of the tree, and recalculate applying the *original* preference ordering to alternative possible continuations of the tree? A yes answer would be recommended by the separability condition which requires that one choose as if one started out with a completely new choice, as if the risks already borne were of no importance. To know how individuals choose in fact is an empirical question. However, the question raised by Machina is to determine what rational normative decision theory *should* prescribe. The following example highlights this point.

If I have time to see *one* film, I prefer [Top Gang 1] to [Top Gang 2], and, if I have time to see *two* films, I would prefer to see the series [Top Gang 1, Top Gang 2] to [seeing Top Gang 1 twice]. These are the fixed preferences. Now, I have 4 hours to go to the movie; so, I walk to the Capitol Theatre at 7 p.m. where Top Gang 1 is playing followed by Top Gang 2 at 9 p.m. At a quarter to 9 p.m., at the end of Top Gang 1, I still have the chance to reconsider what film to watch given the time to watch only one film. According to the principle of separability in deciding which film to see, I would cut off the rest of the decision tree. The choice becomes watching [Top Gang 2] at the Capitol Theatre or [Top Gang 1] at the Versailles Theatre just opposite to the Capitol Theatre. According to the consequentialist reading of the separability condition, I am required to go to the Versailles Theatre and see Top Gang 1, because that choice conforms to my initial preferences when reconsidered only in the light of the options still available and not in the light of what has already occurred. It becomes clear that the separability condition poses not only problems for intertemporal consistency, but choosing this way makes me clearly look crazy.

The preceding example shows that with the separability condition rationality comes in episodes which, at the limit, could be instantaneous. Thus, rationality is not global but atomistic, due to the fact that the criterion of the standard model is independent of time. Such a consequence is inappropriate for an analysis which attempts to treat problems of intertemporal choice. This is a fatal flaw in the traditional analysis, which can be found in the conceptual base of the standard model of expected utility maximization, the belief-and-desire model, which excludes further intentional states. In the context of intertemporal choice, beliefs certainly play a role in linking past and future inasmuch as rational agents must anticipate what the context of future choice will be. But for reconsideration with no additional news, future expectations do not change. With regard to the desire component, the separability condition presupposes, following the consequentialist interpretation, that reconsideration at later points should be forward looking only, without considering, retrospectively, past risks borne, and without incorporating in the evaluation the counterfactual results which could have been reached if another choice had been made. Standards of rationality operate within a time tranche, which is philosophically objectionable from the substantive point of view. On the ontological level concerning the agent, the separability condition slices the agent up into temporally independent episodes without any possibility of relating the multiple selfs who choose without considering the choices made by the former selfs. From a methodological point of view, its seems that it is not possible to treat, in the framework of the standard model which incorporates the separability condition, the intertemporal connections of preferences for dynamic decision problems. There is no way to consider future preferences as *conditional* preferences relative to what happened before future decision points are reached. What is lacking in the bi-linear model of rational choice is a third component whose role is precisely to establish the necessary dependencies of future conditional preferences upon the actual preferences. This role can be played by *intentions*, particularly prospective intentions which figure in the activity of planning.

3. PROSPECTIVE INTENTIONS AND THE PROBLEM OF CONSISTENT PLANNING

The problem of dynamic consistency and the difficulties created by the separability condition are problems which concern the rationality of the *agent* who adopts plans. It appears clearly from the results of the analytical philosophy of action, particularly from the work of Brand (1984), Bratman (1987) and Mele (1992), that the constitutive elements of action plans are prospective intentions contrasted with intentions in action.[7] The thesis to be developed is that prospective intentions exercise the function of intertemporal coordination which rational agents try to realize in the pursuit of plans, and that these intentions are notoriously absent in the conceptual arsenal of standard decision theory. It follows from the principle of dynamic consistency (DC) that rational agents adopting plans must form prospective intentions. A descriptive phenomenology of this kind of intentional states, a phenomenological analysis of the specific intentionality of intentions, tends to conform to this view. To decide to adopt a plan is not only to resolve a decision problem in the light of desires (utilities) and beliefs (probabilities); it comes down to form *in the present* the intention to make further choices and perform actions *in the future*. By adopting prospective intentions, I am committed, until otherwise, to move in a certain direction. It is a sort of promise given to myself, a kind of contract between the actual and the future self, which binds both of us by a common interest. If such an intention is publicly declared, as when I say to my colleagues that I have the (prospective) intention to go to the CPA meeting next month, I authorize, by this illocutionary commitment not as strong as a promise or the signature of a contract, my friends to have certain expectations concerning my future behaviour.

The contribution to be expected from the philosophy of action to the analysis of dynamic choice and its rationality conditions is twofold: (a) it will help to identify the various roles that prospective intentions play and the functions they are supposed to fulfil in the context of deliberation, decision and planning of a course of action over time; (b) it will contribute to identify and make explicit the standards and criteria of adequate performance of the functions the device (i.e. prospective intention) is supposed to fulfil, to identify standards for the accomplishment of the several tasks they are supposed to serve. As the notion of function has normative implications (even for «natural function»), the second step is a crucial prerequisite for normative decision theory; the explication I ask for will lead to further standards and criteria of rational choice, standards which should be integrated in a more adequate conception of *agent* rationality. Brand, Bratman and Mele, among others, have taken the first steps in the direction of a functional characterization of prospective intentions. Here are some indications for a functional analysis.

(1) *The function of inertia and resistance of prospective intentions.* Contrary to intentions in action, prospective intentions do not directly control future decisions and actions. It is precisely the absence of direct control by future-oriented intentions of their execution which is the source of the problem of intertemporal (in)consistency. For example, if prospective intentions could make me extend the hand and put it on the steering wheel, what I have decided now would be performed tomorrow. However, even without direct control, prospective intentions are dispositions which do (and must) persist from the time of their formation until the time of their execution. On the other hand, prior intentions may be revoked. Tomorrow, we could have excellent reasons to reconsider choices made today (in the light of new information about changes in the internal and external environment). Besides the function of inertia and persistence, there is the function of resistance against the unmotivated reconsideration of a prospective intention. This resistance may exist because of the fact that when I form a future-oriented intention, I am committed, to various degrees, to a plan or to an ulterior course of decisions and actions. The fact that I am committed would tend to *block* (not definitively) the possibility of reopening the question every time and to *inhibit* action plans which would be incompatible with the adopted prospective intention. Persistence and resistance are functions of prospective intentions, which must be fulfilled in conformity with the normative principles of rational (non-)reconsideration. For Bratman, they are the principles of the transmission of rationality through time from the adoption of a plan by the formation of a prospective intention at t_0 to the moments $t_1, t_2, \ldots$ when it is possible to reconsider the plan. According to Bratman, such transmission principles take the following form:

(TP)   If at t_0 the criteria of rationality the agent applies recommend to do a at t_1, and if an agent forms then a prospective intention to do a at t_1, and if at t_1 the conditions are identical with those considered at t_0, then it is rational not to reconsider and abandon that intention at t_1.[8]

One could find in Bratman several other more specific principles of rational (non)reconsideration for non-reflective, deliberative and policy based reconsideration. One does not need to give them here. (TP) already illustrates that one cannot find principles concerning the persistence and resistance function of prospective intentions in the standard model of expected utility maximization. Its axioms are about preferences (intensities of desire, i.e. utilities) and beliefs (probabilities)[9] and not about *intentions*.

(2) *The function of monitoring of practical reasoning.* A second task of prospective intentions is to regiment and monitor practical reasoning in the context of extensive deliberation as discussed by Bratman. Prospective intentions pave the way for more specific intentions to solve the problems posed by a plan. The same does not apply to desires. One can have the desire to listen to a concert without specifying whether it will be *Kleine Nachtmusik* or *Kaiserquartett*. But if one has the prospective intention to buy a present for her friend's birthday, this general future-intention must and will generally lead to intention generation, as long as one does not reconsider and abandon it. There are two ways of intention generation and two sorts of principles corresponding to this function of prospective intentions.[10]

(a) Generation of subordinate intentions: My prospective intention to give a talk in Vienna next summer generates in a specific order a series of subordinate intentions such as the intention to register for the conference, to buy the plane ticket, etc., with the generation process extending to the exact moment when I must ultimately form the intentions in action (the operative intentions) to take the flight, arrive in Vienna, walk into the conference room and begin my talk. Subordinate intention generation follows the well-known Kantian principle: «Whoever wants to attain an end is bound also to want to use the necessary means towards its attainment». In this way prospective intentions enter the premises of practical reasoning where they are a measure of incompatibilities, which are intolerable for intentions but not for desires. It is not irrational to have an irresistible desire to spend the next week in the Alps, and desire simultaneously to finish a manuscript of a paper, when both desires cannot be fulfilled at the same time.

(b) Generation of secondary intentions: Prospective intentions, for example a prior intention to go to a concert in Montreal next week, may generate, when certain beliefs and permanent desires are present, the secondary intention to visit my in-laws and, by the fact that I believe that I will be in Montreal next week and will pass by the Laurier subway station, the intention to visit the in-laws may generate a further secondary intention to buy chocolates at the next door. The principles governing this kind of intention generation are much more complex, and there is actually no consensus concerning their formulation.

CONCLUSION

It is by creating inertia and resistance, and by their contribution to intention generation, that prospective intentions fulfil their primary function of intertemporal intrapersonal coordination of action plans, the function of guaranteeing their internal consistency over time, and the function of coordinating more complex activities. It has been shown above that the standard model of expected utility maximization does not provide for intentions in general, nor for prospective intentions in particular. For this reason, the question of the rationality (or irrationality) of (non)reconsideration of a decision in a dynamic context which is crucial for dynamic consistency, and the question of agent rationality, simply cannot be dealt with using this model. The separability condition favours an atomistic conception of rationality of the agent. There are still some philosophers who think that intentions are reducible to a combination of beliefs and desires following the strategy proposed by Audi (1986). But those few will have to fight against a whole army of opponents[11] and there is little hope that they will succeed.

Université du Québec à Trois-Rivières

NOTES

[1]   In conventional economic theory, one has to admit for such problems two (or more) competing independent consumption patterns, which means more than one consumer within the individual, but one cannot specify which one is the sovereign consumer who will make the choice according to the theory of consumer's choice.

[2]   Strotz was followed by several economists among whom are Pollack (1968, 1970, 1976) discussing the existence of a long term utility function, Hammond (1976, 1988) and Yaari (1977) on endogenous preference changes resulting in intertemporal inconsistencies, and finally Machina (1989) and McClennen (1990) who gave to the problem a more philosophical twist.

[3]   The example is McClennen's (1990).

[4]   «For any choice point n_i in a decision tree T, if $D(S)(n_i)$ is not empty and $s(n_i)$ is in $D(S(n_i))$, then $s(n_i)$ is in $D(S)(n_i)$; and if $s(n_i)$ is in $D(S)(n_i)$, then $s(n_i)$ is in $D(S(n_i))$.» (McClennen, 1990, p. 120)

[5]   This property has been explicated by Machina (1989) and discussed in many details for the context of dynamic choice by McClennen (1990).

[6]   The independence axiom stipulates: $X \succ Y$ iff $(pX, (1 - p)Z) \succ (pY, (1 - p)Z$, for any Z and $p > 0$. This axiom implies directly mixture separability, for, with the same probability distribution, p and $1 - p$, X and Y make their contribution to the overall expected utility separately and in isolation of Z. The double application of the axiom gives replacement separability: $[pX, (1 - p)Z] \succ [(pY, (1 - p)Z] \Leftrightarrow X \succ Y \Leftrightarrow [pX, (1 - p)Z] \succ [pY, (1 - p)Z]$. Z may be separately replaced by Z without affecting the value of the overall expected utility. (See Machina, 1989)

[7]   One finds this distinction under various labels in Beardsley (1978, «prospective» and «concurrent intention»), Searle (1983, ch. 3, «prior intention» and «intention in action»), Brand (1984, ch. 5 and 6, «prospective» and «immediate intention»), Bratman (1987, «future-directed» and «present-directed intention»), Mele (1992, ch. 10, «proximal» and «distal intention»), Wilson (1989, «intention for the future» and «present-directed intention»). The distinction was made 50 years ago by one of the distinguished phenomenologists interested in the phenomenology of action, Alfred Schütz (1932, paragr. 9) who distinguished painstakingly intentions whose content is an action qua *actum* apprehended in *modo futuri axacti* and represented as accomplished (action type), and intentions whose content is the acting as an ongoing process conforming to the anticipations of the intentions of the first type, intentions whose content is the *agere* as a particular occurring event.

[8] It is the sort of principle one finds in Bratman (1987) and McClennen (1990). One should systematically examine the following cases:

The decision is	the (non)reconsideration is	retention is
rat at t_0	rat [(non)reconsideration]	rat at t_1
rat at t_0	rat [(non)reconsideration]	¬rat at t_1
rat at t_0	¬rat [(non)reconsideration]	rat at t_1
rat at t_0	¬rat [(non)reconsideration]	¬rat at t_1
¬rat at t_0	rat [(non)reconsideration]	rat at t_1
¬rat at t_0	rat [(non)reconsideration]	¬rat at t_1
¬rat at t_0	¬rat [(non)reconsideration]	rat at t_1
¬rat at t_0	¬rat [(non)reconsideration]	¬rat at t_1

[9] See Schick (1984).
[10] Both forms of intention generation are discussed, for the first time to my knowledge, by Kim (1976).
[11] Bratman (1987), Davidson (1980, ch. 5), Ginet (1990), McCann (1986), Mele (1992) and Searle (1983), to mention only the most prominent.

REFERENCES

Audi, Robert, 1986, 'Intending Intentional Action, and Desire', in Joel Marks (ed.), *The Ways of Desire*, Precedent, Chicago, pp. 17–38.

Beardsley, Monroe, 1978, 'Intending', in Alvin Goldman and Jagwon Kim (eds.), *Values and Morals*, D. Reidel, Dordrecht, pp. 163–184.

Brand, Myles, 1984, *Intending and Acting, Toward a Naturalized Action Theory*, The MIT Press, Cambridge MA.

Bratman, Michael E., 1987, *Intentions, Plans, and Practical Reason*, Harvard University Press, Cambridge MA.

Davidson, Donald, 1980, *Essays on Actions and Events*, Clarendon, Oxford.

Ginet, Carl, 1990, *On Action*, Cambridge University Press, Cambridge.

Hammond, Peter, 1976, 'Changing Tastes and Coherent Dynamic Choice', *Review of Economic Studies* **43**, 159–173.

Hammond, Peter, 1988, 'Consequentialism and the Independence Axiom', in Bernard Munier (ed.), *Risk, Decision, and Rationality*, D. Reidel, Dordrecht, pp. 503–516.

Kim, Jagwon, 1976, 'Intention and Practical Inference', in Juho Manninen and Raimo Tuomela (eds.), *Essays on Explanation and Understanding*, Studies in the Foundations of Humanities and the Social Sciences, D. Reidel, Dordrecht/Boston.

Machina, Marc J., 1989, 'Dynamic Consistency and Non-Expected Utility Models of Choice Under Uncertainty', *Journal of Economic Literature* **27**, 1622–1668.

McCann, Hugh, 1986, 'Rationality and the Range of Intention', *Midwest Studies in Philosophy* **10**, 191–211.

McClennen, Edward F., 1990, *Rationality and Dynamic Choice, Foundational Explorations*, Cambridge University Press, Cambridge.

Mele, Alfred R., 1992, *Springs of Action, Understanding Intentional Behaviour*, Oxford University Press, New York.

Pollack, Robert A., 1968, 'Consistent Planning', *Review of Economic Studies* **40**, 391–401.

Pollack, Robert A., 1970, 'Habit Formation and Dynamic Demand Function', *Journal of Political Economy* **78**, 745–763.

Pollack, Robert A., 1976, 'Habit Formation and Long-Run Utility Functions', *Journal of Economic Theory* **13**, 272–297.

Schütz, Alfred 1932, *Der sinnhafte Aufbau der sozialen Welt, Eine Einleitung in die verstehende Soziologie*, Springer Verlag, Wien, 1960.

Schick, Frederic, 1984, *Having Reasons, An Essay on Rationality and Sociality*, Princeton University Press, Princeton.

Searle, John, 1983, *Intentionality, An Essay in the Philosophy of Mind*, Cambridge University Press, Cambridge.

Strotz, Robert H., 1955/56, 'Myopia and Inconsistency in Dynamic Utility Maximization', *Review of Economic Studies* **23**, 165–180.

Yaari, Menahem Dawns, 1977, 'Endogenous Changes in Tastes: A Philosophical Discussion', *Erkenntnis* **11**, 157–196.

JOCELYNE COUTURE

DECISION THEORY, INDIVIDUALISTIC EXPLANATIONS AND SOCIAL DARWINISM*

INTRODUCTION

There is an important sense in which decision theory[1] is not individualistic. A rational agent will choose the option whose utility, relatively to his own preferences, is the greatest. But the options over which he has to choose and the particular way their utility for him relates to other actor's choices are both part of the context of choice. The rationality of choices, that is, their maximizing character, is then assessed for choices made in a given environment. In interpreting theories framed in a rational choice set up, such an environment will pattern a certain state of the world, a certain social environment, including economic and political contexts shaping, together with individual rationality and preferences, the very outcome of interactive choices.

Among social scientist who favour decision theory there is, however, a very strong tendency to think of the rational agent as being "fully responsible for his choices"; indeed, for many of them, the interest of decision theory consists precisely in the fact that it is, or can readily accommodate, a framework for the individualistic explanations required, or so they believe, in social science. This requirement comes from different horizons and its general formulation is reductionistic; it has been argued for from one or the other of the now usual methodological, semantical or ontological stances.[2] But for the social scientists who favour decision theory, two beliefs seem to mainly inspire the search for individualistic explanations. The first one is the belief that social science is the science of "human action, its course and consequences"[3] embedding, therefore, the science of human intentions, beliefs and desires which, of course, can only be an individual's. In a rational choice framework, social institutions or practices can be accounted for as the result, more likely unintentional, of individual intentional actions. Decision theory here, or so it had been argued, goes hand in hand with evolutionary explanations but in order to contribute to the "science of human action" it has to receive a fully individualistic interpretation. The second belief is that social science, when properly done, can avoid normativity; where it gives an account it should not judge and where it designs working models for society, it has to assess them on normatively neutral grounds.[4] Decision theory makes both possible; explanations can be evolutionary and working models can be assessed as the aggregated outcome of individual choices. But if the assessment of working models is different from social Darwinism, then the individual choices must not be already determined by the context where they take place. As a model of rational choice, decision theory, here again, has to receive a fully individualistic interpretation.

229

M. Marion and R. S. Cohen (eds.), Québec Studies in the Philosophy of Science II, 229–246.
© 1996 *Kluwer Academic Publishers.*

Either the notion of individual rationality as understood in decision theory can play a role in social science or it cannot. If it can, and I think it can, the question remains what role it can reasonably play. I shall argue that reductionist attempts to give that notion of individual rationality a basic role in the social sciences are doomed to be self-defeating; building into individual rationality the constraints imposed by the context of choice, these attempts are likely to end up explaining social behaviour by the impotence of individual maximizing rationality and justifying social choices which are mere restatements of the context where individual rationality has to operate. Trying to make individuals fully responsible for their choices when in fact they are not and cannot be, I shall argue, is the best way to show that individual rationality has very little import in the social sciences.

My principal line of argument will not be from the social sciences but from decision theory itself and from some of the devices which tend to reinforce, in that framework, an allegedly context-free account of rational choice. The use of these devices rests heavily on assumptions concerning individual preference orderings and these assumptions, I shall argue, conflict in various ways with the very axioms of rational choice. I shall first illustrate these assumption at work in two current interpretations of a Prisoner's Dilemma, and show how the "failures of rationality" alleged in both cases rest on a misrepresentation, by way of a putatively context-free representation, of rational choice. I will then argue that when such a representation is carried out by the widely accepted doctrine of revealed preferences the result is that "rational choices" are taken to endorse the very constraints on maximization. I will conclude by pointing out some of the normative consequences of the individualistic interpretations of rational choice for social choice theory.

1. INDIVIDUAL RATIONALITY I: FROM DILEMMAS TO PARADOXES

There is a very simple although not very popular, interpretation of Prisoner's Dilemma (from now on: PD) according to which it is neither a dilemma for a rational agent nor a paradox of individual rationality. Suppose the usual matrix of a PD as represented in Figure 1.

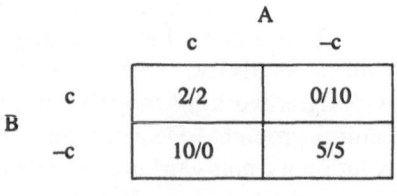

Fig. 1.

Prisoner A and prisoner B, each have to decide whether they will, or will not, confess their joint crime. In order to choose between these two alternatives, each prisoner considers the utilities afforded to him, respectively by confessing and by not confessing. Like most people, including prisoners, both A and B prefer the minimum number of years in jail. Prisoner A is told that if he confesses while B is also confessing then he gets 2 years in jail; but if B is going to confess, A could get 0 by not confessing. A is also told that if he confesses while B does not confess he gets 10 years in jail; but if B is not going to confess, A can get only 5 years by not confessing. A has no way of knowing for sure what B will decide, but A can see that whatever B chooses to do, A is better off with the outcomes of non-confession: that is, 0 year rather than 2 years if B confesses and 5 years rather than 10 years if B does not confess. It is from now on clear to Prisoner A that he will choose non-confession. Prisoner B, being in the same situation also choose to not confess.

Instead of getting what they "prefer most", A and B get what they "dislike least". Should we conclude that A and B, being fully responsible for their choices, either are bad maximizers, or did not really want to maximize? A and B have to choose between two options: confessing or not confessing. Each of these options afford, depending on what the other prisoner does, which they do not know at the time of their choice, a reasonably good and a relatively bad opportunity. They are doing, in that context, exactly what maximizing rationality recommends: they choose the option whose overall utility relatively to their preference ordering is the greatest. They are, in that sense, responsible for their choices but they are not responsible for the fact that the greatest utility they can afford in that set up, is not greater than it is. However, there are other stories about PD situations.

a) A Dilemma

The "dilemma view" pictures the prisoners as Buridan's disciples. This view was advocated, among others, by John Watkins whose narrative goes the following way.[5] In a first step, prisoner A reasons as before and concludes that he is better off by not confessing. But then, so the story goes, A reflects that B will reason the same way and choose to not confess, which for a time, will reinforce A's previous conclusion.

"But dammit," A exclaims to himself, "we *do not* have to resign ourselves to [5] years. We have been offered the chance of two years. Let us grab it. B must surely see this too". But now A has second (*sic*) thoughts: "What if B outsmarts me with [no confession]? [Ten] years! And suppose he does not; then I can outsmart him by [not confessing]. 0 year!". So A veers back toward [non-confession]. But not for long. He sees all too well that if B is likewise veering toward [non-confession], they are back to square one [. . .] "But dammit", A exclaims. . . . [. . .] I am claiming [concludes Watkins] not that rational self-interest does dictate [confession] for A and [confession] for B but only that it does not dictate [non-confession]. The prisoners *are* in a dilemma; there is no determinate solution to their optimisation-problem.

Watkins's prisoners believe that there is something that they "have been offered" and that they "should grab", namely, the chance of 2 years in jail.

Watkins's prisoners here seem to forget two things about their circumstances.[6] First, that it is only through action that they could "grab a chance" whatever it is; whatever is "offered" to A, and also to B, is offered through a choice over exactly two actions: confessing and not confessing. It so happens, and the prisoners have no power to change it, that confessing, which might afford them 2 years in jail, might also afford them 10 years in jail. Either they grab both "chances" by confessing or they do not confess. The second thing overlooked by Watkins's prisoners is that the choice of an action is a choice made by *each* of them separately; neither A nor B can, by his sole choice, bring about an outcome which depends on the choice of both. Each has to choose according to his preferences and on the basis of the utilities he could derive, given the choice of the other prisoner, from confessing and alternatively, from not confessing. They would both choose confession, and then get 2 years in jail, if on the basis of such a calculation, in a context structured as a PD, they will each choose non-confession instead of confession. So, if they are rational maximizers, if they know what interactive choice means, and if they have listened to what the justice of the peace said, how could A and B believe that they have the power to "grab" the outcome of mutual confession? And, if they believe that they are so omnipotent that they could disregard the various constraints to which they are subjected, why are they in a dilemma?

Watkins's prisoners are in a dilemma because they sometimes disregard *and* sometimes take into consideration the environment of their choice. Watkins insists on the fact that 2 years in jail is a chance compared to 5 years. Of course prisoners want to minimise the number of years they spend in jail and for that reason, for any two numbers of years, they prefer the smallest one. Reflecting in that way, they reflect from their own preferences and, looking at the possible outcomes, they discover "chances". But they sometimes remember that in order to get anything they want they have *to choose an action*; if they want 2 years in jail they have to choose to confess which could also give them 10 years. Reflecting that way, they are taking into consideration the environment in which they have to choose.

Prisoners who procrastinate over confessing or not confessing, as in Watkins' story, obviously are confusing ordering their own preferences and choosing an action. Going back and forth from one to the other in trying to decide what to do, they are actually using two different principles of decision: one referring to individual determinants only and the other rightly based on both contextual and individualistic determinants. For them, asking oneself what one prefers the most (without considering whether one *can* have it or not) is on a par, as far as choosing is concerned, with asking oneself which action, given the actual circumstances, affords the greatest utility. Confusing these very different questions, Watkin's prisoners reason as if they were faced with two different utility assignments for the same range of actions. The first assignment is the one made by the justice of the peace, according to which confession affords them either 2 or 10 and non-confession affords them 0 or 5. According to the second assignment, resulting from their wishful thinking, they are offered

2 for confessing and 5 for not confessing. And they are caught: they try to choose between confessing and not confessing, but, considering confession as a prospective choice, for instance, they sometimes have to compare 2 to 0 and 10 to 5 and some other time ("Dammit") they have to compare 2 to 5. No wonder that "rationality dictates neither confession nor non-confession". If rational agents were like Watkin's prisoners, they would be, prisoners or not, in a permanent dilemma. But I think the problem should no longer be ascribed to prisoners.

The dilemma view brings in an interpretation of maximizing rationality according to which maximization is taken to be context free and builds up the alleged dilemma on the conflict between that interpretation and contextualist game-theoretic maximization. Context free maximization, if there were such a thing, would be maximization assessable on the basis of individual parameters only – as here, on the basis of individual preference orderings – and free from the various constraints arising from the environment of choice. It would be maximization where individual rationality is expected to act independently of, if not upon, the very features of the context of choice. To the extent that it supposes maximization to be sometimes assessable on the basis of individual parameters only, the "dilemma view", without assuming it clearly, illustrates the effects of individualistic reduction at any cost; investing individual rationality with the power to act independently of the context of choice, it ends up ascribing to individual rationality the limitations of the context of choices which depends on neither individual preferences, nor will or power.

b) *A Paradox*

The idea that PD shows a paradox of maximizing rationality, another favourite interpretation of PD, also depends on the belief that individualistic determinants play a main role in rational choice. According to the "paradoxical view" the problem is not, as in the dilemma story, that rational agents cannot make up their mind about how to maximize. The alleged problem is rather that utility maximizers, by the very fact that they are utility maximizers, are deprived of the actual capacity to maximize. And the solution is the paradoxical one that in order to maximize, one should not be a maximizer.

Defenders of the paradoxical view will be distressed by the fact that neither A nor B can achieve for himself the outcome of unilateral non-confession that each prefers best (0), for that outcome, they say, may also afford the other what he dislikes the most (10). What they find distressing is that A and B cannot reach the outcome which is better ranked in both A and B's preference ordering (2/2) than the one they actually get (5/5). Are we back to Watkins's idea of grabbing a chance? Not quite. For the paradox of maximizing rationality rather points to the fact that however the utilities of the two prisoners are linked in the joint outcome of their choices, they will fail to satisfy their preferences. So, if their failure is independent of the particular features of their situation, there should be something wrong with maximizing rationality.

To illustrate the reasoning leading to that conclusion, let's modify Figure 1 by putting a name on the joint result of A and B's possible choices as in Figure 2.

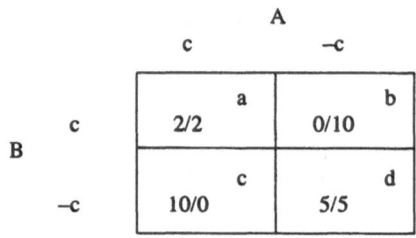

Fig. 2.

Suppose that A and B are asked to rank the four results, according to the utility they afford them. The ranking will tell us how A and B evaluate their own possible performances in the game and will naturally reflect the preference ordering of A and B on the numbers of years that they might spend in jail. A's ranking will be: *b, a, d, c* and B's ranking will be: *c, a, d, b*. The reasoning above assumes that A and B, as maximizers, try to reach their best, that is, respectively, *b* and *c*. The prospect of reaching *d* instead, so goes the reasoning, should be obvious to them as it should be equally obvious that they should aim for *a* which is only their second best, but which is much better for each of them than *d*. Since it is supposed to be obvious for each A and B that their attempt to reach their first best must fail, it should equally be obvious to each of them that his partner is also aiming for his first best; that is a very good reason, for each of them, to *avoid* trying his "second best".[7] But this is not the reason we are given here; we are told here that maximizing rational agents cannot achieve *a*; the presupposition being that, whatever their circumstances are, they will stubbornly never try to reach less than their first best. Only non-maximizing rational agents will try for *a* and, moreover, they will succeed. Why is it that it is rational for non-maximizers to reach for *a* while it is not for maximizers? Is there a different rationality here? According to the standard of maximization referred to in that reasoning, maximizing rationality consists in aiming for one's first best and to continue to aim at one's first best, even when it is *obviously* impossible to reach it, and when it is *obviously* possible under the circumstances to fare better by not going for one's first best. The supposition that maximizing rationality is a special kind of rationality, could make sense if the outcome of a PD was the result of a joint, agreed on, choice.[8] Aiming at one's first best will fail, but what is not said is that it will fail if and only if both partners are aiming at it. It is obvious to one that it will fail when one *knows* what one's partner intentions are. Knowing that and having the opportunity to agree on a different choice, prisoners will indeed have a "special kind" of rationality if they were stubbornly going to stick to non-co-operation; but that will not be maximizing rationality. And to jointly choose co-operation, when they have the opportunity of so choosing is also non-maximization. The rationality of choosing co-operation instead of non-co-operation does not depend on different types of rationality, but on the nature and the amount of the information available to each agent concerning the intentions of his partner and on the possibility of making an agreed choice. In a PD no such information is directly available,[9] and there exists no such opportunity. It is easy to see, in that context, the rationality for each prisoner of playing safe. If we overlook these facts and suppose instead that full information and communication are available to them, then the maximizing rationale for choosing non-confession disappears and with that there arises a "paradox of maximizing rationality".

Reasoning only from the joint outcomes of choices ranked according to individual preference orderings,[10] and supposing, moreover, that rational choices depend on such a reasoning, has that effect of hiding one of the main features of the context of choice represented by a PD. The result is to falsely picture rational agents as choosing solely on the basis of their and their partner's (identical) preference orderings, and entirely free from the limits on communication imposed by the context in which they have to choose. But these constraints do not vanish in thin air; they soon reappear as limits, blamed, this time, on individual rationality.

Nobody to my knowledge ever claimed that prisoners in a PD choose not to confess because they prefer to spend five years in jail. But still, the outcome of a PD continues to appear puzzling in face of the fact that rational agents are maximizers; how could rational agents get 5 years in jail when they "are offered" 2 years? There is a puzzle here when one reflects from the point of view of individual preferences, but the puzzle vanishes when one considers that what is really offered depends on the range of possible actions together with the utilities related to the other agent's choice. Rational agents have definite preferences guiding their choices of action, but their choices are also constrained, in a strict and unambiguous way, by the structure of the situation where they have to choose. The discrepancy between the outcomes of individual choices and individual preference ordering does not indicate that the chooser is in a dilemma or that he does not know how to choose in order to maximize; it simply indicates that a rational chooser is subjected to external constraints, and that he takes them into consideration when he tries to satisfy his preferences.

That situations structured like a PD exist in the real world, is a claim made by not a few social scientists. PD, so the claim goes, can model and then explain, phenomena such as externalities, failure of revolutions, deterrence and so on; it can show that individual rationality plays a decisive role in these phenomena and it can show moreover, what role it plays. The consequence of what has been claimed here is that no individualistic explanation of these phenomena can result from modelling them in a PD. On the one hand, a sensitive interpretation of PD will show that individual rationality plays an important role in the outcome of a PD; but it will also show the no less important role played by the constraints limiting individual maximization such as the limitation of the options of choice, their relative utility, and the foreseeable structural effects of individual choice. Set in that framework, externalities, failure of revolutions, deterrence and so on do not reduce to individual rationality, their explanations have to refer to supra-individual phenomena such as economical, social and political phenomena. On the other hand, an interpretation of a PD which does not fully acknowledge the role played by these contextual constraints and which stresses instead the unique role of individual rationality, will end up showing that the social phenomena it seeks to explain are independent of individual rationality. The two individualistic interpretations of a PD that I have discussed deny that the outcome of a PD is a solution to it, that is, that it is the outcome of a rational choice. The dilemma view, transposed in an explanation of externalities, for instance, will lead to the conclusion that following individual rationality, externalities might or might not arise or at best, that they arise out of a rational vacuum. Not paying attention to the context in which individual rationality operates, it will have no way of showing that individual rationality plays an explanatory role. The paradoxical view will not do better for the "science of human action" in asserting that, but for individual rationality, social interactions will be profitable for each.

2. INDIVIDUAL RATIONALITY II:
FROM REVEALED PREFERENCES TO SOCIAL CHOICE

a) *Revealed Preferences*

In a PD, the discrepancy between what a rational agent gets out of his choices and what he would like to get according to his preference ordering is a very strong reminder of the contextual constraints to which maximization is subjected. As we just saw, one could still deny the existence of those types of constraints by mistakenly evoking the dilemmas and paradoxes of individual rationality; but no one, considering the obvious preference ordering of prisoners, can reasonably claim that, making the choices they make, prisoners simply and unproblematically get what they would like to have in any case.

The standard theory of utility, generally adopted by the theoreticians of social choice, has no such reminder; it assumes that an independent access to the preference ordering of maximizing agents is superfluous and that their choices, and their choices only, reveal their preferences.[11] Making this assumption, the standard theory of utility is not taking for granted that individual choices are uniquely and unproblematically determined by individual preferences and it is not denying that rational agents reveal by their choices only what they can reveal given the contextual constraints to which their choices is subjected.[12] Still, for the view that I am defending, the assumption of revealed preferences is not innocuous. Because of that assumption, I shall claim, the standard theory of utility is likely to become a tool for the individualistic interpretation of rational choice attributing to individual rationality what in fact depends on an important contextual source of constraints on an agent's choice, namely, the composition of the set of options over which an agent has to choose. My line of argument will be the following. Since only choices reveal preferences, what is known of individual preference *orderings* is exactly what is revealed by the choices that an agent has the opportunity to make and that, in turn, depends on the particular options that are made available to him. Unless the complete preference ordering of an agent can be inferred from the various choices he actually had the opportunity to make, which is most improbable, then there is no way to measure, or even to take into consideration, the constraints on maximization resulting from possibly limited ranges of options.[13] For rational choice theory, using the standard theory of utility, the reminder of the contextual constraints to which maximization is subjected, can only consist in a well defined set of options. But such a definition belongs to the interpretation of utility theory and not to the theory itself. For all practical purposes these constraints, in the standard theory of utility, are taken to be nonexistent.

Suppose, to illustrate this point, that Z is asked to choose between playing bridge and playing chess and that he chooses to play chess. The doctrine of revealed preferences tells us that, in that set of options, chess is the one affording Z the greatest utility relatively to his preference ordering. And the doctrine of revealed preferences is right; rational agents, by definition, make maximizing choices. However, it is only, if that ever happens, when Z is asked to choose between playing chess and playing backgammon and chooses to play backgammon that it is "revealed" that the doctrine of revealed preferences being right about Z was only telling half the truth. Z prefers to play chess, but only when he has no opportunity to play backgammon. Z is certainly responsible for choosing to play chess in the first case, but he should not be credited for not choosing to play backgammon when he had no opportunity to choose it.

The assumption that choices reveals preferences has at least two noticeable untoward consequences. First, as shown by the previous example, it generally obscures the fact that the same agent can derive more or less utility from his maximizing choices depending on the set of options over which he can choose. Until Z is presented with a set of options containing both playing chess and playing backgammon, it is impossible to know what is the relative utility for him of playing one or the other. Utilities are not, to that extent, intrapersonally comparable. Second it becomes impossible to take into consideration the fact that two different agents can derive a different degree of utility by making maximizing choices over a same set of options. No one can tell whether B, choosing to play chess with Z, derives more or less utility then Z does by doing it.[14] Utilities are not, to that extent interpersonally comparable.

It is not difficult to see that the standard theory of utility, assuming the doctrine of revealed preferences, might easily introduce tensions in the notion of rationality as it is initially conceived in rational choice theory. In rejecting intrapersonal and interpersonal comparison of utilities, the standard theory of utility allows for a pre-established set of options to impose unnoticed constraints – where there is supposed to be none – on the preferences that rational agents can appeal to when they choose and consequently on the degree of utility that they can derive from their maximizing choices. The pre-established set of options, moreover, even when it is the same for all, imposes constraints which are different for different agents with different preferences. The degree of utility an agent can derive from his choice becomes then a matter of having the *right* preference ordering relatively to the options over which it is possible to choose, that is, the preference ordering in which these very options will be highly ranked whatever other options the feasible set could contain. Within these limits, it remains true that instrumental rationality holds for all, as it remains true that it is equally rational for all to make whatever choices they · make. But if one forgets about the limits, if the context of choice, and most particularly, if the nature of the options over which an agent chooses, is not firmly kept in mind, un-revealed preferences become, for all practical purposes, non-existent and the mediocre outcomes are automatically inputed to the idiosyncrasy of the agent's preferences. Each then appears to be rationally fully responsible for his choices, and his choices appear as the full expression of his free and unconstrained individual desires; but then, the very limits of maximization resulting from the context in which rational choice takes place are simply endorsed by what is called individual rationality. This might turn out to be specially damaging for the scope of "rational choice" and for the underlying notion of rationality when the offered options are modelling, as in social choice theory, frameworks for social interaction.[15]

b) *Social Choice*

When the theoreticians of social choice[16] reflect, as they periodically do, on the very notion of social choice, and on the procedures by which such a choice can be arrived at, they often come to think of it as a function (a social choice function) defined from the orderings, made by individuals over a feasible set, which, as its name makes clear, contains the options socially feasible, or the social states possible, in a given environment. In an intuitive sense, the possibility of defining a social choice function depends then on the convergence of individual orderings.[17] The ordering made by an individual over a given list of social states, is obviously similar to a choice made by an individual according to his preferences in social matters, and based on the utility, or the degree of welfare, afforded to him by each of the options.[18] Social choice so resulting from the convergence of individual choices will then point to the form of society which provides each his greatest utility. But how are we going to assess the exact contribution of individual rationality in social choice so defined? If individual preferences are taken to be revealed by individual choices in the sense understood in the standard theory of utility, then nothing enables us to measure the relative utility afforded to each by the various options they highly rank in their ordering and nothing enables us to measure which utility each would derive from his choice if different social options were possible, that is, were included in the feasible set. Social choice, taken to be defined by individual choices, might just reflect the limited set of options, judged to be socially feasible, over which individuals are given the opportunity to make choices.

Let's suppose, for one moment, that such a social choice function can be defined[19] specifying the nature and type of the institutions which will form the basis of a future society. Let us suppose, then, that starting from the feasible options and the individual orderings over them, we can identify a given social state as being preferred by all. What would that social state be like? My contention is that it will be just a replica of the society in which the rational agents, asked to make such a choice, actually live. And individual rationality will be credited for it.

Consider first the notion of feasible set. Feasibility in a given social environment, depends on what that environment actually is. The social options which are feasible are the ones which can be readily realised by using the resources of that environment, for instance, the social, political, economic and legal institutions, the technology and the modes of production and communication already existing.[20] What is socially feasible, in other words, is a function of the technological, political and economic policies that were made – or not made – in the past.

Faced with the set of feasible options, some individuals will be completely at ease; the options contained in that set correspond to societies, similar in type or even identical, to those they consider as a model of the society in which they would like to live. Some of these options appeal to social interactions similar in type to the ones that they prefer to have and they prefer that type of interaction *because* it is in interactions of that type in which they "realise themselves" better, in which they have the opportunity to use the various talents and capabilities they just happen to have, in which they find, in short, the best opportunities to maximize their utility relatively to their preference ordering. Given that set of options, they have the right talents and therefore the right preference ordering, that is, the preference ordering in which these very options will be highly ranked whatever other options the feasible set could contain.

But, if what is feasible in a given social environment pleases them so much, it is very likely to be because that environment was itself shaped by individuals with similar talents and aptitudes and who also succeed, using *these* talents and aptitudes, to maximize their utility. The feasible set, in other words, contains the options which are such that they favour the flourishing and the multiplication of individuals similar to those who, in the actual society, are the best adjusted. By definition, it contains *only* these options.[21] The individuals inadequately or poorly adjusted to the actual society, are the ones who participate inadequately or poorly in the interactions which take place in the actual society. The non-existence, in the actual society of the ways of life, or frameworks of interaction, which could allow them to use profitably the talents or the capabilities they have is precisely what makes most of them inadequately or poorly adjusted to that society. If the actual environment does not contain the resources allowing for the appearance and sustaining of these ways of life, and if the feasible options are feasible only relatively to the resources of the actual environment, it is clear that the set of feasible options contains only the preferred options of the individuals best fitted for that society, that is, the options allowing for the use of certain competencies to the exclusion of others.

The definition of a social function supposes that *each* individual orders, in view of his preferences, the set of the possible social states. If that set contains only the preferred options of the best adjusted individual's, it is obvious that the individual orderings will converge, if they converge at all, toward these options.[22] The issue of social choice, even if all participate equally in it, will make of individuals, already badly adjusted, individuals less and less adjusted, less and less capable of using the talents that they have, and less and less capable of participating in social interactions since these interactions require increasingly capabilities that they do not have.

An evolutionary explanation could work here very well. Keeping in mind the idea that feasible options are only feasible in a given context, we are provided with an account of the cumulative effect of initial circumstances to which individuals wired in a certain way respond better than others. The fact that certain characteristics are profitable at a given time explains why more and more individuals come to have (adopt or inherit) these characteristics, until the very fact that they are prevalent makes it profitable to possess them. Evolutionary explanations look after the mechanisms of reinforcement, and to the extent that it can represent these mechanisms by a procedure of choices restricted by the environment, where, then, intentional actions have unintentional consequences, rational choice theory constitutes a good evolutionary explanation of our actual social practices.

However the problem of social choice theory is neither to explain the emergence of our actual practices nor to secure the mechanisms by which they could be reinforced. The problem of social choice theory is to define a social order, likely to differ from the actual one by the very fact that it will, or so the claim goes, be *acceptable to each* in the society. To show that that is so, social choice theory has to emphasise the contribution of individual choices to the definition of a social choice function. Trying to do so by relying on the axioms of the standard theory of utility is, as we have seen, risky. When individual preferences are revealed by choices, and by choices only, and when the emphasis is put on individual choices as the unconstrained expression of individual preferences, one is likely to disregard the fact that maximizing choices on the same set of options can be less advantageous for some agents than for others, and the fact that, for the same agent, it will be more advantageous to maximize his utility on a different set of options. Trying to *justify* social choice by individual choices framed in the model of the standard theory of utility, social choice theory will credit individual maximizing rationality for the reinforcement, by the ones who benefit from it, of the actual framework of social interaction, but it will equally credit individual maximizing rationality for the reinforcement of that very framework by the ones who are, and know that they will increasingly be, severely disadvantaged by it.

CONCLUSION

Social choice framed in the standard theory of utility might very well be suboptimal[23] and the outcome of a PD certainly is. From the point of view of utility maximization, sub-optimality indicates, of course, a problem; that it could be, from that very point of view, imputed to maximizing *rationality*, would be a real paradox. It is, I have claimed here, a paradox to which decision theory, used as a framework for individualistic reduction, is committed.

There are two well known contextualist arguments, to neither of which I have appealed, challenging the individualistic character of decision theory. The first one concerns preference formation; it claims that individual preferences are socially acquired and challenges the individualistic conception, standardly ascribed to decision theory, according to which individuals are fully autonomous and socially independent beings. The other argument claims that an account of interactive choices requires a theory of "more than one single person's rationality" and challenges the individualistic conception, standardly ascribed to decision theory, according to which a general theory of rationality reduces to a theory of individual rationality. The individualistic conception that I have challenged here allows for social formation of preferences and for principles governing interactive choices; but it requires that any external contextual causal determinant of rational choice be such that it can itself be ascribed solely to individual rationality.[24] Metaphors for that form of individualism can be found in decision theory, where external causal determinants are sometimes personified as Nature or as Justices of the Peace. But in decision theory, Nature or Justices of the Peace are not thought to make use of maximizing rationality and, moreover, they do not often play an explanatory role in social science. The contextual determinants relevant in social science are part of a social environment, including economic or political contexts, and to show that they can be ascribed solely to individual rationality requires more than a claim that they *emerge* as the *unintentional* results of human action. It will at least require that these contexts are themselves shown to be the intentional results of a rational (maximizing) choices. But how could they be, since they are themselves constraints on further maximization of utility? This is the paradox of individualistic reductions of decision theory. Individual rationality, taken to have no constraints external to individual rationality, is assigned full responsibility for the very constraints

on maximization and therefore, for suboptimal choices. The science of human
action, construed on this model, amounts to little more than the repetitive story
of a self-defeating rationality.

Université du Québec à Montréal

NOTES

* This paper had the financial support of the Social Sciences and Humanities Research Council
of Canada.
[1] Decision theory is used here in a broad sense, so as to include formal theories of rational
agency, and, in particular, the theory of games, utility theory and social choice theory.
[2] For anthologies, see for instance: O'Neil (1973) and Broadbeck (1968).
[3] Weber (1978), p. 4
[4] The idea that individual choices can constitute a normatively neutral ground for social choice,
as the more general idea that social science must avoid normativity, can be, and indeed, has
been challenged (e.g.: in Sen, 1979). Although they both seem to me obviously wrong, I will
not directly challenge any of these ideas here. As far as they are concerned, my argument
rather points to the inconstancy between the idea of neutrality as argued for in social science
and the individualistic interpretations of rational choice. The substantive normative import of
these individualistic interpretations has been discussed in Couture (1993).
[5] Watkins (1985). Watkins's argument is that the dilemma becomes more obvious as the difference between the outcomes decreases. My criticism of the account he gives of the PD is,
however, untouched by a change in the value of the outcomes (provided, of course, their ratio
remains the same) and I took the liberty to restate, in the quotation below, the values I already
use in Figure 1.
[6] What they forget, in short, is the difference between their preference ordering and the set of
options over which they have to choose. If A prefers w to x, x to y to z, then his preference
ordering is $\langle w, x, y, z \rangle$. Suppose that the options over which A has to choose are a, b, c, d, and
suppose moreover that the utility afforded by a is w, the utility of b is x, the utility of c is y
and the utility of d is z. Relatively to his preference ordering, the option a is the one affording
A the greatest utility and that is the one he will rationally choose. If the preference ordering of
A and the utilities afford by each option are unchanged but that only the options c and d are available, A maximizes his utility by choosing c. Watkins' prisoners reason as if, in that case, it
will be relevant to the choice they have to make *between c and d*, to reflect that they prefer w
and x to z.
[7] This points out to a first way (cf. also note 8 and 9) of formulating the discrepancy between
the game-theoretic perspective and the account described here, where agents are reasoning as
if they where choosing an outcome. In the game-theoretic perspective, where each prisoner
has to separately choose *an action*, each is always considering the two outcomes (including
the non-symmetrical ones) related to each action. Each prisoner will consider that his trying
his first best might fail, but also, might *not* fail, depending of what his partner does. In the second
perspective (choosing an outcome) it is irrational not to reach for one's second best, but in the
first perspective, to talk of a as the second best, of both prisoners is, of course, begging the
question. The relevant observation about A and B's above ranking is rather that while A and
B each rank something before a, they also both rank d before something else. So d and not a
is the second best of both A and B. And that is what they get.
[8] The hypothesis that prisoners communicate in view of making an agreement immediately rules
out any of the possible joint outcomes of a PD but one: the outcome resulting from mutual
confession. That hypothesis has, to that extent the symmetrical effect sometimes seen as a consequence of equal rationality (cf. note 9).
[9] Indeed, it had been argued that equal rationality implies identical behaviour; a rational agent

knows that his partner is equally rational and should therefore know that he will choose as he himself will. Following that reasoning it could be possible for one of the partners, to secure the outcome *a*. But that violates the condition on the independence of choice.

[10] Re: "in order to maximize one should not be a maximizer". What looks like a mere semantical ambiguity on "maximization" turns out to be a real confusion on the proper way to assess the maximizing character of choices. In the view criticised here, a choice is maximizing if its outcome equals or approaches the greatest utility relative to one's preference ordering and not, as it should be, if it affords the greatest possible utility relatively to one's preference ordering *given the context of choice and available options*. Hence the *maximizers*, on one hand, pictured here as agents trying to have their "first best", and failing to "maximize", because they get instead their second worse and, on the other hand, *rational* agents, pictured as non-maximizers since they only try to have their second best relative to their preference ordering. In the same fashion, recipes to achieve maximizing choices have portrayed the latter as agents making counter-preferential choices, or as maximizing on a cryptic preference ordering, or as having the "character" of being a co-operator.

[11] By "standard" theory of utility I mean a formal theory of utility making the assumption of revealed preference. That assumption was already made by Fisher (1892) and, independently, by Pareto (1896). It has received a canonical formulation in Samuelson (1932) and is characteristically adopted by orthodox economists of the present day. The standard theory of utility has indeed been challenged (e.g. Elster and Roemer, 1991) but it still has, or so many believe, no serious rival in social science.

[12] Samuelson is often considered as the father of the contemporary doctrine of revealed preferences and he first conceived of it as a tool for a theory of demand based on the consistent behaviour of consumers rather than on their preferences. For him, and his disciples, the very notion of preference should then be superfluous: "If an individual's behaviour is consistent, then it must be possible to explain that behaviour without reference to anything other that behaviour". But as Sen (1973) remarks, quoting the previous sentence of Little (1949), "Faith in the axioms of revealed preference arises [. . .] not from empirical verification, but from the intuitive reasonableness of these axioms interpreted precisely in terms of preference." (p. 63)

[13] To contribute to a theory of *consistent* behaviour the doctrine of revealed preference should minimally presuppose transitivity and allow therefore for inferences concerning individual preference orderings. Suppose that A chooses *a* when presented with *a* and *b*; that he chooses *b*, when presented with *b* and *c* and chooses *d* when presented with *d* and *b*. One can then infer that he will choose *a* when presented with *a* and *c* and choose *d* when presented with *d* and *c*. But it is impossible to tell which of these two choices afford A the greatest utility since, from A's previous choices, one cannot infer anything concerning what he would choose when presented with *a* and *d*.

[14] Unless B's and A's preference orderings have been inferred (cf. note 13) from their previous choices. This is a different ground for establishing a point initially made by Harsanyi (1955): interpersonal comparison of utility is a special case of intrapersonal comparison.

[15] Preferences will then be interpreted as preferences for results of interactions. In the previous example, for instance, Z and B will both rank in decreasing order all the possible results of playing a game, from winning gloriously, to losing lamentably with, in between, winning comfortably, barely loosing, etc. Relatively to an identical preference ordering, the utility afforded to each by playing chess, backgammon and bridge is obviously linked to B's and Z's talents for playing each of these games. When Z chooses to play chess because backgammon is not available, he might actually choose to play a game that he is likely to lose, that is, he chooses to use talents which does not allow him to satisfy his preferences. It will be rational for a maximizer to act in such a way. Z, finally losing the game, one should conclude, not that Z is bad at chess – no one can compare how he would have fared (the utility he could have derived) by playing backgammon – but that he is a bad player, period. A limited set of alternatives put constraints on preferences that it is possible to reveal, but when the alternatives are different frameworks for interaction, the constraints are constraints on the talents that it is possible to use.

[16] My main reference here will be Arrow, specially: (1963) and (1978)

[17] Suppose a society to be the set $N = \{1, \ldots, i, j, \ldots, N\}$ of individuals. The cardinality of that set is denoted by N and is finite.

An ordering R is a complete, transitive and reflexive relation on X, the set containing at least 3 possible social states.

For Arrow, social choice is defined by a function g of the list of individual orderings $\langle R_i \rangle_{i \in N}$ where R_i is an ordering on X which reflects the welfare of the individual i in different states. If for two orderings R and R' $xRy \leftrightarrow wR'z$ and $yRx \leftrightarrow zR'w$ then $R: \{x, y\} = R': \{w, z\}$.

[18] Actually, most formal theories of aggregation are formulated in terms of rational choice. See, for example: Blair and Muller (1983), Bordes (1976), Elster and Hylland (1986), Suzumura (1983), Sen (1977) and (1982), and Gibbard (1973).

[19] Arrow has shown that such a function (cf. g, in the note 16) cannot satisfy simultaneously the four following conditions: (Relatively to R, P and I denote respectively the strict preference and indifference)

1.   *Unrestricted Domain (U)*
        All $\langle R_i \rangle_{i \in N}$ logically possible are in the domain of g.

2.   *Indpendence of the irrelevant alternatives (I)*
        For all ordered pairs $A \in X \times X$, if $\langle R_i \rangle$, $\langle R'_i \rangle$ are two lists such that:

        $\forall i \in N: R_i: A = R'_i: A$

        then

        $R: A = R': A$ where $R = g(\langle R_i \rangle)$ and $R' = g\langle R'_i \rangle)$.

3.   *Pareto (weak) condition (P)*
        For all pair $x, y \in X$

        $[\forall i \in N: xP_i y] \rightarrow xPy$.

4.   *Non-dictatorship (D)*
        There is no individual d such that for each element in the domain of g, $\forall x$, $y \in X$:

        $xP_d y \rightarrow xPy$.

POSSIBILITY THEOREM. *If an ordering satisfies* (U), (I), *and* (P), *then, there is a $d \in N$ such that for all $x, y \in X$: $xP_d y \rightarrow xPy$.*

[20] Arrow's second condition (condition I), stipulates that only the options feasible in that sense are part of X. "Since the chosen element from any environment is completely defined by knowledge of the preferences as between it and any other alternative in the environment, it follows that the choice depends only on the ordering of the elements of that environment. In particular, the choice made does not depend on preferences as between alternatives which are not in fact available in the given environment, nor – and this is probably more important – on preferences as between elements in the environment and those not in the environment. It is never necessary to compare available alternatives with those which are not available at a given moment in order to arrive at a decision." Arrow (1978), p. 218.

[21] It should be noticed that that does not contradict the first of Arrow's condition (condition U), since it only concerns the domain of the social function which corresponds to the set of individual orderings. I am talking here of the domain of individual choices, which corresponds to the set of social states. In view of the previous note, it is clear that the condition I, which admits a restriction of the domain of individual choices, also has the effect of restricting the domain of the social choice function. It is nonetheless true that the domain of individual choices includes "all the logically possible individual orderings" (condition U).

[22] In that perspective, it is not surprising that the three first "intuitive" Arrow's conditions

are not compatible with the fourth one. The situation I am describing here is the situation where there is, not one individual, but a group of individuals "whose preferences are automatically society's preference independent of all other individual" Arrow (1978), p. 226. This is not strictly speaking a dictatorial choice, but rather an oligarchic choice.

[23] This follows from the previous section: social choice, defined from a different set of alternatives might offer individuals more, without decreasing the welfare of the others, than it does on the feasible set considered. Moreover, the Pareto condition (cf. note 19) applies only when the preference relation derived from it is complete; which might not be the case, even when individual preference orderings are.

[24] This individualistic conception is discussed in Nozick (1977).

REFERENCES

Arrow, K. J., 1963, *Social Choice and Individual Values*, John Wiley and Sons, New York.
Arrow, K. J., 1978, 'Values and Collective Decision-Making', in P. Laslett and W. G. Runciman (eds.), *Philosophy, Politics and Society*, 3e series, B. Blackwell, Oxford, pp. 215–232.
Blair, D. H. and Muller, F., 1983, 'Essential Aggregation Procedures on Restricted Domains of Preferences', *Journal of Economic Theory* 21, 186–194.
Bordes, G., 1976, 'Consistency, Rationality and Collective Choice', *Review of Economic Studies* 43, 447–457.
Broadbeck, M., 1968, *Readings in the Philosophy of Social Sciences*, University of Minnesota Press.
Couture, J., 1993, 'Sélection rationnelle', to appear in: *Les Cahiers scientifiques de l'ACFAS*, Montréal.
Elster, J. and Roemer, J. (eds.), 1991, *Interpersonal Comparison of Well-Being*, Cambridge University Press, Cambridge.
Elster, J. and Hylland, A., 1986, *Foundations of Social Choice Theory*, Cambridge University Press, Cambridge.
Fisher, 1892, *Mathematical Investigation in the Theory of Value and Prices*, in: Transactions of the Connecticut Academy of Sciences. Re-publication: Yale University Press, New Haven, 1925.
Gibbard, A., 1973, 'Manipulation on Voting Schemes: A General Result', *Econometrica* 41, 587–601.
Hammond, P., 1976, 'Equity, Arrow's Conditions, and Rawls's Difference Principle', *Econometrica* 44, 793–804.
Harsanyi, J. C., 1955, 'Cardinal Welfare, Individualistic Ethics and Interpersonal Comparison of Utilities', *Journal of Political Economy* 63, 309–321.
Kalai, E., 1985, 'Solutions in the Bargaining Problem', in L. Horwitz, D. Schmeider and H. Sonnenschein (eds.), *Social Goals and Social Organisation*, Cambridge University Press, pp. 77–105.
Little, I. M. D., 1949, 'A Reformulation of the Theory of Consumer's Behaviour', *Oxford Economic Papers* 1.
Nozick, R, 1977, 'On Austrian Methodology', *Synthese* 36, 353–392.
O'Neil, J. (ed.), 1973, *Modes of Individualism and Collectivism*, Heineman, London.
Pareto, V., 1906, *Manuel d'Economie Politique*, trad. 1927 (2e ed.).
Roberts, K. W. S., 1980, 'Possibility Theorems with Interpersonally Comparable Welfare Levels', *Review of Economic Studies* XLVII, 409–420.
Samuelson, P. A., 1938, 'A Note on the Pure Theory of Consumers Behaviour', *Economica* 5, 61–71.
Sen, A., 1976, 'Welfare Inequalities and Rawlsian Axiomatics', *Theory and Decision* 7, 243–262.
Sen, A., 1979, *Collective Choice and Social Welfare*, North-Holland.
Sen, A., 1973, 'Behaviour and the Concept of Preference', in J. Elster (ed.), *Rational Choice*,

New York University Press, New York, 1986, 60–81. Reprinted from *Economica* **40**, 1973, 241–259.

Sen, A., 1977, 'Social Choice Theory: A Re-Examination', *Econometrica* **45**, 53–89.

Sen, A., 1982, *Choice, Welfare and Measurements*, MIT Press, Cambridge Mass.

Suzumura, K., 1983, *Rational Choice, Collective Decision and Social Welfare*, Cambridge University Press, Cambridge.

Watkins, J., 1985, in R. Campbell and Lanning, Sowden (eds.), *Paradoxes of Rationality and Co-operation: Prisoner's Dilemma and Newcomb's Problem*. University of British Columbia Press, Vancouver.

SUSAN DWYER

DISPOSITIONS TO EXPLAIN

> "it is usually a mistake to suppose that the micro-descriptions 'explain' the macro-phenomena. If someone tries to correlate or identify dispositions like intentions or beliefs with 'micro-phenomena' like neural synapses, or tries to explain the former in terms of the latter, then of course he is likely to lose . . . the explanatory benefit (Wilkes [1989], p. 165)."

1. As inveterate "why?"-askers, human beings are disposed to seek for explanations of phenomena, exotic and familiar alike; and ironically, the familiar often proves the most intractable for explanation. Thus, while we are all native speakers of some human language, the possession and exercise of this common ability is notoriously difficult to account for. In this paper I investigate the prospects for a dispositional explanation of linguistic competence. Such an explanation is well-motivated; however, dispositions, and thus alleged explanations that invoke them, have a vexed history in the philosophy of science. But while it is possible to explicate a notion of dispositions that invests them with genuine explanatory currency, such modifications threaten to make them inappropriate for the explanation of linguistic behavior. I consider this threat, and suggest that it is not as serious as it appears. It is plausible to attribute to speakers linguistic dispositions; indeed, such ascription is a part of the best explanation of their linguistic behavior. In short: speakers of natural languages have a disposition to explain, and some of their dispositions *do* explain.

2. Intuitively, speakers[1] are ripe for epistemic characterization. A speaker of a natural language has manifold abilities: she is able to converse successfully and impart information; she can lie, deceive, tell the truth, make distinctions, ask questions and answer them. Typically, a speaker is able to tell when two sentences are incompatible, when two sentences mean the same, and when one sentence follows from another. She is also able to make judgments of truth and falsity, revealing an ability to ascertain relations between her language and the world. In doing all these things a speaker exploits the *meaningfulness* of expressions in her language and, correspondingly, her and her interlocutors' *understanding* of those expressions. At the very least, to understand a sentence is to know what that sentence means, and so it is natural to say that a speaker's various linguistic capacities are to be characterized, at base, as *knowledge* of meaning, or semantic knowledge (see, e.g., Dummett, 1991, pp. 101–102; Larson, 1990).

These intuitive reasons for attributing linguistic knowledge to speakers

M. Marion and R. S. Cohen (eds.), *Québec Studies in the Philosophy of Science II*, 247–263.
© 1996 *Kluwer Academic Publishers*.

receive considerable theoretical support in the literature. Broadly speaking, we can distinguish two types of argument for the claim that we ought to attribute speakers with linguistic knowledge. Dummett (1978, 1991, pp. 88–92), foremost proponent of what we can call the *a priori* argument, focuses on the idea that speaking and understanding a language are rational activities carried out by agents with intention and purpose, and he argues that linguistic knowledge must be attributed to speakers as a way of making such activities intelligible. Alternatively, driven by the "logical problem of language acquisition," theoretical linguists advance more *a posteriori* arguments to the effect that we best explain adult speakers' attainment of a mature capability with language, if we posit that children come into the world equipped with a considerable amount of knowledge about humanly possible languages (see, e.g., Chomsky, 1986; Lightfoot, 1991, ch. 1; Pinker, 1989).[2]

From both perspectives, descriptive and normative observations suggest that linguistic knowledge is rather special: (1) Ordinary speakers are unable to state (or, often, to recognize) the theory or rules they are alleged to know (Devitt and Sterelny, 1987, p. 138; Kirkham, 1989, pp. 211–212; Platts, 1979, p. 232; Stich, 1971); (2) they are not aware of deploying this information when they routinely speak and understand their language; and (3) it is only on pain of gross circularity that one could insist that speakers' linguistic knowledge is explicit, i.e., represented and available to them in their native tongue (Devitt and Sterelny, 1987, pp. 139–40; Harman, 1967, pp. 75, 79). Thus it is argued that linguistic knowledge not only is, but *must* be implicit, or tacit (Dummett, 1991, p. 94; also Dummett, 1973, p. 217; 1976, p. 70; 1977, p. 373; Higginbotham, 1990; Crispin Wright, 1986, p. 209). Furthermore, proponents of both types of argument recognize that a crucial feature of linguistic behavior that requires explanation is the apparent unboundedness of speakers' linguistic capacities. It is a fact about competent speakers that they are able to understand and produce a potential infinity of novel (well-formed) sentences in their language. For this reason, it is argued that the content of linguistic knowledge must be (a) a finitely axiomatized, compositional theory of meaning for the language in question, and/or (b) the (suitably parameterized) rules and principles of Universal Grammar.[3]

Without further argument, then, I will take it that there are sufficiently compelling reasons to credit speakers with implicit linguistic knowledge (henceforth, ILK). My real interest here devolves on what kind of account we are able to provide of such knowledge, and especially on whether that account can be made consistent with its *a priori* motivations, i.e., with the idea that its attribution explains linguistic behavior *as* rational behavior. This is a matter of pressing importance. The notion of ILK is not without its critics: McDowell (1977) and Searle (1990) contend that it is incoherent; Platts (1979), Quine (1972), and Wright (1981) argue that the attribution of ILK to speakers is vacuous; and, relatedly, Foster (1976) and Kirkham (1989) claim that the only respectable construal of the attribution to speakers of ILK is instrumentalistic; speakers don't really have such knowledge, they merely behave 'as if' they do. And it is not clear that we can formulate a substantive account of ILK which is both (1) immune to these challenges, and (2) amenable to the kind of explanatory service we arguably require of it.

The central difficulty here hinges on the notion of implicitness. If, as Dummett and others (e.g., Montefiore, 1989, Wilkes, 1989) have argued, linguistic behavior is paradigmatically *intentional* behavior (speaking and understanding a language are rational activities carried out by purposive agents), any adequate explanation of it must advert to the intentional states of the agents whose behavior it is. But we have seen that the relevant intentional states must be tacit. Instrumentalism can be very tempting with respect to the attribution of intentional states *simpliciter* (e.g., Dennett, 1991), and it might be thought that there is more reason to think of the attribution of *implicit* intentional states in this light. It is agreed that ordinary speakers will not acknowledge the theory implicit knowledge of which theorists attribute to them; hence, it is only speakers' unreflective linguistic behavior which serves to verify the ascription of ILK. And so the following difficulty – 'Quine's challenge' – presents itself. As we will see below, it is easy to construct two (or more) extensionally equivalent theories of meaning (say, T1 and T2) for a language L, where T1 and T2 are extensionally equivalent if they yield the same meaning-specifications for expressions of L. Thus it becomes a fair question how the ILK theorist is to justify the attribution to a speaker of implicit knowledge of T1 *rather than* implicit knowledge of T2 (see Quine, 1972). Ordinary speakers' behavior will not, it appears, be sufficient to warrant the ascription of a *particular* theory; hence, it is better to construe any ascriptions of ILK instrumentalistically.

However the proponent of ILK cannot leave things here. The attribution of ILK to speakers is, recall, primarily motivated by *explanatory* concerns, but to 'go instrumental' about it commits one to merely re-*describing* speakers' behavior (Dummett, 1981; Wright, 1981).[4] Hence, a more robust notion of ILK must be specified. Moreover, anyone who holds out explanatory hope for ILK will have to say how the knowledge in question is responsible for the possession and manifestation of the range of abilities had by competent speakers of a natural language.

Both of these desiderata can be met quite straightforwardly if we provide a *physicalistic* characterization of ILK; that is, if we say that to implicitly know a theory of meaning for L just is to be in a particular brain state (Chomsky, 1986, ch. 4; 1988a, p. 4; 1988b, p. 3). But herein lies the rub for the *a prioristically* inclined theorist. Perhaps he can embrace the consequence of an empirical investigation that establishes some identity between the intentional states that comprise a speaker's linguistic knowledge and her neurophysiology; but he cannot, on pain of inconsistency, accept from the outset a physicalistic construal of ILK. For he argues that linguistic behavior, because of the type of behavior it is, necessitates *intentional* explanation. More subtly, he will be sensitive to plausibility constraints on the notion of ILK. For should it turn out that the only coherent account of ILK *requires* positing particular physical states or processes in speakers, he will be unable to appropriate the notion for his explanatory purposes.

Thus, the insistence that linguistic knowledge be implicit gives rise to a real tension in the project of explaining linguistic behavior via its attribution. At the one end, in order to be a candidate for a genuine *explanation*, the attribution of ILK must be more than instrumental; it must be realistic. But paradigmatic realistic explanations work by positing physical entities and events, and mapping their causal interactions. As Quine puts it, "[I]f we are limning the true and ultimate structure of reality, the canonical scheme for us is the austere scheme that knows . . . no propositional attitudes but only the physical constitution and behavior of organisms (1960, p. 221)." But at the other end, if we take seriously the idea that linguistic behavior is intentional behavior, and thus demands intentional explanation, brute physical explanations just seem to be explanations of the wrong sort. The challenge for the proponent of ILK, then, is to say how the attribution of *implicit* linguistic knowledge to speakers can be (1) genuinely explanatory, i.e., he must respond to Quine's challenge and say how possession of ILK is causally efficacious, and (2) thoroughly intentional. To put the point another way, what is needed is an ontologically neutral, non-instrumental account of ILK.

It is for these reasons that a dispositional account of ILK appears attractive. Intuitively, the appeal lies in the following two considerations: first, intentional notions readily admit of dispositional characterizations; e.g., if S believes that p, she is disposed, all things equal, (i) to judge that p, (ii) to assert that p, (iii) to act as if p were true, (iv) to deny that not-p, and so on; second, it is now widely held that dispositions, if they are to be explanatorily useful, must be understood in terms of their bases (dispositional or categorical).[5] And it is arguable that we can specify the bases of linguistic dispositions in terms of *something* (I leave matters deliberately vague for the moment) inside the speaker's head, thereby putting ILK somewhere where it can do its causal work in the production and comprehension of language. With a proper understanding of dispositions in hand, we can claim that to attribute a speaker with ILK is to ascribe to her a set of structured dispositions, viz., dispositions to judge the meaning of expressions in her language. The pressure is then, clearly, to provide this proper understanding of dispositions. And in particular, we will want to see whether the bases of linguistic (semantic) dispositions can be specified in a way that is acceptable to the *a priori* defender of ILK.

3. It will facilitate discussion of these issues to have a concrete proposal before us. In his provocative (1981) paper, Gareth Evans presented a sketch of a theory of ILK that was not vulnerable to the full force of Quine's challenge, and that was intended to respond to another serious challenge due to Crispin Wright (1981). Wright's query was whether the structure-reflecting, or compositionality requirement on a theory of meaning can be justified. That requirement

finds expression in the idea that an adequate semantic theory for a language L will not only deliver a meaning specification for each well-formed expression in L, but will do so in such a way as to reveal how the meanings of complex expressions depend on the meanings of their constituents and the manner of their composition. Wright (1986) develops the point and argues that the strongest support for this requirement is forthcoming from the semantic theorist's ambition that a theory of meaning for a natural language ought to explain how its speakers are able to produce and understand novel expressions in the language. As we have seen, it is tempting to provide an explanation of the unboundedness of speakers' linguistic capacities in terms of their possessing ILK; the further move here is simply to specify that the object of such knowledge be a finitely axiomatized, compositional semantic theory. Thus, Wright argues, the justification for the structure-reflecting requirement is tied to the possibility of constructing a plausible, realistic (as opposed to instrumentalistic) account of ILK.

Clearly, Wright's worry is related to Quine's. The provision of a realistic, empirically respectable account of ILK depends on meeting the problem of extensionally equivalent theories. And if Quine's challenge cannot be met, then there will be no justification for attributing to speakers implicit knowledge of any particular theory, let alone one that is structure-reflecting. Wright's challenge is an important one. For on the one hand, as he himself points out (*modulo* some qualifications), "should it emerge that there is absolutely nothing to made of the notion of implicit linguistic knowledge . . . Davidson's project will prove to have been a waste of time (1986, p. 208)." Thus, much in the recent history of the philosophy of language appears to be hostage to a favorable prognosis for ILK. On the other hand, and more to our present purposes, Wright's focus on the notion of *structure* alerts us to something that appears to be pivotal in the explanation of how the possession of ILK can be causally efficacious in speakers' behavior. But more of this below.

Evans' proposal was this: that "we construe the claim that someone tacitly knows a theory of meaning as ascribing to that person a set of dispositions – one corresponding to each of the expressions for which the theory provides a distinct axiom (1981, p. 124)." These dispositions will be dispositions to judge the meaning of each expression of the language as being such-and-such. Evans invites us to imagine a finite language which contains 10 names $(a, b, c, . . .)$ and 10 predicates $(F, G, H, . . .)$, and thus 100 well-formed expressions, or sentences (e.g., Fa, Gb, Hc . . .). Now consider two theories of meaning for L: T1 (the listiform theory) consists of 100 distinct axioms of the form 'Fa is true iff John is bald,' and T2 (the articulated theory) consists of 21 axioms – 10 of the form 'a denotes John', 10 of the form 'an object satisfies F iff it is bald,' and 1 compositional axiom of the form: 'A sentence coupling a name with a predicate is true iff the object denoted by the name satisfies the predicate.' T1 and T2 are extensionally equivalent because they each specify the same meanings for the sentences of L.

Now if we take it that an agent is a speaker of L just in case she is disposed to make correct semantic judgments concerning the expressions of L, it is far from obvious how Evans' proposal speaks to Quine's challenge. An L-speaker's semantic judgments concerning the expressions of L are correctly described by *both* T1 and T2. And so it would appear that the semantic dispositions of an L-speaker who is credited with implicit knowledge of T1 will be identical to those of an L-speaker who is ascribed implicit knowledge of T2. Moreover, it might be doubted that the attribution to speakers of semantic *dispositions* is explanatory of speakers' competence. The ascription of dispositions appears to do no more than *report* on an L-speaker's linguistic behavior; it does not appear to explain it.

Quine's challenge and the worry about the explanatory force of dispositions are in fact two sides of the same coin. If we say that to implicitly know a theory of meaning is just to be disposed to make correct judgments concerning the meaning of expressions in L, then it will be very difficult to detect a difference in the linguistic behavior of any two L-speakers sufficient to warrant the attribution of implicit knowledge of the listiform theory to the one and of the articulated theory to the other. Thus the appeal to dispositions to characterize ILK, far from making that notion 'empirically adequate,' as Quine would say, seems to compound the problem. However, there is a way out for the (Evansian) friend of ILK, the explication of which will require a short detour.

4. Dispositions (dispositional properties, dispositional predicates and dispositional explanations) have had, until recently (e.g., Cartwright, 1989; Sober, 1983) rather bad press in the philosophy of science, and it does not seem to be much of an advance to press them into service in the characterization of ILK and thus in the explanation of linguistic behavior. But in investigating what is thought to be problematic about dispositions it is useful to distinguish between (1) questions having to do with dispositions *simpliciter*, and (2) questions having to do with the appeal to dispositions in the explanation of linguistic behavior in particular.

A familiar worry about dispositions is due to Feigl (1958) who characterized dispositional statements as "promissory notes," and emphasized by Pap (1962) who writes, "in subsuming an observed regularity under a disposition concept one anticipates an explanation in terms of the intrinsic structural microproperties of the things involved (p. 282)." The worry here is best seen by considering the distinction between (1) *bare* dispositions and (2) *full-blooded* dispositions. Bare dispositions involve no more than "the truth or falsehood of certain 'could' or 'would' propositions and certain other particular applications of them (Ryle, 1949, p. 46)." Thus, they are simply counterfactuals facts of the following form:

BD1: Given conditions C, if the creature were to receive input I, then the creature would produce output O.

BD2: Were this piece of salt to be put in water, it would dissolve.

Thus, while it is agreed that dispositional statements signal the existence of potentially interesting regularities, ascriptions of bare dispositions cannot be genuinely explanatory, because they are silent on the basis or the cause of those regularities. Weimer (1984) suggests, "[D]ispositional analyses are incomplete in that they cannot stand alone as explanations: they invariably point beyond themselves to a more penetrating and rarefied level of analysis (p. 162)." So, notoriously, on a bare dispositionalist account we are unable to explain why an object credited to possess a disposition (perhaps, in virtue of the kind of object it is) fails to manifest that disposition. Put in terms of the above schema, the bare dispositionalist cannot say why, given conditions C and input I, the creature does not produce output O; or why a particular piece of salt placed in a beaker of water does not dissolve. Since, on this view, dispositions are simply counterfactual facts, the theorist is able only to *report* that the expected behavior did not occur.

Two (not obviously distinct) moves can be made to salvage for dispositional ascriptions some explanatory currency. In the first place, in order to allow that the relevant subjunctive conditionals can be true of the objects they are about *even when* those objects fail to manifest the specified dispositions given relevant initial conditions, the dispositionalist can insert a *ceteris paribus* clause. For example, **BD2** becomes: *Ceteris paribus*, were this piece of salt to be put in water, it would dissolve. Thus, when the alleged disposition fails to manifest – when a quantity of salt does not dissolve in water – the ascription of the disposition is not thereby falsified. Rather the theorist is directed to seek for what, in the given circumstances was 'not equal'. But taking to heart Pap's remark above, one might go further.

Modifying the relevant subjunctive conditionals with *ceteris paribus* clauses allows that they can come out true, but we might ask after what *makes* them true when they are. Hence the dispositionalist might specify some occurrent, or steady state of the object or creature whose presence is (causally) responsible for the manifestation of the disposition in question, and to which we can point in accounting for failures of manifestation. So we might opt for a full-blooded account of dispositions, thus:

FBD1: There is some state of the creature, S, such that given conditions C and input I, *ceteris paribus*, S together with I serves to produce output O.

FBD2: Salt has a molecular structure, X, which interacts with the molecular structure of water, such that *ceteris paribus* when salt is placed in water it dissolves.[6]

The specification of these states is to be given in terms of "whatever state of the object scientists find to be responsible for manifestation of the disposition when a suitable initiating cause acts upon the object. In the case of brittleness, for instance, the state will be a certain sort of bonding of the molecules of the brittle object (Armstrong, 1973, p. 13)."[7] Complete explanations of the creature's (object's) behavior will therefore make ineliminable appeal to something 'inside' it. Dispositional ascriptions can therefore be understood in a way that makes them explanatorily respectable. But it might be argued that nonetheless there is something problematic about the use of dispositions in the explanation of linguistic behavior.

One reason to think so is captured (but not endorsed) by Andrew Wright (1991): "A cowardly person can act bravely, if suitably motivated. Animate dispositions are not sure-fire but, to some degree, under the control of their owners, especially in the case of capacities (p. 48)." But this allusion to the 'control' that animate creatures exercise over the manifestation of their dispositions does not present a *special* problem for linguistic dispositions. For it is clear that few, if any, inanimate dispositions are sure-fire: it is not the case that salt *always* dissolves in water. And we have seen that dispositional attributions can be modified to deal with this fact. The defender of animate, including linguistic, dispositions can appeal to the same strategy. He can add a *ceteris paribus* clause to the conditional used to ascribe the disposition, and/or specify some abiding property of a speaker which provides a causal explanation of both the manifestation and the failure of the manifestation of the relevant dispositions.

Kripke (1982) has also been critical of the appeal to linguistic dispositions, especially in the context of responding to the Wittgensteinian sceptic. The issue here is essentially the same, though Kripke frames it in terms of normativity, rather than speaker's 'control'. Anyone who professes to provide an account of understanding, or knowledge of meaning must be sensitive to the fact that meaning is essentially normative. In particular, the linguistic dispositionalist will need to provide an account of error. The specification of a coherent competence/performance distinction will require being able to determine of a given piece of behavior whether it is (1) the imperfect manifestation of the disposition D1, or (2) the perfect manifestation of a different disposition D2. Graeme Forbes (1983) has investigated the plausibility of transferring the explanation offered above of 'dispositional malfunction' in the case of inanimate dispositions to the case of speakers' semantic errors. The strategy is initially appealing; but the question arises as to precisely *which states* of a speaker we need appeal in making out an explanatory account of linguistic dispositions. We will return to this crucial issue below.[8]

5. With these remarks in hand, let us return to Evans' proposal. Recall that we left off our discussion wondering whether Evans could vindicate the notion of ILK against Quine's challenge. Evans recommends that we construe the attribution of implicit knowledge as the ascription of a certain set of dispositions, and indeed insists that

it is essential that the notion of a disposition used in these formulations be understood in a full-blooded sense. . . . When we ascribe to something the disposition to V in circumstances C, we are claiming that there is a state S which, when taken together with C, provides a *causal* explanation of all the episodes of the subject's V-ing (in C). So we make the claim that there is a common explanation to all those episodes of V-ing (1981, p. 125).

T1 and T2, the extensionally equivalent semantic theories for L, differ with respect to how much structure in the expressions of L they discern; T1, with its 100 axioms, treats each expression as an unstructured whole, whereas T2, with 21 axioms, treats each as articulated. We have noted that Evans suggests that we ascribe to a speaker of L just as many dispositions as the number of "expressions for which the theory provides a distinct axiom (p. 124)." In light of the discussion of the last section, then, we can imagine the attribution of implicit knowledge of T1 or T2 as involving the ascription to the relevant L-speaker of 100 distinct *states*, or 20 distinct *states*, respectively.[9] So, for example, when we attribute implicit knowledge of T1 to a speaker, we explain her understanding of any one sentence of L (e.g., Fa) in terms of some *one* state of her which corresponds to the relevant axiom: "Fa is true iff John is bald.' And of the speaker who is said to implicitly know T2, we say (i) that her understanding of Fa is underpinned by two distinct states, viz., that to do with the predicate F and that to do with the name a; and (ii) that her understanding of Fa involves a state that is also involved in her understanding of Ga; viz., that to do with the name a. Thus while the linguistic dispositions of a speaker who implicitly knows T1 will be underpinned by a single state, those of a speaker who implicitly knows T2 will always involve two distinct states, and crucially many of her linguistic dispositions will have states in common, e.g., all her dispositions involving the name a will involve the state that corresponds to the axiom governing a.

As Evans would have it, when we attribute dispositions to a speaker we *thereby* attribute a set of causal-explanatory states to her. Seeing things this way allows us to formulate empirically testable hypotheses about, for instance, patterns of decay which would reveal which theory an L-speaker implicitly knows. Thus, suppose an L-speaker loses facility with the expression Fa. If she also loses facility with all expressions in L involving that proper name and that predicate – i.e., if we witness her failing to judge the meanings of Fb . . . Fj, and of Ga . . . Oa – then we have evidence in favor of ascribing to her implicit knowledge of T2, the articulated theory. However, if her semantic deficit is localized to the expression Fa, then we ought to conclude that, of the two theories, she implicitly knows T1, the listiform theory. For that theory contains 100 *distinct* axioms, and we would not expect to see trickle down effects of the same order (Evans, 1981, pp. 127–129). Thus Evans' proposal appears to provide the friend of ILK with a response to Quine's challenge (cf. note 9.)

It will be clear that what is doing the real work in Evans' account is the notion of causal explanatory states in speakers; and this returns us both to our central question concerning the nature of these states, and to the question of whether Evans has answered Wright's challenge. About the latter, while

it might be true that speakers' linguistic abilities are underpinned by significant causal structure, it is still not clear why that fact warrants the imposition
of the structure-reflecting requirement on a theory of meaning. As Davies
(1987) puts it, "why does *causal* structure justify articulation in a *semantic*
theory? (p. 448)." And about the former we need to ask if we must understand these causal explanatory states as *neurophysiological* states. If the answer
to this question is "yes," the prospects for the *a priori* theorist's appropria-
tion of the dispositionalist account of ILK look grim indeed.

The idea of causal explanatory states of speakers finds its home in a rather
abstract constraint on theories of meaning. This constraint is implicit in
Dummett (1976), p. 109; and is made explicit in Davies (1981) as the "Mirror
Constraint" (hereafter, MC). The MC posits a relation between causal-explanatory states in the speaker and the structure, the *derivational* structure, of the
theory of meaning she is said to implicitly know. The idea is this: If a speaker
is able, on the basis of her understanding of $s_1, s_2 \ldots s_n$ to understand some
arbitrary (new) sentence s, then it must be true of the semantic theory, knowledge of which we attribute to her, that the "resources used in derivations of
meaning specifications for $s_1 \ldots s_n$ are jointly sufficient for the derivation
of the meaning specification for s (p. 446)." (See also Evans, 1985, pp. 25–6.)
Thus, "the derivational structure in a semantic theory should match the causal-
explanatory structure in actual speakers (Davies, 1987, p. 447)."

The MC certainly appears to lend to the notion of tacit knowledge real
empirical bite, for it insists that we credit a speaker only with having inter-
nalized a set of rules that are connected with her psychological states. So
attributions of ILK are anchored in the actual agent we are considering. But
two issues arise. First, the MC (as stated) is very weak; perhaps so weak as
to warrant only instrumental ascriptions of ILK. For consider precisely
what the MC is: 'On the outside' we have a theory of meaning which has a
particular derivational structure, and 'on the inside' we have something or other
going on in the agent's brain or psychology. What is proposed is that the
agent tacitly knows a certain theory, if the derivational structure of that theory
mirrors the causal-explanatory structure in the speaker. But any number of pairs
of semantic theory and causal-explanatory structure can be considered to stand
in this relation. Hence there is a danger that the attribution of ILK will be
decidedly non-empirical. The solution demands a detailed and empirically
testable specification of the causal-explanatory structure in the agent. And it
is here, secondly, that the temptation to 'go physicalistic' is particularly strong.
Indeed, while Evans maintains – as surely he must, for it is his very point –
that "there is a clear empirical difference between these two models of competence, and hence between tacit knowledge of T1 and T2," he says that
"[T]he decisive way to decide which model is correct is by providing a
causal, presumably neurophysiologically based, explanation of comprehension
(p. 127)."

Putting to one side the obvious worry that "physical states do not explain
exercises of *understanding* – they explain at most the physical events in

which that exercise is embodied (Forbes, 1983, p. 236, my emphasis)," it is not clear that *all* the neurophysiological states and processes causally responsible for a speakers' competence are semantically relevant. In an ingenious thought-experiment, Davies (1987) has shown how some of the states that are properly part of the relevant set of a speaker's causal explanatory states have no plausible counterpart in a semantic theory. Thus imagine that the neural site underlying the specific sub-competence of a speaker with the proper name *a* ceases to receive some vital nutrient, causing it to malfunction. The manifest effect for the speaker will be an inability to recognize the meaning of that term. But her failure to comprehend here is not due to anything semantic. What is going on in an agent's neurophysiology may have nothing to do with meaning, even though it might underpin the physical processes that make comprehension of meaning possible.

It is not obvious that Evans intended his account of ILK to be a prolegomenon to the reduction of that notion to neurophysiology. And in any case, it would appear that the *neurophysiological* individuation of the speaker's causal-explanatory structure will be too coarse grained, or simply irrelevant to the project of explaining her *semantic* competence. A more important philosophical consequence of reduction must also be noted. If we are in possession of knowledge concerning the neural states that causally underpin speakers' semantic abilities, then it might be argued that we can do away with the dispositional characterization of linguistic competence altogether. And insofar as the positing of such neural states is required to explicate the notion of ILK, we will have 'explained' ILK by explaining it away.

The proponent of ILK might, of course, just reject talk of the MC, and of the concomitant causal-explanatory states in speakers. But this would not be advisable. For she must advert to such states in order to (1) render the dispositions she ascribes explanatorily respectable, (2) answer Quine's challenge, and (3) vindicate the requirement that theories of meaning must be structure-reflecting. The more satisfactory strategy, if Evans' account is to meet Wright's challenge and be of use to the *a priori* proponent of ILK, is to reject a physicalistic understanding of the relevant causal-explanatory states of speakers.

It is by no means obvious that the 'descent' to the causal level necessarily entails a descent to the neurophysiological. That is, causal explanations need not be physical explanations. It is open to us to think of the causal-explanatory structure mirrored by a semantic theory, which adequately describes a speaker's linguistic competence, as *psychological*. This is a strategy that will hold obvious appeal for the theorist looking for an intentional explanation of linguistic behavior. However, even if the causal-explanatory states required by the dispositional account of ILK can be construed as in some sense psychological, there is reason to doubt that these states that ground a speaker's linguistic competence are genuine intentional states of the type sought after by the *a priori* theorist.

6. Evans claims that while his account

leads one to the thought of a correspondence between the separable principles of a semantic theory
and a series of internal states of the subject, dispositionally characterized, . . . to regard these
states as states of knowledge or belief, that is to say, states of the same kind as are identified
by the ordinary use of these words, is wrong and capable of leading to confusions (1981, pp.
130–131).

For while there are analogies between tacit knowledge and ordinary cases of
belief, these are "very far from establishing tacit knowledge as a species of
belief (p. 131)." For Evans, "it is of the essence of a belief state that it be at
the service of many distinct projects, and that its influence on any project
be mediated by other beliefs (p. 132)." Thus, for example, my belief that the
food before me is poisonous can, in conjunction with various of my desires
and other beliefs, manifest itself in indefinitely many ways; if I wish to
debilitate my colleague, I will offer him a taste; if I wish to innure myself
to the effects of the poison, I might take a little of the food every day, and
so on. However, "possession of tacit knowledge is exclusively manifested in
speaking and understanding a language; the information is not even potentially
at the service of any project of the agent, nor can it interact with any other
beliefs of the agent (p. 133)." Hence, Evans recommends that we think of
the states which ground a speaker's semantic dispositions as 'subdoxastic.'

If Evans is right, then the *a priori* defender of ILK is faced with the following dilemma. If he wants a speaker's semantic dispositions to be genuinely
explanatory, he must specify the relevant causal-explanatory states which
ground them. These he may specify (a) physicalistically, or (b) psychologically. If he opts for (a), any hope of providing an *intentional* explanation of
linguistic behavior is gone; if he opts for (b), the psychological states in
question will not be genuine intentional states. Neither horn is appealing;
but since (a) is not really a live option, we should ask what the *a priori* theorist
can make of (b).

It seems to me that the disanalogies between ILK and ordinary cases of
belief indicate, to the extent that they indicate anything, *not* the rejection of
ILK as consisting in genuine intentional states, but rather a need to adjust
our ordinary understanding of belief. *If* linguistic behavior is the rational
behavior *par excellence*, as Dummett suggests; and *if* the account of ILK
here presented provides the best explanation of that behavior; and *if* the
states which constitute a speaker's ILK do not appear to be 'genuine' intentional states, then perhaps we need to rethink our notions of belief and
knowledge. Of course, there is a whiff of dogmatism on both sides, as is always
the case when we impose what we take to be conceptual constraints on explanations and theories. So it is worth investigating which set of these *a priori*
constraints we find the most plausible. (Notice, I do not take "*a priori* constraints" to refer to those that are accessible by the clear light of reason;
rather they are conceptual constraints that we discover as operative in our
practice of inquiry.)

Evans' worry above concerns what it is to be a belief. But suppose we give up talk about the essence of *beliefs* and focus instead on what we know about *believers*. Certainly if I believe that p, then given what else I believe, and desire, and hope for, etc., my believing that p will have ramifications for what I do, how I understand my options, what else I come to believe, and so on. But beliefs get their purchase just to the extent that their attribution explains an agent's behavior;

> Representational mental states should be understood primarily in terms of the role that they play in the characterization and explanation of action. What is essential to rational action is that the agent be confronted, or conceive of himself as confronted, with a range of alternative possible outcomes of some alternative possible actions (Stalnaker, 1984, p. 4)."

This "pragmatic" construal imposes next to no constraints on what counts as *a* belief. Indeed, the pragmatist is not really very interested in that question, if it is taken to be an *ontological* question. The pragmatist's emphasis is on explanation, and on his view, "[B]eliefs are conditional dispositions to act." Thus when we attribute beliefs to an agent, we are ascribing a complex of interlocking dispositions, where these are "real causal properties of persons, and not simply . . . patterns of actual and possible behavior (Stalnaker, 1984, p. 82; p. 16)."

This is a gross oversimplification of the pragmatic picture, but my point in sketching it is to show that there is an alternative conceptual story to be told about belief, which has the consequence that linguistic knowledge (beliefs), understood as a set of dispositions, is not obviously ruled out as a candidate for a genuine intentional state(s). Indeed, Stalnaker (1984) claims that

> [L]inguistic action . . . has no special status. Speech is just one kind of action which is to be explained and evaluated according to the same pattern (p. 4). . . . If the strategy can be carried out, then we will have a foundation for the kind of semantic theory that explains the meaning and content of linguistic expressions in terms of the intentions and conventions of language users, and which explains what it is to understand a language in terms of the capacities of speakers and hearers to use the language to serve their needs and desires (p. 21).

The pragmatic strategy, then, will appeal to the *a priori* defender of ILK for the following reasons: (1) it allows him to maintain a dispositional account of ILK, that is (2) explanatorily respectable, and (3) does not commit him to any particular physical thesis about the grounding states of a speaker's semantic dispositions. Thus it would seem to make possible the non-instrumental, ontologically neutral account of ILK he was seeking.

There is, however, a final worry to address. I said earlier that the real sticking point about formulating an account of linguistic knowledge is its alleged implicitness, and it might be wondered whether the pragmatic account can handle *implicit* intentional states, or dispositions. I have argued elsewhere (1992) that there is no theoretically interesting distinction between what we are wont to call implicit and explicit intentional states. Fundamentally, two things are at issue when we attribute beliefs to an agent of which she appears to be unaware (and linguistic beliefs certainly are such): (1) the kind of access an agent has or *needs* to the content of those beliefs, and (2) the degree of awareness an agent has to those contents themselves.

Let us distinguish between agent-access and agent-awareness in the following way: an agent has access to a proposition, *p*, if she is in a mental state whose content is that proposition; an agent is aware of a proposition, *p*, if she is in a *different* mental state, viz., a state whose content is *p* and which has the special property of her being aware of its content. The robust phenomenon of first person authority is relevant here. It appears to be the case (a) that we *are* aware of the content of some of our beliefs, and (b) that, individually, we are more likely to be right about the content of those beliefs than are those others who observe us. Now let us mark off a type of mental state – called a special belief – in terms of the following two conditions: (i) its content is a particular proposition, and (ii) if the .agent has such a state, she is aware of that proposition. So, for example, those beliefs about which we are thought to be first person authoritative are special beliefs. Linguistic beliefs are clearly *not* special beliefs in this sense. Consider the following:

(1)   John expects to feed himself.

(2)   I wonder who John expects to feed himself.

As competent speakers of English we know that in (1) "John" and "himself" must refer to the same person, viz., John, whereas in (2) we know that they *cannot* be coreferential. Linguists explain these facts on the basis of our tacitly knowing a particular grammatical rule,

> C: An anaphor must have an appropriate c-commanding antecedent.

Thus the attribution of ILK appears to involve the attribution of many mental states the contents of which are propositions like those expressed by (the *sentence*) C. However, as we have noted, speakers are not in general aware of the contents of these mental states (i.e., of these propositions). On the other hand, speakers do have access to the contents of their linguistic beliefs. For our linguistic behavior, our understanding of our language, *antecedes* any explicit knowledge of the rules of syntax we might acquire. If C is a rule of English, then the mental state whose content is the proposition expressed by C is causally implicated in every English speaker's linguistic behavior, whether or not she is aware of the relevant proposition.

The relation between agent-access and agent-awareness with respect to linguistic beliefs is perhaps best grasped by considering the English-speaking linguist. In virtue of being a speaker of English such a person has access to the proposition expressed by C, while in virtue of being a linguist, she is aware of that proposition. Of course, ordinary speakers too can become aware of the relevant propositions, as anyone who has taken a syntax (or philosophy of language) class will attest. But the point to note is that while familiarity with linguistic or semantic theory often results in the increased salience of some features of one's language, linguists and philosophers of language do not speak their native language differently than their non-expert compatriots. Thus, with respect to a linguistic belief, access to the proposition that is its content is necessary for awareness of that proposition, but agent-access does not entail agent-awareness. And there seems to be no principled reason why this does not hold for all our beliefs. Certainly we have beliefs of which we are unaware, but this is not sufficient to warrant a distinction between *kinds* of belief. Far from implicit beliefs (i.e., non-special) being problematic, the difficult question really faces the sceptic about implicit intentional states. Once we take account of (a) the phenomenon of first person authority, and (b) the fact that access to a proposition does not entail awareness of it, it becomes a fair and obvious question why we are aware of *any* of our beliefs at all.

This last point made, the *a priori* proponent of linguistic knowledge can safely help himself to the dispositionalist account of ILK. Indeed, we would expect him to have a disposition to do just that.

McGill University

NOTES

[1] Throughout, when I use the term "speaker" I intend it to denote an agent who is able both to produce and interpret untterances in a natural language.

[2] I deliberately omit mention of Davidson's work for the simple reason that he has been, and remains resolutely non-committal concerning the question of whether speakers genuinely *know* a theory of meaning for their language. And here I am interested in the prognosis for a *realistic* rather than an *instrumentalistic* account of linguistic knowledge.

[3] We need not think of these objects of knowledge as mutually exclusive; for there is some reason to think that a marriage of sorts can be effected between Davidsonian style theories of meaning and linguistic theory (Higginbotham, 1986).

[4] Instrumentalism about ILK will appear attractive to the theorist who thinks the sole purpose of constructing a theory of meaning is to reveal the semantic machinery of the language in question. For then it would simply be beside the point whether speakers of the language actually know the theory or not. But this position, which is Foster's (1976), "generates an intolerable gap between the concepts of meaning and understanding (Wright, 1986, p. 210)." See Wright for further remarks.

[5] I do not wish to enter into the debate about the correct characterization of the bases of dispositions here. See Prior (1985) for thorough coverage of the relevant positions; and more recently, Cartwright (1989) and Blackburn (1990).

[6] The 'cp-modifier' and the 'base-specification' strategies are related and in rather interesting ways, discussion of which is beyond the scope of this paper. Specifying the base of a disposition and modifying the relevant subjunctive conditional used to attribute the disposition appear to have the same effect of rendering dispositional talk explanatory respectable. For (1) the specification of a base attributes some causally powerful state to the object, and (2) adding the cp clause allows the subjunctive conditional to be true, even when the object fails to manifest the disposition ascribed to it. But whether or not one thinks these strategies amount to the *same thing* will depend, ultimately, on what one's views about causal laws are.

[7] See also Dummett (1979), p. 31; Evans (1980), pp. 102–103; Quine (1960), p. 34, p. 223.

[8] It might be responded that there is a *special* kind of normativity at work in language which is, strictly speaking, orthogonal to the competence/performance distinction as it is cashed out in Forbes' account or in theoretical linguistics (see, e.g., Bourdieu, 1991). But if so, then it must be pointed out that the terms of the debate have been changed.

[9] Martin Davies (1987) has pointed out that Evans' initial proposal stands in need of minor revision here. Reporting a subsequent objection of Wright's (1986), Davies writes: ". . . a subject who has the interlocking dispositions that Evans describes – one for each of the twenty atomic expressions of L – will *ipso facto* be disposed to assign the correct meanings to whole sentences. But, in the theory T2, the twenty axioms for those names and predicates are not sufficient to yield meaning (truth conditions) specifications for whole sentences (p. 444)." The question is how to incorporate the force of the compositional axiom into the relevant states of the speaker who is credited with implicit knowledge of an articulated theory. Davies' suggestion is that we imagine yet another theory of meaning for L, viz., T3, which "contains an axiom for each name of L – indeed, the very same axiom as in T2. But the axioms for the predicates are different. For the predicate 'F', we have [:] a sentence coupling a name with a predicate 'F' is true iff the object denoted by the name is bald. . . . What T3 does is to parcel out the content of T2's compositional axiom among the predicates of the language (p. 445)." This has the effect of making the case that Evans' proposal will not allow us to distinguish between *every* pair of extensionally equivalent theories.

REFERENCES

Armstrong, D. M., 1973, *Belief, Truth and Knowlege*, Cambridge University Press, Cambridge.
Blackburn, S., 1990, 'Filling in Space', *Analysis* **50**, 62–65.
Bourdieu, P., 1991, *Language and Symbolic Power*, in J. B. Thompson (ed.), trans., Gino Raymond and Mathew Adamson, Polity Press, Cambridge.
Cartwright, N., 1989, *Nature's Capacities and their Measurement*, Clarendon Press, Oxford.
Chomsky, N., 1986, *Knowledge of Language*, Praeger, New York.
Chomsky, N., 1988a, 'Language and Interpretation: Philosophical Reflections and Empirical Inquiry', ms.
Chomsky, N., 1988b, *Language and the Problems of Knowledge. The Managua Lectures*, The MIT Press, Cambridge, Ma.
Davies, M., 1981, *Meaning, Quantification, Necessity*, Routledge & Kegan Paul, London.
Davies, M., 1987, 'Tacit Knowledge and Semantic Theory: Can a Five Per Cent Difference Matter?', *Mind* **96**, 441–462.
Dennett, D., 1991, 'Real Patterns', *Journal of Philosophy* **87**, 27–51.
Devitt, M. and Sterelny, K., 1987, *Language and Reality*, The MIT Press, Cambridge, Ma.
Dummett, M. A. E., 1973, 'The Philosophical Basis of Intuitionistic Logic', reprinted in *Truth and Other Enigmas*, Harvard University Press, Cambridge, Ma.
Dummett, M. A. E., 1976, 'What is a Theory of Meaning? II', in G. Evans and J. McDowell (eds.), *Truth and Meaning*, Clarendon Press, Oxford.
Dummett, M. A. E., 1977, (with the assistance of R. Minio) *Elements of Intuitionism*, Oxford University Press, Oxford.
Dummett, M. A. E., 1978, 'What Do I Know When I Know a Language?', Lecture at the Centenary Celebrations of Stockholm University.
Dummett, M. A. E., 1979, 'Comments', in A. Margalit (ed.), *Meaning and Use*, Reidel, Dordrecht.
Dummett, M. A. E., 1981, 'Objections to Chomsky', Review of *Rules and Representations*, *London Review of Books*, 3–16 September, 1981.
Dummett, M. A. E., 1991, *The Logical Basis of Metaphysics*, Harvard University Press, Cambridge, Ma.
Dwyer, S., 1992, 'Doxastic Capacity and Implicit Belief', *Cahiers d'Epistemologie*, No. 9226, Montreal.
Evans, G., 1980, 'Things without the Mind – A Commentary upon Chapter Two of Strawson's *Individuals*', in Z. Van Straaten (ed.), *Philosophical Subjects. Essays Presented to P. F. Strawson*, Clarendon Press, Oxford.
Evans, G., 1981, 'Semantic Theory and Tacit Knowledge', in S. Holtzman and C. Leich (eds.), *Wittgenstein: To Follow A Rule*, Routledge & Kegan Paul, London.

Evans, G., 1985, *Collected Papers*, Clarendon Press, Oxford.

Feigl, H., 1958, 'The "Mental" and the "Physical"', in H. Feigl, M. Scriven and G. Maxwell (eds.), *Concepts, Theories, and the Mind-Body Problem*. Minnesota Studies in the Philosophy of Science, vol. II. University of Minnesota Press, Minneapolis, pp. 370–497.

Forbes, G., 1983, 'Scepticism and Semantic Knowledge', *Proceedings of the Aristotelian Society*, New Series, vol. 84, pp. 223–237.

Foster, J., 1976, 'Meaning and Truth Theory', in G. Evans and J. McDowell (eds.), *Truth and Meaning*, Clarendon Press, Oxford.

Harman, G., 1967, 'Psychological Aspects of the Theory of Grammar', *Journal of Philosophy* **64**, 75–87.

Higginbotham, J., 1986, 'Linguistic Theory and Davidson's Program in Semantics', in E. LePore (ed.), *Truth and Interpretation*, Basil Blackwell, Oxford.

Higginbotham, J., 1990, 'Philosophical Issues in the Study of Language', in D. Osherson and H. Lasnik (eds.), *Language*, The MIT Press, Cambridge, Ma.

Kirkham, R., 1989, 'What Dummett Says About Truth and Linguistic Competence', *Mind* **98**, 207–224.

Kripke, S., 1982, *Wittgenstein on Rules and Private Language*, Harvard University Press, Cambridge, Ma.

Larson, R., 1990, 'Semantic Knowledge', in D. Osherson and H. Lasnik (eds), *Language*, The MIT Press, Cambridge, Ma.

Lightfoot, D., 1991, *How to Set Parameters: Arguments from Language Change*, The MIT Press, Cambridge, Ma.

McDowell, J., 1977, 'On the Sense and Reference of a Proper Name', *Mind* **86**, 159–168.

Montefiore, A., 1989, 'Intentions and Causes', in A. Montefiore and D. Noble (eds.), *Goals, No-Goals, and Own Goals: A Debate on Goal-Directed and Intentional Behaviour*, Unwin Hyman, London.

Pap, A., 1962, *An Introduction to the Philosophy of Science*, Free Press, Glencoe, Ill.

Pinker, S., 1989, *Learnability and Cognition: The Acquisition of Argument Structure*, The MIT Press, Cambridge, Ma.

Platts, M., 1979, *Ways of Meaning*, Routledge & Kegan Paul, London.

Prior, E., 1985, *Dispositions*, Aberdeen University Press, Aberdeen.

Quine, W. V. O., 1960, *Word and Object*, The MIT Press, Cambridge, Ma.

Quine, W. V. O., 1972, 'Some Methodological Reflections on Current Linguistic Theory', in D. Davidson and G. Harman (eds.), *Semantics of Natural Language*, Reidel, Dordrecht.

Ryle, G., 1949, *The Concept of Mind*, Hutchinson, London.

Searle, J., 1990, 'Consciousness, Explanatory Inversion, and Cognitive Science', *Behavioral and Brain Sciences* **13**, 585–596.

Sober, E., 1983, 'Dispositions and Subjunctive Conditionals; or, Dormative Virtues are No Laughing Matter', *The Philosophical Review* **91**, 591–596.

Stalnaker, R., 1984, *Inquiry*, The MIT Press, Cambridge, Ma.

Stich, S., 1971, 'What Every Speaker Knows', *The Philosophical Review* **80**: 476–518.

Weimer, W. B., 1984, 'Limitations of Dispositional Analysis of Behavior', in J. R. Royce and L. P. Mos (eds.) *Annals of Theoretical Psychology: Vol I*, Plenum Press, New York, pp. 161–198.

Wilkes, K., 1989, 'Representation and Explanation', in A. Montefiore and D. Noble (eds.), *Goals, No-Goals and Own Goals: A Debate on Goal-Directed and Intentional Behavior*, Unwin Hyman, London.

Wright, A., 1991, 'Dispositions, Anti-Realism and Empiricism', *Proceedings of the Aristotelian Society*. New Series, vol. lxxxxi, 1990/1991, pp. 39–59.

Wright, C., 1981, 'Rule-Following, Objectivity and the Theory of Meaning', in S. Holtzman and C. Leich (eds.), *Wittgenstein: To Follow a Rule*, Routledge & Kegan Paul, London.

Wright, C., 1986, *Realism, Meaning and Truth*, Basil Blackwell, Oxford.

CLAUDE PANACCIO

BELIEF-SENTENCES:
OUTLINE OF A NOMINALISTIC APPROACH*

What I call a nominalistic analysis of belief-sentences is an explication of the meaning of sentences of the form 'A believes that p' that does not commit the analyst to the existence of special abstract entities such as sets, Fregean senses and propositions, or universal properties and relations. I take it, without discussing the point here, that *if it is feasible*, such an analysis is *prima facie* philosophically preferable to a realistic one for general reasons of clarity and ontological economy. But the question, of course, is: is it, in some reasonable way, feasible? Standard nominalistic approaches to belief-sentences all face well-known and crucial difficulties. The quotational theory, according to which the object-sentence '*p*' is quoted or mentioned within the belief-sentence 'A believes that p', fails the Langford–Church translation test.[1] The metalinguistic analysis, on the other hand, according to which the 'that'-clause within the belief-sentence is, in every case, a primitive metalinguistic predicate fails to satisfy Davidson's learnability requirement.[2] And Davidson's own approach, the so-called paratactic theory, according to which 'A believes that p' means simply 'A believes this. p' fails, for all its virtues, to accommodate some important illocutionary and logical characteristics of belief-sentences.[3] These negative results from the last four or five decades of debates led more than one phi-losopher to conclude drastically that nominalism just can't cope with belief-sentences in any satisfactory way. What I would like to show here is that there is still at least one more line of investigation for nominalism in this endeavour. Space being severely limited, I will be content to sketch the general features of this hitherto unexplored approach and to list some of the motivations for it as well as some of its theoretical advantages.

The basic idea of the theory is that what is required for a satisfactory analysis of belief-sentences – and of the other indirect contexts as well – is not a special category of *designata* for the 'that'-clauses considered as names or as singular terms of some sort, but rather an enrichment of our conception of the *logical syntax* of natural languages. According to this view, it is not the 'that'-clause as a whole that is in need of a special explanation, but the sole connective 'that'. The thesis will be that 'that' is in such contexts an abbreviation for a complex and heterogeneous expression, which I will try to spell out.

1. IN SEARCH OF AN INDEXICAL CONNECTIVE

Let us start with this: if we renounce, as I think we should, both the quotational and the metalinguistic analysis of indirect contexts and if we're not ready to countenance abstract entities as *designata*, then the only way left for us will be to admit that the words occurring within the scope of 'that' in a

265

M. Marion and R. S. Cohen (eds.), *Québec Studies in the Philosophy of Science II*, 265–277.
© 1996 *Kluwer Academic Publishers*.

belief-sentence are being used by the speaker in their plain and ordinary sense and with their normal reference. In 'Plato says that Socrates is a philosopher', 'Socrates', then, will neither be mentioned nor turned into a special metalinguistic expression or part of such an expression. It will simply be the ordinary proper name 'Socrates', the *designatum* of which is Mr. Socrates himself. And similarly 'philosopher' in such a context will be nothing but a token of the familiar predicate which you and I like to apply to ourselves and to our masters of the past.

Considerations of this sort were crucial to Donald Davidson's solution of the riddle. Davidson breaks 'Plato says that Socrates is a philosopher' into two sentences: 'Plato says that' and 'Socrates is a philosopher', and although the second of these is not taken to be asserted by the speaker, all the words that occur in it have, according to this view, nothing but their normal semantical functions. The problem, though, with this paratactic analysis is precisely that it is paratactic, that it involves, that is, the breaking down of the original statement into two distinct juxtaposed complete sentences. This is the direct source of all the problems in Davidson's theory. It is, in particular, what makes it difficult for him to account for quantification from the outside of the second sentence into it, and for the embedding of 'that'-clauses into one another (as in 'Aristotle says that Plato says that Socrates is a philosopher'). It is also this paratactic feature which forces into the first sentence an improbable metalinguistic reference – accomplished by the use of 'that' seen as a demonstrative – to a particular subsequent utterance of the person attributing the belief.[4] If this diagnosis is correct, what is needed, then, is a non-paratactic analysis of belief-sentences which would nevertheless preserve the virtues of Davidson's theory. And this goal can only be reached by seeing the 'that' of a belief-sentence. as being (or, at least, incorporating) a sentential connective, an approach which involves that neither the word 'that' in such a context nor the 'that'-clause taken as a whole are singular terms.

But what sort of connective would it be? One important feature we will have to take into account in answering this question is the obvious fact that the sentential clause that follows it is not in general asserted by the speaker. And it will be relevant, therefore, to recall that there is a whole class of logical connectives which also possess this very same feature, the familiar conditional connectives: 'if', 'only if', 'if and only if' (thereafter: 'iff'). A second obvious characteristic of a belief-sentence that should at this point enter into consideration is the fact that the left member of it ('A believes') does not itself constitute a complete sentence. From which it can straightforwardly be concluded that 'that' in such contexts is not *merely* a sentential connective since, superficially at least, it does not connect two sentences with one another.

Now, the one way I can see to keep all this together is to say that 'that' in a belief-sentence is, for all its superficial simplicity, an abbreviation for a complex and heterogeneous expression incorporating at least a neutral complement for the verb 'believe' (in order to give a sentential form to the left member) and a sentential connective of the conditional type. The hypothesis which then suggests itself is to read:

–   Noah believes that it rains

as:

–   Noah believes something which is true iff it rains.

This tentative analysis would account for the fact that the words 'it rains' are in such a context used in their ordinary way while, nevertheless, the corresponding sentence is not asserted by the speaker. And it would centrally accommodate Davidson's idea that what is in fact going on in the utterance of a belief-sentence is that the speaker attributes a belief to some agent and informs the hearer of the content of this belief by uttering himself in the last part of the belief-report a sentence of his own which he claims to have the same content as the belief he is currently attributing.

That's all very well. But the problem, of course, for this attractive solution is to specify the exact nature of the equivalence connective 'iff' involved in the analysis and to specify it in such a way that it accounts for the well-known logical and semantical peculiarities of indirect contexts. What we are looking for is a connective that would capture the exact sort of content-equivalence we want to posit between the belief that is talked about and the last clause of the belief-report. It cannot be, for example, the usual truth-functional 'iff' since, obviously, the equivalence in actual truth-values does not suffice for these purposes: 'Noah believes that it rains' cannot simply convey the information that Noah believes something which has, at the moment of utterance, the same truth-value as the sentence 'it rains'. And nor is strict equivalence sufficient here, at least not in every context: attributing to somebody the belief that two plus two are four is not merely crediting that person with an unidentified belief that is true in all possible worlds; it does precisely specify, among all necessarily true beliefs, which one it is that the person is supposed to have, and the strict 'iff' would not account for that.

In order to find out forms of equivalence that are stronger even than strict equivalence, what has to be taken into consideration is the internal characteristics of the relevant sentences or representations. Thus the French sentence 'David est malade' is more exactly translated in English by 'David is ill' than by 'David has some illness' because the former English sentence more exactly corresponds to the original French one with respect to its internal structure (same logical form, same number of terms, coreferential subjects, coextensive predicates, etc.). What is involved in this case is closely akin to what Carnap has called 'intensional isomorphism'.[5] But there are many possible degrees of isomorphisms. And to each one of these degrees there will correspond a special form of equivalence, each one of which will be stronger than strict equivalence. Some of these will depend, among other things, on the correspondence between the propositional connectives of the sentences: 'il pleut et il ne neige pas', for example, is more exactly translated by 'it rains and it does not snow' than by 'either it rains or it snows, but it does not snow' although both translations are strictly equivalent to one another. Some will depend, as in the 'David is ill' example, on the internal structure of the atomic sentences involved, and among them some will depend on the connotations of the terms as well as on their extensions, and so on. Any feature, in fact, that can correctly be considered as relevant, at one level or another, for the fine-grained syntactical or semantical analysis of discourses and representations will determine, in some context, at least one particular form of equivalence which will be stronger than strict equivalence.

Which one, then, among all these equivalences, do we need to capture for our analysis of belief-sentences? Once the question is thus formulated, it becomes obvious that the answer is: it depends! If the belief which is reported is about mathematical matters, for example, we might want to resort to one of the strongest possible forms of equivalence available. But in other contexts, only a much rougher brand will often suffice. The historian of philosophy who attributes some beliefs to, say, William of Ockham or Martin Heidegger is not always expected to reproduce, in so doing, the exact structure of the original sentences of his author. A few lines summary of hundreds of pages of the original text can even, in some contexts, be accepted as a completely accurate report. In other cases, only a relation of coreferentiality will be needed between some of the original terms and some of the reporter's. And so on. The form of equivalence which is required for the analysis of belief-reports widely varies with the contexts of such reports.

Does that striking variability irremediably compromise the type of analysis we have been considering of 'A believes that p' as 'A believes something which is true iff p'? I don't think so. We are already quite familiar with linguistic expressions which are context-sensitive with respect to their semantical import, aren't we? We call them 'indexicals'. Why, then, shouldn't the required 'iff' be such an indexical expression? Why, after all, shouldn't there exist *indexical connectives*? Connectives, that is, which are such that their precise contribution to the meaning of the complex sentences in which they occur would vary with the context according to some determinate rule? Such a rule would correspond for these connectives to what David Kaplan calls the 'character' of an indexical expression.[6] Think, for example, of the complex connective 'now that' (as in 'now that it rains, we are all wet'): its precise import depends on a determinate contextual feature, the moment of the utterance; it is, therefore, an indexical connective. What I am suggesting is that the connective we are looking for belongs to the same category.

Admittedly, its character would be much more complicated than that of 'now that', and formulating the relevant rule in a precise way is something I won't even try to do here. But the requirements, at any rate, are clear. One would have first to list (maybe in a recursive way) all the differents brands of equivalence that could be relevant for belief reporting, and then to specify the contextual features, if any, which in each case determine which one of these equivalences will be expressed by our indexical connective. This might look

like a formidable task. But whatever our approach is to belief-sentences, it
is a task that, under one guise or another, simply cannot be avoided. The
relevant phenomenon, here, is that the conditions of accuracy for a belief-report
are context-dependent. This is something that any philosopher interested in
such reports will have to deal with eventually, whether he or she is a nomi-
nalist or not. And dealing with it can only be done by identifying the different
relevant forms of equivalence and the contextual conditions under which
each one of these forms enters into consideration. With respect to such a
task, the only particularity of the approach I am advocating is that it resorts
to a special propositional connective. But this does not make the task more
difficult to any degree, on the contrary. Of course, it might turn out that there
is no razor-sharp rule involved here, but only very imprecise ones. But
this would not jeopardize the approach either. It would only show that the
connective we are looking for is a fuzzy one, which, after all, wouldn't be
very surprising.[7] The important point here is that all the well-known – and
contextually variable – limitations to the principle of extensionality that
are associated with belief-sentences will be accounted for in this theory
by resorting not to special *designata* for the names and predicates which
occur in the 'that'-clauses, but to a special context-sensitive equivalence
connective.

2. TANTUMSI

Unfortunately, the simple analysis we started from does not work for another
reason. It unfolds, remember, '*A* believes that *p*' into '*A* believes something
r and *r* is true iff *p*. But 'iff', however variable, can only mark an equivalence relation: symmetrical, that is, reflexive, and transitive. And if we
look at things a little more carefully, we will realize that the relation we
need in this position, between '*r* is true' and '*p*', is neither reflexive nor
symmetrical.

Consider, first, the following rather curious argument. From:

(1)   *A* believes something *r*,

it can be inferred by standard logical rules that:

(2)   *A* believes something *r* and *r* is true iff *r* is true,

for any sort of 'iff', however strong or however weak, as long as it marks
a reflexive relation. And according to our proposed analysis, (2) should
authorize:

(3)   *A* believes that *r* is true,

since the analysis contends that in such contexts 'that' is nothing but an
abbreviation for 'something *r* and *r* is true iff'. But the derivation of (3)

from (1) is surely unacceptable. In the first place, (3) is not even a complete sentence since it incorporates a free variable r. And even if we read it as:

(4) A believes that something is true,

we wouldn't want it to be derivable from (1). We should, in effect, be able to say that an agent has some belief without automatically being committed to attributing to this agent the belief that something *is true* (we might want to say, for example, that this particular agent doesn't even have the concept of 'truth').

This seems to prove that no reflexive connective will do the job here, at least not without some more explicit limitations as to the conditions of applicability of the proposed analysis. But the real trouble, in fact, is that the logical relation we need to mark in the considered position is not a symmetrical one. The left and right members of the 'r is true iff p' part of the *analysans* are simply not interchangeable with one another. The left member should *always* incorporate the truth-predicate and be of the form 'r is true', while the right member can be any closed sentence whatsoever. Of course, all this could be accounted for by keeping the equivalence connective and specifying some explicit constraints on the analysis as a whole and on its conditions of application rather than by resorting to a special non-symmetrical connective. But it will prove much easier, I think, in accordance with the politics we have been following so far, to have the connective itself do the largest part of the job. Admittedly, this will require, as we shall see, a hitherto unheard of sort of connective. But since the kind of linguistic constructions we are ultimately dealing with here are so frequent in practice (they might very well include, as we shall see below, all the so-called intensional constructions), the introduction of a single special connective for all these constructions is not, I hope, something that will look intolerably *ad hoc*.

Let us reflect at this point that when we acknowledged earlier the need for a strong sort of equivalence (strong isomorphism, for example) in order to account for (at least some) belief-reports, the equivalence we were talking about was supposed to hold between the content of the attributed belief r and that of the 'p'-clause of the belief-report, and not between a sentence of the form 'r is true' and the 'p'-clause. Suppose, for example, that the original agent A sincerely uttered in French '*deux plus deux font quatre*'; we could correctly say 'A believes that two plus two are four', and the strongest possible condition of accuracy for such a report is that 'two plus two are four' be exactly (syntactically and semantically) isomorphical with '*deux plus deux font quatre*', and not, of course, with anything like 'r is true', or even with " '*deux et deux font quatre*' is true".

So, if we want to pursue on the same tack we have been following so far, what we need in place of our original 'iff' is an asymmetrical and irreflexive conditional connective, something in fact which will be more akin to 'only if' than to 'iff', our proposed analysis of 'A believes that p' thus becoming something like: 'A believes something r and r is true only if p'. But, of course, this newly introduced 'only if' has to be very special. Not only does it have to be context-sensitive, as we already saw, but its correct conditions of use should also incorporate somehow a reference to another logical relation, a relation of equivalence this one, holding, in such case, between a certain *value* of the 'r'-variable and 'p' itself. I will conventionally mark this connective by the Latin word '*tantumsi*', and will now provide, in a sketchy way, a *metalinguistic* explication of its normal conditions of use.

The procedure I will follow for the introduction of this special connective will be to characterize both syntactically and semantically what I will call its basic use, and then to extend its acceptable uses beyond this basic one with the help of a single standard principle of existential generalization. Let us first stipulate, then, a rather restrictive rule of formation for the sentences in which '*tantumsi*' will be admitted according to its basic use. The rule I am thinking of for this purpose has three components:

(1)   the right member 'p' of a complex phrase of the form '– *tantumsi p*' is a closed sentence;

(2)   the left member of the same complex phrase normally has the form 'r is true', where 'r' is a bound variable, the sole acceptable values of which are representations capable of truth and falsity (sentences, mental representations, and so on);[8]

(3)   the whole 'r is true *tantumsi p*'-clause is within the range of a quantifier binding the variable 'r'.

Secondly, truth-conditions for sentences of the type just characterized can be metalinguistically given according to the following schema:

–   A sentence of the form '$(\exists r) \ldots r$ is true *tantumsi p*' is true iff there is at least one admissible value of 'r' which is *equivalent* to 'p'.

This locates the equivalence where we want it, that is: between the original representation we are talking about (the value of 'r') and 'p', rather than between 'r is true' and 'p'.

But now, let us remember that '*tantumsi*' is supposed to be an indexical connective. How is this idea cashed out in the truth-conditions we have just given? Well, as we have seen, what is expected to vary with the context of utterance is precisely the type of equivalence that is required between 'p' and the value of 'r'. Let us suppose, therefore, that there exists a denumerable list of such types of equivalence, and that each one is identified by a numerical subscript: $equivalence_0$, for example, could be material equivalence; $equivalence_1$ strict equivalence; and so on for all these stronger forms of equivalence I have alluded to earlier. '*Tantumsi*' could then be derivedly broken down into exactly as many forms, each one of which being identified by the numerical subscript corresponding to the type of equivalence its basic truth-conditions require. Thus, we would have:

–   A sentence of the form '$(\exists r) \ldots r$ is true $tantumsi_1\ p$' is true iff there is at least one admissible value of 'r' which is equivalent$_1$ to 'p'.

The general formulation of the truth-conditions for such sentences would then become:

–   A sentence of the form '$(\exists r) \ldots r$ is true $tantumsi_n\ p$' is true iff there is at least one admissible value of 'r' which is equivalent$_n$ to 'p'.

And the character (in Kaplan's sense) of the general indexical connective $tantumsi_n$ will then be the rule – whatever it is – according to which the context of utterance determines – however imprecisely – the numerical value of the subscript 'n'.

So much for the basic use. We can now, starting from there, extend the acceptable linguistic contexts for '$tantumsi$' by allowing in *some cases* for existential generalization from within the right member of a '$tantumsi$'-sentence of the basic sort, so that we could validly, in these cases, move, for example, from a basic sentence of the form '$(\exists r) \ldots r$ is true $tantumsi_n\ A$ is F' to a new '$tantumsi$'-sentence of the form '$(\exists r)(\exists x) \ldots r$ is true $tantumsi_n\ x$ is F'. Which cases exactly will that be? Well, I won't attempt to characterize them with any precision here since that would bring us much too far into general considerations about reference and quantification. But one thing I want to say is that these will be exactly the same kind of cases as those where existential generalization is warranted from within any normal *conditional* clause. Consider, for example, the following conditional sentence:

If Peter comes, Mary will be happy.

It is clear that for *some* tokens of the proper name 'Peter' in such a context – let us call them 'directly referential occurrences' –, the inference will be good from such a sentence to:

$(\exists x)$ if x comes, Mary will be happy.

The relevant principle of existential generalization, then, will be something like:

If a term has a directly referential occurrence within a conditional clause, existential generalization is warranted with regard to that term; otherwise not.

Of course, we should, in order to be complete, specify what exactly a directly referential occurrence of a term is, but in whichever way we do it, the important point now is that this very same principle will straightforwardly apply to '$tantumsi$'-clauses since '$tantumsi$', however special it is, is nevertheless a conditional connective.

All acceptable cases of existential quantification from the outside into belief-contexts will be accounted for in this manner. In other words: no such quantification will be considered as yielding a true affirmative sentence unless this sentence could, at least in principle, be inferred, with the help of this very same rule of existential generalization for conditional clauses, from a true '*tantumsi*'-sentence of the basic sort. As to the question when exactly is a directly referential occurrence of a term admissible within a basic '*tantumsi*'-clause, that will utterly depend, in accordance with the already stated truth-conditions for the basic '*tantumsi*'-sentences, on the brand of equivalence which is required in the particular case under consideration. The important point, for the moment, is that the whole phenomenon of quantifying into indirect contexts thus appears as just a special case of quantifying into conditional contexts, a point which has not often been made in the literature.

On the other hand, our characterizations of both the basic and the extended uses of '*tantumsi*' make it clear that this peculiar connective would never occur but in the immediate neighborhood of the truth-predicate. And this suffices to explain why the whole 'something which is true *tantumsi*' part of unfolded belief-sentences is at the surface level of ordinary languages abbreviated into one single, preferably short, word ('that' in English or '*que*' in French, for example): there can simply be no other use for '*tantumsi*' apart from such surroundings and, therefore, nothing would be gained in the general economy of ordinary languages by independently lexicalizing it.

3. PROSPECTS AND ADVANTAGES

The proposed analysis, then, is now to read '*A* believes that *p*' as:

$(\exists r)$ *A* believes $r \wedge r$ is true $tantumsi_n$ *p*.

Admittedly this provides no explication at all of what believing is, except that it is supposed to be a relation between some agents and things, such as sentences or representations, that can have truth-values. Since what we are exploring here is, from the outset, a nominalistic approach, these sentences or representations should always be singular tokens rather than types, and this, of course, will have important consequences on the ways according to which it will be admissible to quantify over them. But what exactly this belief-relation is is something I won't go into here. Whether, for example, the relevant representational token should be – or should be able to be – in some way produced by the agent, or whether it could under some circumstances be utterly independent from him or her is a question I will leave open for the moment. The task I have undertaken was solely to elucidate up to a certain degree – and from a stringently nominalistic standpoint – the *logical form* of belief-sentences, and this does not require at this point any detailed explication of the psychological or physical nature of the central belief-relation that is called for.

The only further refinement that might be advisable with respect to this belief-relation in the context of our analysis will be to replace the ordinary verb 'believes' by a conventionally fabricated phrase: 'believes-true'; so that the *analysans* of '*A* believes that *p*' will now become (and this will be its final form):

$(\exists r)$ *A* believes-true $r \wedge r$ is true *tantumsi$_n$ p.*

It will thus be more conspicuous that the verb we need in this position does not have the same grammar as the corresponding ordinary one (it does not, in particular, accept 'that'-clauses as complements) and that the required relation crucially has something to do with the truth-value of the representational token that terminates it.

Above all, this last amendment opens the way to a welcome generalization of our analysis to other types of attitude ascriptions. '*A* desires that *p*', for example, can now, in the same vein be read as:

$(\exists r)$ *A* desires-true $r \wedge r$ is true *tantumsi$_n$ p.*

And '*A* intends that *p*' will become:

$(\exists r)$ *A* intends-true $r \wedge r$ is true *tantumsi$_n$ p.*

And so on. Even illocutionary verbs will be tractable along the same lines. '*A* orders that *p*' will be analyzed as:

$(\exists r)$ *A* orders-true $r \wedge r$ is true *tantumsi$_n$ p,*

and '*A* promises that *p*' as:

$(\exists r)$ *A* promises-true $r \wedge r$ is true *tantumsi$_n$ p.*

Such a generalization could not very easily have been accomplished had we decided to keep the corresponding ordinary verbs into the *analysans*, because while it might sound plausible to say that the complements of the ordinary verb 'to believe' normally denote representations even when they do not have the forms of 'that'-clauses, this is certainly not so with the verbs 'to desire', 'to intend', 'to order' and so on. The value of the variable '*x*' in '$(\exists x)$ *A* desires *x*' (a banana, for example) or in '$(\exists x)$ *A* orders *x*' (an assassination, let's say) is normally not a representation at all, let alone a representational token. The introduction into the *analysans* of a special '*X*-true' verb for every relevant ordinary verb *X* smooths away this superficial difficulty and makes it conspicuous that what is needed in each case is, just like in the belief case, a particular relation of the agent to some sentence or representation. Once this is done, the generalization of our proposed analysis is completely natural.

We might even, along the same tack, consider a further generalization of this type of analysis to all intensional contexts. Modal predicates such as 'necessary' or 'possible' will then give place to corresponding phrases such as 'necessarily-true' and 'possibly-true'; and the corresponding modal sentences such as 'it is necessary that *p*' or 'it is possible that *p*' will respectively be interpreted as:

$(\forall r)$ if r is true *tantumsi$_n$ p,* then r is necessarily-true, and:

$(\forall r)$ if r is true $tantumsi_n$ p, then r is possibly-true.

The logical form of these new sentences differ from that of the previous ones by the use of universal quantification instead of existential quantification. The reason for this is that once we have accepted that the values of r are tokens rather then types, we have, accordingly, to interpret sentences such as 'it is necessary that p' as being about all tokens, if any, of a certain sort rather than about one particular existing token as in the case of belief or desire attribution. This nominalistic analysis of such modal sentences thus avoids being committed to the actual existence of particular representational tokens of any sort. Universal quantification, in this case as in any other, calls for a subsequent *conditional* clause 'if . . . , then', and thus nothing is asserted by such sentences about the actual existence of anything.

As far as I can see, this sort of analysis satisfies all the usual requirements in the field. It accounts in principle (although I have not spelled out the details) for the well-known limitations to substitutivity in indirect contexts and, at the same time, for the contextually sensitive variations of these limitations. It also shows how to deal with quantification from the outside into indirect contexts. Since it involves no explicit mention of the sentence or representation which is referred to (the one which is supposed to be the value of the variable r), it passes without problems the Langford–Church translation test. Since, on the other hand, it does not require special primitive metalinguistic predicates, it also satisfies Davidson's learnability condition and it most naturally complies with the principle of compositionality. And it easily allows, as well, for the embedding of indirect contexts into other indirect contexts.[9]

As to the inferences involving belief-sentences (or, for that matter, other intensional sentences) that we want to consider as valid, they will, as far as I can see, turn out to be correctly admitted as such under this analysis. Consider, for example, the following inference, which, according to some authors, should force us to treat the 'that'-clauses of belief-sentences as singular terms:[10]

– A believes whatever B believes
– B believes that it rains
_____

–   Therefore: A believes that it rains.

Taking into account normal constraints for quantification over particular tokens, this inference should be represented under our approach as:

– $(\forall r)$ If B believes-true r, then $[(\exists r')$ r' is equivalent$_n$ to $r \wedge A$ believes-true $r']$
– $(\exists r)$ B believes-true $r \wedge r$ is true $tantumsi_n$ it rains
_____

–   therefore: $(\exists r')$ A believes-true $r' \wedge r'$ is true $tantumsi_n$ it rains.

And this will certainly turn out to be a valid inference although the complete proof of it would require some intermediate steps involving the (uncontroversial) principle that:

If $(\exists r)(\exists r')$ such that r is equivalent$_n$ to r', then if r is true *tantumsi$_n$ p*, r' is also true *tantumsi$_n$ p*,

a principle, in fact, which is easily derived from the very truth-conditions I have given above for '*tantumsi*'-sentences.

The prospects, then, are quite good for this sort of nominalistic approach. If its generalized version should ultimately turn out to be tenable, it would efficiently neutralize, for example, both George Bealer's argument that nominalism in general fails because it cannot account for quantification into intensional contexts,[11] and Stephen Schiffer's argument that compositional semantics in general is impossible because it cannot account for belief-contexts.[12] The one crucial premise of both these arguments that is here abandoned is the idea that the 'that'-clauses of indirect and intensional contexts should be in any promising compositional semantics, whether nominalistic or not, interpreted as singular terms. A way has been shown here of doing away with this premise. The procedure has a price, of course: the introduction of a new sort of (unlexicalized) indexical logical operator such as '*tantumsi*'. But if the manoeuvre succeeds, this might not be too high a price to pay, considering all the philosophical jobs this single special connective is expected to accomplish.

Université du Québec à Trois-Rivières

NOTES

* This paper stems from a research project supported by the Canadian Social Sciences and Humanities Research Council and by Quebec's FCAR Fund, both of which I want to express my gratitude to. Thanks are also due to Élizabeth Karger, Grzegorz Malinowski, François Récanati, Philippe de Rouilhan, and Daniel Vanderveken for helpful remarks and criticisms on a previous version of the paper.
[1] On the Langford–Church translation test and its critical application to quotational approaches of belief-sentences, see Alonzo Church, 'On Carnap's analysis of statements of assertion and belief', *Analysis* 10 (1950), 97–99.
[2] The metalinguistic approach to belief-sentences has been proposed by Israel Scheffler in 'An inscriptional approach to indirect quotation', *Analysis* 14 (1954), 83–90. For its criticism from the point of view of the learnability requirement, see George Bealer, *Quality and Concept* (Oxford, Clarendon Press, 1982), 28–29.
[3] The paratactic account has first been proposed by Donald Davidson as a theory of indirect contexts in 'On saying that', *Synthese* 19 (1968), 130–146 (reprinted in the author's collection of essays *Inquiries into Truth and Interpretation*, Oxford, Clarendon Press, 1984, 93–108). It has been extensively discussed since then, for example, by J. A. Foster in 'Meaning and truth theory' (in *Truth and Meaning*, ed. by G. Evans and J. McDowell, Oxford, Clarendon Press, 1976, 1–32), Brian Loar in 'Two theories of meaning' (also in *Truth and Meaning*, 138–161), and Stephen Schiffer in chapter 5 of *Remnants of Meaning* (Cambridge, Mass, The MIT Press, 1987, 111–138). It seems that the theory cannot without importantly being revised account for

typical logical phenomena such as quantification from the outside in indirect contexts (think of a sentence such as 'Galileo said of a certain person that she baked terrific lasagna', an example adduced by Schiffer in *Remnants of Meaning*, p. 125), or the embedding of a 'that'-clause into another one ('Laplace said that Galileo said that the earth moves', another example from Schiffer's book, p. 132). The theory, though, has many extremely interesting features which I will try to recapture in my own approach.

[4] Stephen Schiffer, in particular, in *Remnants of Meaning* (*op. cit.*, 133–137) has insisted on an important drawback of supposing the intervention of such a metalinguistic demonstrative reference. It is a consequence of this approach, he explains, that the hearer of a belief-sentence could perfectly well understand what is being *asserted* by the speaker without understanding the content that is being attributed by him to the belief he is reporting. Suppose the hearer of 'Sam said that flounders snore' fails to understand the 'flounders snore' part, she would nevertheless, in Davidson's theory, understand everything that is asserted by the speaker, since she would correctly understand the 'Sam said' part and also correctly identify the referent of the demonstrative 'that' (while she wouldn't understand the words 'flounders snore', she would nevertheless correctly identify them as what the first sentence – the only one, remember, which is asserted by the speaker – is about). But this consequence, Schiffer rightly remarks, is unacceptable.

[5] See Rudolf Carnap, *Meaning and Necessity*, 2nd ed., Chicago, The University of Chicago Press, 1956, 56–59.

[6] See David Kaplan, 'On the logic of demonstratives', *Journal of Philosophical Logic* 8 (1978), 81–98.

[7] Stephen Stich, for example, has forcefully insisted upon the impreciseness of the conditions of accuracy for belief-reports in *From Folk Psychology to Cognitive Science. The Case Against Belief*, Cambridge, Mass., The MIT Press, 1983.

[8] This rules out from the left member of the sentence metalinguistic descriptions of sentences or representations (cases, for example, such as 'Janet's last sentence is true *tantumsi p*'). This restriction should ideally be relaxed. But for the sake of simplification, I will completely neglect this eventuality here.

[9] Thus 'Aristotle believes that Plato believes that Socrates is a philosopher' will be read as: '$(\exists r)$ [Aristotle believes-true $r \wedge r$ is true $tantumsi_n$ (($\exists r'$) Plato believes-true $r' \wedge r'$ is true $tantumsi_n$ Socrates is a philosopher)]'.

[10] See, for example, G. Bealer, *Quality and Concept*, Oxford, Clarendon Press, 1982, 23–25; and S. Schiffer, 'Belief ascription', *Journal of Philosophy* 89, 10, Oct. 1992, 504–505.

[11] This is the main theme of Bealer's recent paper, 'Universals', *Journal of Philosophy* 90, 1, Jan. 1993, 5–32.

[12] This is one of the central theses of Schiffer's book, *Remnants of Meaning*, Cambridge, Mass., The MIT Press, 1987.

MARTIN MONTMINY

VERIFICATIONISM AND THE
MOLECULAR VIEW OF LANGUAGE*

1. MEANING AND METAPHYSICS

In Section 5 of 'Two Dogmas of Empiricism', W. V. Quine considered the possibility of drawing the analytic-synthetic distinction by means of a verificationist model of meaning. Quine regarded such an approach to meaning as more promising than the ones he had examined in the first four sections, since it explicitly aimed at giving an empirical sense to the concepts of meaning and analyticity, by grounding these concepts on the observation of how actual theories are related to experience. Under the verificationist model, the meaning of a statement lies in the observations that would support or refute it. Quine's "countersuggestion" against the possibility of defining strict meanings by this approach was *holism*, namely that "our statements about the external world face the tribunal of sense experience not individually but only as a corporate body" (Quine, 1951, p. 41).

Michael Dummett's verificationist theory of meaning might be seen as a reply to Quine's criticisms: not only does Dummett claim that we can make sense of the concepts of meaning and analyticity within such an approach, but he also contends that the elaboration of a meaning-theory for our language would make it possible to provide answers to questions about the type of logic we should adopt and to metaphysical problems concerning the ontological status we should attach to certain classes of entities or events.[1] Actually, Dummett proposes that instead of addressing metaphysical questions directly (in which case we are confronted with various relatively vague rival metaphors between which, lacking any precise criterion, we are not in a position to choose), we first try to settle the question of what is the correct model of meaning for the statements of the disputed class.[2] For Dummett, the task of constructing a meaning-theory may in principle be taken on without metaphysical prejudices: a theory succeeds inasmuch as it provides an account which accords with the actual use of the language by speakers.[3]

The meaning of a statement determines in which circumstances it may be correctly asserted (or denied): hence, meaning determines in which types of inference the statement may be correctly used. A disagreement concerning the truth of statements of a certain form or the validity of certain types of inference thus reveals a deep disagreement concerning the kinds of meaning possessed by the statements involved.[4] A deep disagreement about meaning arises when one (or each) of the parties claims that the meaning ascriptions to statements proposed by the other party are incoherent. Hence, this kind of disagreement does not merely concern the question whether a given meaning is, as a matter of fact, attached to a certain expression according to its standard

279

M. Marion and R. S. Cohen (eds.), Québec Studies in the Philosophy of Science II, 279–293.
© 1996 *Kluwer Academic Publishers.*

use (such a disagreement would simply be verbal), but rests on divergent conceptions of what meaning is.

For Dummett, a theory of meaning is a theory of understanding: it must account for what a speaker knows when he knows the meanings of sentences and expressions of his language. It must thus give an explicit account of what knowledge of the language consists in. Of course, the knowledge a given speaker has of his language may not consist in the explicit knowledge of the propositions of a meaning-theory: his knowledge of these propositions is rather, in most of the cases, implicit, and is manifested by a practical ability.[5] The task of a meaning-theory is thus to give a theoretical representation of a complex practical ability which anyone must master in order to understand and speak the language.

A meaning-theory has two parts.[6] The first consists of a semantic theory (or theory of reference) and a theory of sense. The semantic theory has to specify, for each sentence of the language, the conditions of application of the notion taken as central.[7] It also has to specify what types of semantic value the different expressions of the language possess.[8] The theory of sense should lay down in what a speaker's knowledge of the semantic theory consists, by associating specific practical abilities with certain propositions of the semantic theory. To know the sense of an expression is to know everything that is necessary in order to determine its semantic value.[9] The second part of a meaning-theory is the theory of force. The theory of force describes what a speaker must know, in addition to the semantic theory, in order to understand all the relevant aspects of use. It must give an account of the various types of linguistic act which may be performed by uttering a given sentence. A sentence may be used to make an assertion, give a command, ask a question, etc. The aspect of a sentence which serves to indicate which type of linguistic act is being performed is the *force* of that sentence.[10]

Dummett subscribes to the principle that *meaning is use*, that is, meaning depends entirely on use: a speaker's understanding of a sentence must be capable of being fully manifested by his linguistic practice. Dummett's criticism of realism rests essentially on this principle. Let me briefly present this criticism, which concerns more precisely the realist view which identifies the meaning of a sentence with its truth conditions, and according to which the central notion of a theory of meaning is the classical concept of truth. For the realist, every (unambiguous) sentence possesses a definite truth-value independently of our capacity to discover it. In other words, the realist adheres to the principle of bivalence, which says that any sentence, decidable[11] or not, is determinately either true or false. Dummett points out that it is not possible to explain the knowledge of the truth conditions of undecidable sentences in terms of the ability to recognize that those conditions of undecidable sentences in terms of the ability to recognize that those conditions are fulfilled or not, since such a recognition is in principle impossible. In consequence, the realist model (according to which each sentence has a meaning given by its truth conditions and a determined truth value independent of our knowledge) violates the principle that meaning is use, since it cannot state what constitutes a speaker's knowledge of the condition for the truth of every sentence, and is therefore seriously deficient.[12]

2. DUMMETT'S VERIFICATIONISM

Dummett adheres enthusiastically to the verificationist model of language proposed by Quine (1951), according to which language forms an articulated structure of interconnected sentences upon which experience impinges only at the periphery. Dummett endorses Quine's criticism concerning the logical positivists' type of verificationism: the verification of a non-peripheral (or theoretical sentence) cannot be represented by a simple sequence of sense-perceptions. Dummett's verificationist theory takes account of the fact that the process of establishing a staement as true does not always consist in a sequence of bare sense experiences, and emphasizes the existence of inferential links between sentences.

A verificationist theory takes as central the notion of verification: to know the meaning of a sentence is to be in a position to recognize the circumstances in which it is conclusively verified.[13] In a verificationist theory, the semantic theory and the theory of sense somewhat merge: the condition for the application of the central notion is such that a speaker can directly manifest his knowledge of it by the way he actually uses the language.[14] Let us note however that Dummett has also proposed to account for meaning through a pragmatist theory, which takes as central the notion of *consequences*: to know the meaning of a sentence consists in knowing what difference believing in its truth would make, from the speaker's point of view.[15] Actually, Dummett regards the verificationist and pragmatist approaches not as rival theories, but as complementary aspects of the same enterprise.[16] It is however fundamental to him that the two features of meaning (namely verification and consequences) should not be mixed within the same theory.

Let us come back to the verificationist theory. According to this approach, the meaning of a given statement is to be explained in terms of its verification conditions, which must be accessible to any competent speaker. In general, what establishes a sentence as true or false consists in a mixture of observation and inference (deductive or inductive). In some limiting cases, the verification consists in making an observation alone: understanding an observation sentence consists in the ability to recognize the experience or sequence of experiences which confirms it. In other cases, only inference is needed to establish the statement as true: for Dummett, the truth of a mathematical statement is established by purely linguistic operations, and does not involve, even indirectly, any observation. In general, understanding a given sentence S requires the ability to recognize the validity of an argument whose premises are observation sentences and whose conclusion is S. To know the meaning of the sentence 'The Earth is spherical' for example, does not consist in associating with it a complex sense experience which would constitute its verification; it consists rather in understanding the means by which we might in practice be led to accept such a sentence as conclusively established. Therefore, with the exception of observation sentences, sentences are not correlated with experiences individually, but collectively, along with other sentences with which they are connected.

Dummett describes his model as "molecular", since the understanding of a sentence may depend on the understanding of a rather important fragment of the language to which the sentence belongs.[17] The model is neither *atomistic* nor *holistic*. Let me explain how an atomistic conception of meaning correlates with each axiom of the theory giving the meaning of a word a specific ability which would constitute the knowledge of the meaning of this word. For Dummett, the basic semantic unit is the sentence, and knowledge of an axiom giving the meaning of a word is rather manifested in the use of whole sentences:[18] understanding a word consists in the ability to understand how it contributes to the meaning of *certain* sentences in which it appears. For every expression, one must distinguish between two kinds of sentences containing it: sentences whose understanding *precedes* that of the expression (understanding the expression thus consists in the ability to understand those sentences) and sentences whose understanding *requires* that of the expression (an understanding of the expression, combined with an understanding of the other constituents of the sentence and of the sentence structure yields an understanding of the sentence).

The model proposed by Dummett is not holistic, since it does not deny that for every individual sentence, there is a specific ability constituting the knowledge of its meaning. Each sentence of the language possesses its own meaning, and such a meaning may be grasped without a knowledge of the entire language: on the molecular view, for each sentence, the knowledge of a determinate fragment of the language suffices for a complete understanding of that sentence. The molecular model of meaning is incompatible with the rejection of the analytic-synthetic distinction: the verification conditions of a sentence determine in which circumstances it may rightly be judged as true (or false), and provide a clear criterion for separating factual matters from semantic matters.[19]

The molecular view of language imposes a hierarchical structure on the sentences of the language. Dummett sometimes talks of a gradation of sentences in terms of their *complexity*: the language fragment whose understanding is needed in order to understand a given sentence should contain only sentences of lower or equal complexity.[20] It should be noted that complexity here is not logical complexity, since the understanding of a sentence of a certain (apparent) logical complexity may depend on the understanding of sentences of superior logical complexity: "We therefore cannot assess complexity solely in terms of overt logical complexity, but must extend it so as to count one sentence as more complex than another if an understanding of it depends upon an understanding of the other" (1987, p. 249). The notion of complexity should be construed as a general notion defined in terms of dependance relative to understanding. According to this notion, the sentences of lowest complexity are observation sentences.

To ensure the coherence of his hierarchical model, Dummett must distinguish between two types of verification, *direct* and *indirect* verifications. In a meaning-theory of a verificationist type, a direct verification provides the meaning of the verified sentence, and our understanding of the sentence consists in the capacity to recognize such a verification of it. The direct verification of a sentence is established in accordance with the composition of the sentence, and must always proceed from less complex premisses to a more complex conclusion. In the case of mathematical statements, Dummett uses the expression 'canonical proof' to designate a direct verification. All the valid arguments are not constitutive of the meaning of their conclusion, since some are *indirect* means of obtaining it. An indirect verification is an argument whose understanding is not directly associated with the meaning of the verified statement. We rather say that it is the validity of such an argument which flows from the meanings of the expressions which appear essentially in it.

For a systematic account of our language to be possible, Dummett writes, our linguistic practice must be coherent: more precisely, there must be *harmony* between the different aspects of use, that is, between the conclusions reached by direct verifications and those obtained by indirect means. Such a requirement may be formulated as follows: the language as a whole must be a conservative extension of the fragment of it which is necessary for the direct verification of any sentence. That is, take a fragmentary language containing only the expressions which occur in a given sentence and others whose understanding is necessary to the understanding of these expressions: it must not be possible to establish with the whole language any sentence of the fragmentary language which could not already be established in that fragmentary language.[21]

Clearly, because of the constraints that it imposes on use, Dummett's view on meaning has a revisionist character. All the aspects of use are not sacrosanct and some may be criticized: "An existing practice in the use of a certain fragment of language is capable of being subjected to criticism if it is impossible to systematise it, that is, to frame a model whereby each sentence carries a determinate content which can, in turn, be explained in terms of the use of that sentence" (1973b, p. 220). A particular linguistic practice is justified (and therefore immune from criticism) if it is faithful to the meanings of the sentences involved, otherwise it must be revised. For example, if we conceive the meanings of logical constants as given by their contribution to certain specific forms of inference, it might be the case that other forms of inference (for example, ones that are accepted in classical logic but rejected by intuitionistic logic) do not conform to these meanings, and should therefore be abandoned.[22]

3. TWO DIFFICULTIES

Dummett's approach raises several difficulties. In the following, I shall present two of them that I find particularly noteworthy, and in the final section I will try to show why Dummett's view does not provide us with any general means or method for drawing an analytic-synthetic distinction. The two difficulties I want to bring up concern the question of how to determine when a given statement may be considered true or correctly asserted. According to a "radical" version of verificationism, a statement is true or may be correctly asserted only if we have a *direct* and *conclusive* verification of it (or some effective method for obtaining it). The two difficulties concern respectively these two conditions for a statement to be true or correctly assertable. Let me start with the conclusive character of the verification. The problem is that there are many statements which are (generally) admitted and play an important role in science or in practical reasoning, but which are not susceptible of being established conclusively. Admitting that some of these statements might nevertheless be correctly asserted would greatly complicate Dummett's theory: different degrees of verification should then be allowed. Dummett is aware of this problem: "[A verificationist theory of meaning] must operate, not with a simple notion of direct verification, but with a more general notion of canonical grounds, qualified by what, if anything, is counted as overthrowing them" (1987, p. 284).[23]

But such a general notion of verification is problematic. The situation could be put this way. Concerning a given kind of sentences for which there is no conclusive evidence, two alternatives should be considered: either the notion of conclusive verification should be relaxed so that these sentences could be correctly asserted and serve in inferences, or the use of these sentences should be revised to make it agree with the meanings attached to them by the meaning-theory. In order to decide between these two alternatives, Dummett proposes to regard a statement as conclusively established (in some new sense of the term) if, given a verification of it, "we could not conceive of subsequently encountering *stronger* counter-evidence" (1987, p. 284). However, this new notion of conclusive verification is unsatisfactory, appealing as it does to our intuitions (that is, to what we *conceive*) concerning eventual data. In other words, the distinction between sufficiently and insufficiently conclusive verification may vary a lot from one speaker to the other.

Besides, some *a priori* criterion for drawing such a distinction would appear highly questionable: I do not see how the new notion of conclusive verification could be formulated without taking into consideration the methods actually used in contemporary science for establishing theoretical statements. For example, among the present-day competing models in Cosmology or Quantum Field Theory, none may be said to have been, strictly speaking, conclusively established. New subtle and hitherto unthought-of experimental devices are constantly designed in order to back up or refute some of these models. Consequently, the question of how well supported a given model or theoretical hypothesis is, relatively to eventual counter-evidence, is highly contextual and subject to change, given that new practically possible ways of testing it may always be discovered. This would seem to have unfortunate consequences for Dummett's view, since it would tend to make unclear and partially arbitrary in their application the notions of truth and assertability conditions.[24]

Let us now examine the second difficulty. It may be the case that a given sentence is established by some indirect means, while it is practically impossible to establish it by a direct verification.[25] For example, there are many ways to establish as true a sentence like 'The dog was in the house at noon' (it may be done by inspecting some clues, from the knowledge of the owner's habits, etc.), which do not constitute a direct verification of it, that is, the observation of the dog in the house at noon. In some circumstances, only certain indirect methods may be available, while a direct verification is impossible. This seems to go against the constraint of harmony, which says that it should not be possible to establish indirectly a sentence which cannot be established directly. These considerations force Dummett to relax the constraint of harmony: "[A] sufficient condition for [the correctness of an assertion] is that there exist effective means by which, at the relevant time, someone appropriately situated *could have* converted observations that were actually made into a verification of the statement asserted" (1991, p. 268).

Such a relaxation on the demand of harmony implies a "stretching" of the notion of truth, and a certain concession towards realism, since we now admit that there might be a certain gap between truth and the recognition of a direct verification. The "new" notion of truth is explained in counterfactual terms, that is, in terms of what *would have been observed* if one had been in a position to do so. If we take those remarks seriously, it becomes difficult strictly to maintain, within a molecular view of meaning, the principle that meaning is use. Indeed, we may now attribute to a given speaker the knowledge of the meaning of a certain sentence (since he may correctly assert it in a given circumstance) despite his not being in a position to manifest his knowledge by a specific ability. In the case of obsrevation sentences (or sentences close to the periphery), this should not be a great worry, since the speaker might be able to manifest his knowledge of the meanings of those sentences in *other* circumstances, that is in recognizing, in such circumstances, the experiences which directly establish their truth. However, when it comes to theoretical sentences (or sentences further from the periphery), the problem is more acute: it might very well be the case that a speaker (or even a community of speakers) is *never* in a position to manifest his knowledge of the meaning of the sentence (since a direct verification of it is, for him or anyone, out of reach),[26] even though, it seems, he may still correctly assert it (since by hypothesis, the sentence can be established by indirect means and could have been directly verified).

This criticism is analogous to Dummett's criticism of realism. Remember that Dummett criticizes the realist view for not being able to account for the meanings of individual statements in terms of speakers' specific abilities. The main culprit is the principle of bivalence or, if one prefers, the classical concept of truth. In place of this concept, Dummett introduces a concept of truth which is defined in terms of the central notion of the meaning-theory, namely direct verification. But, as we have seen, since we cannot simply identify the truth of a sentence with its having been directly verified, truth must be explained in terms of what has or could have been directly established if one were appropriately situated. However, by allowing 'S is true', for some sentences S, to be defined in terms of counterfactual conditionals, Dummett's notion of truth becomes vulnerable to the same kind of criticism as the classical one. The difficulty is simply that the connection between the speaker's grasp of the conditions for the truth of S and the actual means he uses to recognize it as true or false is severed, precisely when the antecedent of the corresponding conditional remains unfulfilled.

4. THE ANALYTIC-SYNTHETIC DISTINCTION

We now come to the question of whether Dummett's approach enables us to draw an analytic-synthetic distinction. The possibility of establishing such a distinction by means of a verificationist theory rests essentially on the possibility of distinguishing, for any statement of the language, a direct verification of it from an indirect one. Unfortunately, Dummett gives no indication of how to establish such a distinction in general. He even admits that he is not in a position to specify the distinction for statements of scientific theories:

A historical conjecture may commend itself as neatly explaining what would otherwise be puzzling: but there seems a reasonable distinction between this case and finding direct historical evidence for the conjecture; if so, we may maintain that only the latter goes to determine its content. With scientific theories, on the other hand, the distinction appears to vanish. I have to confess that I do not know how a verificationist theory of meaning ought to treat scientific statements. (1987, pp. 284–285)[27]

It would nevertheless be interesting to examine the way Dummett deals with the meanings of logical laws and constants, since he has written a lot more on this topic. In Chapters 11–13 of *The Logical Basis of Metaphysics*, he presents a model for the meanings of logical constants. According to Dummett, let us recall, the justification of logical laws must ultimately rest on the meanings of logical constants. Hence, the grounds for adopting a logic (that is, a class of laws and rules of inference) consist essentially in the feasibility of a systematic and coherent account of that logic in terms of the meanings of its logical constants. For Dummett, the meanings of logical constants are specified by their introduction and elimination rules. More precisely, according to a verificationist theory, the introduction rules provide the meanings of logical constants, while for a pragmatist theory, the meanings are represented by the elimination rules.

Dummett imposes certain general requirements on introduction and elimination rules: these rules must comply with uniqueness[28] and stability. The notion of stability is defined in the following way. Let us suppose that we adopt a given set I of introduction rules, and take it as giving the meanings of the logical constants. From these rules, we may propose a characterization of what is a valid argument,[29] and we may determine a set E of elimination rules which come out as valid under this characterization (that is, the rules which are in harmony with the introduction rules). Next, by regarding the set E of elimination rules as giving the meanings of the constants, we may define what is a valid argument, and determine a set I' of introduction rules which are valid under this definition. If the two sets I and I' come out as identical, then the requirement of stability is satisfied.[30]

Certainly a logical system which complies with stability presents genuine advantages, like symmetry and elegance. However, there is no *a priori* reason to think that our logic should absolutely satisfy this requirement. In other words, I do not see any reason why the demand for stability should be the ultimate one to consider (apart from uniqueness). One may think of pragmatic constraints like simplicity, convenience, etc., and, of course, agreement with our actual practices. Hence, it may very well be the case that the different demands we try to meet are antagonistic to the point that we are forced to renounce the *complete* satisfaction of each. Besides, nothing shows that conformity with stability is related to the *meanings* of logical constants. Depending on the logical constant, our intuitions oscillate between associating its meaning with its introduction rule or with its elimination rule. However, there is no *a priori* reason why we should assume that the logical constants should get their meanings entirely from introduction (or elimination) rules.[31] It would be pertinent at this point to recall Quine's criticism of what he called the *linguistic doctrine of logical truth*. Instead of claiming, as this doctrine does, that the fundamental laws of logic rest on meaning, one could contend that they rather reflect certain properties that everything possesses:

Consider [. . .] the logical truth 'Everything is self-identical', or '$(x)(x = x)$'. We *can* say that it depends for its truth on traits of the language (specifically on the usage of '='), and not on traits of the subject matter; but we can also say, alternatively, that it depends on an obvious trait, viz., self-identity, of its subject matter, viz., everything. The tendency of our present reflections is that there is no difference. (Quine, 1960, p. 113)

In other words, since we do not possess any criterion to decide between these two views, both lose their alleged explanatory power.

A major difficulty of Dummett's approach is that it seems to presuppose a *transcendental* notion of meaning.[32] In order to claim, as he does, that a meaning-theory would give us the means to determine what is the adequate logic, and would justify some of our linguistic practices and criticize others, one must appeal to an objective notion of meaning which is, up to a certain point, independent of our actual use. Hence, for Dummett, if certain speakers, be they members or non-members of our linguistic community, have linguistic practices which are incompatible with the meaning-theory, it is not because their words have a different meaning but because their use is incoherent: this is, I believe, the way to understand Dummett's revisionism.

Should we then say that the meanings of logical and mathematical statements, for example, are fixed for ever, and should not be affected by the evolution of our knowledge in these fields? It seems that Dummett is not ready to go as far as to claim that: "As mathematics progresses, so the relevant notion of a canonical proof will change, and hence the meanings of our mathematical statements are always, to some degree, subject to fluctuation" (1977, pp. 402–403).[33] But this remark is in a way very surprising given the general objectives that Dummett links with the theory of meaning: if we admit that logical statements (and logical constants) can change their meanings, we must then admit that there is no *unique* model of their meanings. In consequence, the metaphysical question which a meaning-theory is expected to settle do not have a determinate and permanent answer, since such a theory is subject to evolution.

The dilemma that Dummett faces is the following: either he claims that meaning is fixed "for good" by the meaning-theory independently of the specificities of use of certain periods of history or linguistic groups, or he concedes that meaning may evolve with the development of science and mathematics. If he opts for the first alternative, he may consistently claim that a univocal answer might be given to the various metaphysical questions which concern us, but for that, he must appeal to a transcendental notion of meaning which seems to be contrary to his approach[34] and to the principle that meaning is use. In any cases, Dummett does not give us any indication of how to characterize such a notion. On the other hand, if he chooses the second alternative, Dummett allows a better correspondence between the meaning-theory and linguistic practice, but then he certainly cannot hope to reach the objectives he attaches to the theory of meaning: the resolution of various metaphysical questions and the justification of our logic can only receive contextual answers, relative to the use of the relevant linguistic community.

However, some of Dummett's remarks point to what *could* be a way out of the dilemma. The fact that our linguistic practices evolve shows that a *static* meaning-theory is inadequate; we should rather account for meaning by means of a *dynamic* theory,[35] which is supposed to take into consideration the changes in use. For example, in a dynamic theory, what counts as a canonical proof of a statement may change with the development in mathematics. A dynamic account has first to specify the meanings of statements at a given time, on the basis of their use of speakers, and second it has to specify the changes of meaning resulting from changes in use.

The dynamic meaning-theory should, among other things, account for the fact that some analytic statements may eventually be rejected. The rejection of an analytic statement may occur only if there is a change of meaning such that a canonical proof of the statement (which does not appeal to observation) is no longer possible. The difficulty of course is to distinguish between theoretical changes and changes of meaning, and this does not seem possible without appealing (again) to some transcendental criterion. In order to make it possible for a dynamic theory to establish the relevant distinctions, there must exist a strict and universal criterion allowing us to determine, from any evolving linguistic practice, what is the appropriate notion of direct verification (or canonical proof) at any given state in its evolution, and how such a notion must be adjusted according to changes in that practice. Otherwise, different specifications of meaning would be possible given a certain use (there would then be an indeterminacy of meaning), and there would be different ways of accounting for the changes of meaning related to changes in use. In consequence, the appeal to a dynamic theory merely moves the task of specifying a transcendental semantic notion to a different level. The feasibility of a dynamic account of meaning rests on the possibility of providing a transcendental criterion that enables us to determine what we should regard as a direct verification of a statement given *any* possible use of this statement. Unfortunately, Dummett does not give us any clue as to how we could formulate such a criterion.[36]

The criticisms that I have raised with regard to Dummett's views aim at his molecular conception of meaning: in my opinion, Dummett has not shown in any convincing way how a molecular meaning-theory may be constructed. In other words, he has not given a satisfactory response to Quine's objections to the analytic-synthetic distinction, since he has not proposed any criterion enabling us to draw such a distinction. Actually, several of Dummett's remarks indicate that he views his molecularism as a *methodological hypothesis* that one is forced to adopt in order to construct an adequate meaning-theory.[37] Without such a hypothesis, Dummett writes, we would have to make do with a holistic model of meaning which cannot be used for criticizing our linguistic practice, since within such a model sentences do not have determinate meanings.[38] My own view is that given the link that Dummett establishes between the construction of a meaning-theory and the solving of metaphysical disputes, molecularism cannot simply be accepted as a hypothesis but must rather be argued for in a convincing way, which we are still awaiting.[39]

Université de Montréal

NOTES

* I am grateful to the Social Sciences and Humanities Research Council of Canada for an award which helped make this paper possible. I owe thanks to Daniel Laurier for stimulating discussions on the topics dealt with here. I would also like to thank Lawrence Deck for helping me with the English version of this paper.
[1] Dummett writes: "Any justification for adopting one logic rather than another as the logic for mathematics must turn on questions of *meaning*" (1973b, p. 215); "[T]he whole point of my approach to [. . . the various disputes concerning realism] has been to show that the theory of meaning underlies metaphysics" (1978, p. xl). See also 1991, introd., and 1993a. Dummett uses the phrase 'the theory of meaning' to designate a branch of philosophy which we usually

call 'the philosophy of language', whereas 'a meaning-theory' would designate "a complete specification of the meanings of all words and expressions of one particular language" (1991, p. 22). In this paper, all references are to Dummett's writings, unless otherwise specified.

[2] The disputed class of statements might comprise, for example, statements about physical reality, mathematical statements, statements about the past, the future, mental states, etc.

[3] Dummett is very clear on the fact that a model of meaning ought to conform with our actual linguistic practice. See 1973a, p. 310; 1973c, pp. 381–382; 1976, p. 116; 1981, p. 591; 1991, pp. 13–14, 340.

[4] Dummett characterizes the dispute between realists and anti-realists as a dispute about the principle of bivalence and the forms of argument which are based on it.

[5] Actually, for Dummett, knowledge of a language should not be assimilated to a simple practical knowledge, like knowledge of how to swim for example: "[Knowledge of a language] is an acquired ability to engage in a practice of such a kind that one cannot know what engaging in it consists in until one has acquired the ability to do so" (1991, p. 94). Contrarily to a simple practical knowledge, knowledge of a language is only marginally a "know how" (as for instance the ability to produce and recognize certain sounds): learning a language concerns mainly what to do, since it is only in learning how to speak an dunderstand a language that one can come to know what it is to speak it. In other words, one can know what swimming is (that is, what one must do in order to swim) without being able to swim, but one cannot know what it is to speak a language if one does not speak it. Knowledge of a language is thus in between purely practical knowledge and explicit theoretical knowledge. See 1987, pp. 262, 283; 1991, pp. 93–97; 1993c, pp. 94–96.

[6] See 1976, pp. 74–75, 82, 127; 1991, pp. 24–25.

[7] In his writings, Dummett uses somewhat interchangeably the expressions 'statement' and 'sentence' (see 1976, p. 67, note). In what follows, neither will I establish any important distinction between these expressions.

[8] The semantic value of an expression may be, for instance, its reference or extension.

[9] Hence, for Dummett, sense determines semantic value, since the semantic value of an expression follows from its sense and certain facts concerning external reality. See 1991, p. 123.

[10] The sense of a sentence is thus only one ingredient of its meaning. The other ingredients are its force and its *tone*. (See 1991, pp. 113–122.) The tone comprises different aspects of meaning which do not belong to sense or force. For instance, the tone concerns the difference in meaning between 'and' and 'but', 'dead' and 'deceased', 'enemy' and 'foe', etc. In what follows, I will focus on the first part of the theory of meaning, and, to avoid cumbersome turns of phrase, I will simply pretend that the meaning of an expression boils down to its sense.

[11] A decidable sentence is a sentence for which there exists an effective procedure which would, in a finite time, put us into a position in which we could recognize whether or not its truth conditions obtain.

[12] I could not, without considerably modifying the course of this exposition, review the different possible replies to Dummett's argument against realism. I nevertheless invite the interested reader to take a look at McGinn (1980), Loar (1987), Wright (1987) and Laurier (1991a). Two remarks deserve to be made though. First, it should be noted that an essential premise of Dummett's argument is that a meaning-theory should be *molecular*, that is, it should associate with each sentence, taken individually, a specific meaning. (I will come back to this notion in a moment.) Second, it must be clear that a repudiation of Dummett's argument does not necessarily lead to endorsing realism. For example, the possibility of a non-molecular meaning-theory combined with an anti-realist conception, should not be excluded a priori. I will not discuss this issue any further here.

[13] See however 1976, where Dummett proposes to account for meaning in terms of *falsification*, and even claims (pp. 123–126) that the notion of falsification is, in the order of explanation, prior to that of verification.

[14] See 1976, p. 127; 1983, p. 122.

[15] See 1983, pp. 102, 118–122; 1991, pp. 211–214.

[16] The two theories are complementary if the requirement of *stability* (I will explain this

notion in a moment) is satisfied. See 1987, pp. 279–280; 1991, pp. 320–321; 1993b, pp. 162–163.

[17] In certain places however, Dummett associates the meaning of a sentence, not with its complete verification from the periphery (or its complete canonical proof in the case of mathematical statements), but with the understanding of the *last step* in establishing it (or of what its *immediate* consequences are). See 1983, p. 117; 1987, pp. 233, 272–273.

[18] See 1976, p. 72; 1991, pp. 100–101, 224; 1993c, p. 100.

[19] See 1975, pp. 118–119.

[20] We may admit that in certain cases, the understanding of a sentence could only be possible in conjunction with the understanding of other sentences of equal complexity. In such cases, we would say that a grasp of these sentences can only be acquired simultaneously.

[21] Another way of expressing this constraint would be the following: when an expression is added to the language (or to a fragment of the language), the rules for its use should fix its meaning, but its addition should not affect the meanings of the expressions which already belonged to the language (or fragment).

[22] We find here, presented from a different perspective, Dummett's argument against the realist conception which admits bivalence: certain constraints are imposed on a meaning-theory (that is, meaning is use, molecularism, etc.) and we find out that the forms of inference depending on the principle of bivalence are devoid of justification in such a theory.

[23] See also 1991, p. 317; 1993a, p. 473; and Prawitz (1987, p. 145).

[24] In the final section I will come back on the importance, from Dummett's point of view, of having strict and well-defined semantic notions.

[25] See 1973a, p. 316; 1973b, p. 222; 1991, pp. 219–220, 267–279, 309–310.

[26] For example, an empirical statement which is directly verifiable now may subsequently become only indirectly verifiable. Let me also mention the case of statements for which a direct verification would be too costly in terms of energy, time, etc., so that it is never performed (but could have been) and the speaker contents himself with an indirect verification.

[27] As far as I know, in his later writings, Dummett is neither more explicit nor more optimistic on that question.

[28] A criterion for uniqueness is given in 1991, pp. 246–247.

[29] See 1991, pp. 260–261.

[30] The demand for stability for the whole language could be expressed in the following way: "This condition is that we do not demand more, in justification [i.e. verification] of a statement, than is required for it to carry the commitments [i.e. consequences] we take it to bear; conversely, we should not understand it as carrying fewer commitments than would be warranted by what we require as a justification of it" (1993b, p. 163).

[31] Actually, Dummett (1991, pp. 272–277) admits that the meanings of the conditional and the universal quantifier cannot be entirely given by their introduction rules, since they also depend on the non-deductive principles which are admitted into the subordinate deductions (in the first case, the principles which allow us to infer the consequent whenever we assume the antecedent; in the latter, the inductive principles allowing to infer the quantified statement from the free-variable one).

[32] For a similar criticism, see Wright (1987, sect. 8).

[33] See also 1991, pp. 301–303.

[34] For Dummett, let us recall, a meaning-theory must conform with our actual linguistic practices.

[35] See 1973c, pp. 410–416; 1991, pp. 243–244.

[36] It is important to recall that for Dummett, it is not merely the meanings that the speakers themselves ascribe to their statements by their use of these statements, or by the explanations they give of this use, which must be determined (insofar as one could make sense of such a notion of meaning); it is *the* adequate model of meaning, which would justify, among the set of linguistic practices, only those that are coherent and allow a systematic account, that must be determined. The dynamic character of the theory no doubt ensues from the fact that *certain* changes in the linguistic practice demand a revision of the model of meaning. Hence,

according to the dynamic account of meaning, a multiplicity of adequate models of meaning is possible, given the potential evolution of ourl inguistic practice. We thus encounter again the problem raised before: the various metaphysical questions and the question of the justification of logic can only receive contextual answers, relative to the use of a given linguistic community at a given time. Hence, far from resolving the dilemma presented above, the dynamic account *combines* the difficulties that are peculiar to each of its horns.

[37] Dummett writes: "[M]y own preference is, therefore, to assume as a methodological principle that holism is false" (1975, p. 121). See also 1973a, p. 309; 1975, pp. 132–133; 1983, pp. 117–118; 1987, pp. 234, 249–251; 1991, pp. 241–242.

[38] This is not the only objection that Dummett raises against holism. For a more thorough examination of the issue, see Loar (1987), Tennant (1987) and Laurier (1991a).

[39] Dummett sometimes calls his view a "research programme", hoping for leniency I suppose, not to regard it as if it were a sharply defined and articulated philosophical system. However, I think that despite the programmatic character of his enterprise, Dummett should provide us with some convincing arguments that it is feasible, and since molecularism is at the heart of this enterprise, it would be especially important to have a positive argument to the effect that the molecular view of language is the correct one.

REFERENCES

Devitt, Michael, 1984, *Realism and Truth*, Princeton University Press, Princeton.

Dummett, Michael, 1963, 'Realism'. Reprinted in *Truth and Other Enigmas*, Harvard University Press, Cambridge (MA), 1978, pp. 145–165.

Dummett, Michael, 1973a, 'The Justification of Deduction'. Reprinted in *Truth and Other Enigmas*, Harvard University Press, Cambridge (MA), 1978, pp. 290–318.

Dummett, Michael, 1973b, 'The Philosophical Basis of Intuitionistic Logic'. Reprinted in *Truth and Other Enigmas*, Harvard University Press, Cambridge (MA), 1978, pp. 215–247.

Dummett, Michael, 1973c, 'The Significance of Quine's Indeterminacy Thesis'. Reprinted in *Truth and Other Enigmas*, Harvard University Press, Cambridge (MA), 1978, pp. 375–419.

Dummett, Michael, 1974, 'The Social Character of Meaning'. Reprinted in *Truth and Other Enigmas*, Harvard University Press, Cambridge (MA), 1978, pp. 420–430.

Dummett, Michael, 1975, 'What Is a Theory of Meaning?', in S. Guttenplan (ed.), *Mind and Language*, pp. 97–138.

Dummett, Michael, 1976, 'What Is a Theory of Meaning? (II)', in G. Evans and J. McDowell (eds.), *Truth and Meaning*, pp. 67–137.

Dummett, Michael, 1977, *Elements of Intuitionism*, Clarendon Press, Oxford.

Dummett, Michael, 1978, 'Preface', in *Truth and Other Enigmas*, Harvard University Press, Cambridge (MA), pp. ix–lviii.

Dummett, Michael, 1979, 'What Does the Appeal to Use Do for the Theory of Meaning?', in A. Margalit (ed.), *Meaning and Use*, pp. 123–135.

Dummett, Michael, 1981, *Frege: Philosophy of Language*, 2nd ed., Duckworth, London.

Dummett, Michael, 1983, 'Language and Truth', in R. Harris (ed.), *Approaches to Language*, pp. 95–125.

Dummett, Michael, 1987, 'Replies to Essays', in B. Taylor (ed.), *Michael Dummett*, pp. 219–330.

Dummett, Michael, 1991, *The Logical Basis of Metaphysics*, Harvard University Press, Cambridge (MA).

Dummett, Michael, 1993a, 'Realism and Anti-Realism', in *The Seas of Language*, Clarendon Press, Oxford, pp. 462–478.

Dummett, Michael, 1993b, 'Truth and Meaning', in *The Seas of Language*, Clarendon Press, Oxford, pp. 147–165.

Dummett, Michael, 1993c, 'What Do I Know When I Know a Language', in *The Seas of Language*, Clarendon Press, Oxford, pp. 94–105.

Evans, Gareth and John McDowell (eds.), 1976, *Truth and Meaning*, Oxford University Press, Oxford.

Gabbay, D. and F. Guenthner (eds.), 1986, *Handbook of Philosophical Logic*, Vol. 3, D. Reidel Publishing Company, Dordrecht.

Guttenplan, Samuel (eds.), 1975, *Mind and Language*, Clarendon Press, Oxford.

Harris, Roy (ed.), 1983, *Approaches to Language*, Pergamon Press, Oxford.

Kirkham, Richard L., 1989, 'What Dummett Says about Truth and Linguistic Competence', *Mind* **93**, 207–224.

Laurier, Daniel, 1991a, 'Comprendre ou interpréter', in D. Laurier (ed.), *Essais sur le sens et la réalité*, pp. 101–131.

Laurier, Daniel (ed.), 1991b, *Essais sur le sens et la réalité*, Bellarmin/Vrin, Montréal/Paris.

Laurier, Daniel, 1993, *Introduction à la philosophie du langage*, Mardaga, Liège.

Loar, Brian, 1987, 'Truth Beyond All Verification', in B. Taylor (ed.), *Michael Dummett*, pp. 81–116.

Margalit, A. (ed.), 1979, *Meaning and Use*, D. Reidel Publishing Company, Dordrecht.

McGinn, Colin, 1980, 'Truth and Use', in M. Platts (ed.), *Reference, Truth and Reality*, pp. 19–40.

Platts, Mark (ed.), 1980, *Reference, Truth and Reality*, Routledge & Kegan Paul, London.

Prawitz, Dag, 1987, 'Dummett on a Theory of Meaning and Its Impact on Logic', in B. Taylor (ed.), *Michael Dummett*, pp. 117–165.

Quine, W. V. O., 1951, 'Two Dogmas of Empiricism', in *From a Logical Point of View*, Harper and Row, New York, pp. 20–46.

Quine, W. V. O., 1960, 'Carnap and Logical Truth'. Reprinted in *Ways of Paradox and Other Essays*, Harvard University Press, Cambridge (MA), 1976, pp. 107–132.

Sundholm, Göran, 1986, 'Proof Theory and Meaning', in G. Gabbay and F. Guenthner (eds.), *Handbook of Philosophical Logic*, pp. 471–506.

Taylor, Barry (ed.), 1987, *Michael Dummett. Contributions to Philosophy*, Martinus Nijhoff Publishers, Dordrecht.

Tennant, Neil, 1987, 'Holism, Molecularity and Truth', in B. Taylor (ed.), *Michael Dummett*, pp. 31–58.

Wright, Crispin, 1987, 'Dummett and Revisionism', in B. Taylor (ed.), *Michael Dummett*, pp. 1–30.

NOTES ON THE AUTHORS

PAUL BERNIER, born in Québec City, holds a B.A. (1986) and a M.A. (1988) in Philosophy from the Université Laval, and a Ph.D. in Philosophy (1993) from the Université de Montréal. He is currently working as Post-Doctoral Fellow at the Graduate Center of the City University of New York. His main interests are in Philosophy of Mind, Cognitive Science and Philosophy of Language. He has recently published 'Narrow Content, Context of Thought and Asymmetric Dependency'.

MURRAY CLARKE, born in Winnipeg (1956), received his M.A. from Dalhousie University and his Ph.D. from the University of Western Ontario. He is Associate Professor of Philosophy and Graduate Program Director at Concordia University. His primary interests are in Contemporary Epistemology and the History and Philosophy of Science. He has recently published in *Philosophical Studies, Synthese, Philosophy of Science*, and his work has been anthologized.

JOCELYNE COUTURE is Docteur en Philosophie (Logic and Philosophy of Science) from the University of Aix-Marseille I (1982). She is Professor of Philosophy (Ethics and Political Philosophy) in the Department of Philosophy at the Université du Québec à Montréal. Her fields of interest are Methodology, Decision Theory and Contemporary Contractarian Ethics. She is editor of *Éthique et rationalité*, co-editor of *Éthique sociale et justice distributive* and *Métaphilosophie/Reconstructing Philosophy*? She has published papers in Ethics, Logic and Philosophy of Mathematics.

FRANÇOIS DUCHESNEAU, born in Montréal (1943), holds an Agrégation de Philosophie (France) and a Doctorat-ès-Lettres in Philosophy from the Université de Paris I. He started his teaching career at the University of Ottawa. A former editor of *Dialogue*, he is Professor of Philosophy and Associate Dean of the Faculty of Arts and Sciences at the Université de Montréal. His primary interests rest with History of Science and Methodology from 17th to 19th Century, and with Philosophy of Biology. He authored five books: *L'empirisme de Locke, La physiologie des Lumières: empirisme, modèles et thèories, Genèse de la théorie cellulaire, Leibniz et la méthode de la science, La dynamique de Leibniz.*

PAUL DUMOUCHEL is professor in the Department of Philosophy at the Université du Québec à Montréal. He is the author of numerous articles in

295

M. Marion and R. S. Cohen (eds.), Québec Studies in the Philosophy of Science II, 295–298.
© 1996 *Kluwer Academic Publishers.*

Philosophy of Science and Philosophy of Mind. His book, *Le corps social: essai sur les émotions* is due off the press in 1994.

SUSAN DWYER was born in Kuala Lumpur (1958), holds a B.A. (Hons) from the University of Adelaide (1986), and a Ph.D. in Philosophy from the Massachusetts Institute of Technology (1991). She is an Assistant Professor in Philosophy at McGill University. Her primary interests are in the Philosophy of Language and Moral Epistemology. She is also the editor of *The Problem of Pornography* (1994).

DENIS FISETTE, born in Sherbrooke, Québec (1954), was educated in Germany (Berlin) and Canada (Ph.D. Université de Montréal, 1986) and worked as a Post-Doctoral Fellow at Stanford University (1987–1989). He is Professor of Philosophy at the Université du Québec à Montréal. His primary interests are Phenomenology, Philosophy of Mind, Psychology and Action Theory. He has published several papers on these topics and has been guest editor of a special issue of *Lekton*, of which he is the current editor, on Daniel Dennett and of *Philosophiques* on phenomenology and intentionality. He is the author of *Husserl et Frege* (1994) and is writing a book on phenomenology and psychology.

J. NICOLAS KAUFMANN, born in Switzerland, studied Psychology and Philosophy in Belgium. He holds a Ph.D. in Philosophy, specializing in Philosophy of Science, from the University of Louvain (1978), He is Professor of Philosophy and Graduate Program Director at the Université du Québec-à-Trois-Rivières, and past President of the Canadian Philosophical Association. He published a number of articles in his fields of interest: Philosophy of Social Sciences, Philosophy of Mind and Action, Phenomenology.

MAURICE LAGUEUX, born in Montréal (1940), holds a Ph.D. (3e cycle) in philosophy from the Université de Paris-Nanterre (1965) and a M.A. in economics from McGill University (1970). He is a Professor of Philosophy at the Université de Montréal. His primary interests are Philosophy of History, Methodology of Economics and Philosophy of Architecture. He is the author of *Le marxisme des années soixante* and of numerous articles.

DANIEL LAURIER was born in 1951 in Montréal. He studied philosophy at the Université du Québec à Montréal and the Université de Provence Aix-Marseille I, where he obtained his Ph.D. in 1983. He was a post-doctoral fellow at York in 1983–84 and at Cambridge in 1984–85. He is professor of philosophy of language and analytical philosophy at the Université de Montréal since 1987. He is the editor of *Essais sur le sens et la réalité* (1991), co-editor of *Essais sur le langage et l'intentionalité*, author of *Introduction à la philosophie du langage* (1993) and of several articles in the Philosophy of Language and the Philosophy of Mind. His main interests are in the theory of meaning

and intentionality. He is currently working on a monograph on radical interpretation.

JAMES McGILVRAY was born in India in 1942. He received his Ph.D. in
philosophy from Yale in 1968, and has been at McGill University for twenty
years. Except for a two-year period, he was Chair of the Department of
Philosophy from 1984 to 1994. He has worked on issues of time and tense,
the semantics of natural languages, and issues of color perception. A recent
book, *Tense, Reference and Worldmaking* (1991) combines issues of tense with
the semantics of natural languages. An article defending a subjectivist approach
to color is forthcoming in *Synthese*. Prof. McGilvray continues to work on both
issues of color perception and the semantics of natural languages.

MARTIN MONTMINY, born in Sept-Îles, Québec (1963), studied Physics
at Université Laval, where he obtained a B.Sc. and a M.Sc. He is currently
working on his Ph.D. dissertation in Philosophy at the Université de Montréal.
His dissertation bears on the Foundations of Semantics of Natural Languages,
a topic on which he has already published papers.

ROBERT NADEAU, born in Montréal (1944), holds a Doctorate in Philosophy
from the Université de Paris X (1973). He wrote his dissertation under Paul
Ricoeur's supervision on Ernst Cassirer's Philosophy of Language. Since 1971,
he has been a Professor of Philosophy at the Université du Québec à Montréal,
where he teaches Epistemology, the Philosophy of the Social Sciences and,
for the last twelve years, the Methodology of Economics. He has edited three
books, co-authored one and published numerous articles and chapters in edited
books. He recently co-edited a special issue on Nelson Goodman of the *Revue
Internationale de Philosophie* (Vol. 46, No. 185, 2/3/1993) and he is now
preparing a special issue of *Dialogue* on Philosophy and Economics (Vol.
34, No. 3, Summer 1995). He is also completing a monograph on Hayek's
methodological writings. He is presently director of the *Research Group on
Rationality and Social Sciences* (funded by the Québec Government) and he
is the managing editor of the long-standing working papers series *Cahiers
d'épistémologie* (as of March 1994 more than 190 issues have been published).
Finally, he is a former President of the *Canadian Philosophical Association*
and from October 1991 to June 1994 served as President of the *Canadian
Federation of Humanities*.

CLAUDE PANACCIO, born in Montreal (1946), studied Philosophy and
Medieval Sciences at the Université de Montréal (Ph.D. 1978). He is Professor
of Philosophy at the Université du Québec à Trois-Rivières and, since 1991,
co-editor of *Dialogue*, the journal of the Canadian Philosophical Association.
His main fields of interest are Philosophy of Mind, Philosophy of Language,
Ontology, and Medieval Philosophy. As well as many articles, he has authored
Les mots, les concepts et les choses (on William of Ockham and today's

Nominalism) and co-edited *Philosophie au Québec* (with P. A. Quintin) and *L'idéologie et les stratégies de la raison* (with C. Savary).

MICHEL SEYMOUR, born in Montréal (1954), holds a Ph.D. from the Université du Québec à Trois-Rivières (1986). He was a Research Fellow at the University of California at Los Angeles (1988–90). He is Associate Professor in the Department of Philosophy at the Université de Montréal. He has published articles in areas such as Philosophy of Language and Philosophy of Mind. He is the author of *Pensée, langage et communauté. Une perspective anti-individualiste* (1994).

EVAN THOMPSON, born in Ithaca N.Y. (1962), holds an A.B. in Asian Studies from Amherst College and a M.A. and Ph.D. in Philosophy from the University of Toronto. From 1992–94 he was an Assistant Professor in the Department of Philosophy at Concordia University. He is currently an Assistant Professor in the Department of Philosophy and the Center for the Philosophy and History of Science at Boston University. His primary areas of interest are Cognitive Science, Philosophy of Biology, Philosophy of Mind and Comparative Philosophy. He is the author of *Colour Vision: A Study in Cognitive Science and the Philosophy of Perception* (1994), and (with Francisco J. Varela and Eleanor Rosch) of *The Embodied Mind: Cognitive Science and Human Experience* (1991). His articles have also appeared in *Behavioral and Brain Sciences, Synthese, Philosophical Studies, Metaphilosophy*, and *Philosophy East and West*.

ALAIN VOIZARD, born in Montréal (1961), B.A. and M.A. in philosophy (Université de Montréal, 1982 and 1984, respectively), D.É.A. and Doctorat Nouveau Régime (Université de Paris I, Panthéon-Sorbonne, 1985 and 1991, respectively), and a Post Doctoral Fellow at the Université de Montréal (1991–93). He is presently Professor in the Department of Philosophy of the Université du Québec à Montréal. He has written a number of papers and articles on subjects ranging from Wittgenstein, matters of Philosophy of Language, Theories of Truth, to Epistemology. His main interests are General Epistemology and Theory of Knowledge, Analytical Philosophy of Language, Probability Theory and Decision Theory.

NAME INDEX

Boston Studies in the Philosophy of Science

123. P. Duhem: *The Origins of Statics*. Translated from French by G.F. Leneaux, V.N. Vagliente and G.H. Wagner. With an Introduction by S.L. Jaki. 1991
ISBN 0-7923-0898-0

124. H. Kamerlingh Onnes: *Through Measurement to Knowledge*. The Selected Papers, 1853-1926. Edited and with an Introduction by K. Gavroglu and Y. Goudaroulis. 1991
ISBN 0-7923-0825-5

125. M. Čapek: *The New Aspects of Time: Its Continuity and Novelties*. Selected Papers in the Philosophy of Science. 1991
ISBN 0-7923-0911-1

126. S. Unguru (ed.): *Physics, Cosmology and Astronomy, 1300-1700*. Tension and Accommodation. 1991
ISBN 0-7923-1022-5

127. Z. Bechler: *Newton's Physics on the Conceptual Structure of the Scientific Revolution*. 1991
ISBN 0-7923-1054-3

128. É. Meyerson: *Explanation in the Sciences*. Translated from French by M-A. Siple and D.A. Siple. 1991
ISBN 0-7923-1129-9

129. A.I. Tauber (ed.): *Organism and the Origins of Self*. 1991
ISBN 0-7923-1185-X

130. F.J. Varela and J-P. Dupuy (eds.): *Understanding Origins*. Contemporary Views on the Origin of Life, Mind and Society. 1992
ISBN 0-7923-1251-1

131. G.L. Pandit: *Methodological Variance*. Essays in Epistemological Ontology and the Methodology of Science. 1991
ISBN 0-7923-1263-5

132. G. Munévar (ed.): *Beyond Reason*. Essays on the Philosophy of Paul Feyerabend. 1991
ISBN 0-7923-1272-4

133. T.E. Uebel (ed.): *Rediscovering the Forgotten Vienna Circle*. Austrian Studies on Otto Neurath and the Vienna Circle. Partly translated from German. 1991
ISBN 0-7923-1276-7

134. W.R. Woodward and R.S. Cohen (eds.): *World Views and Scientific Discipline Formation*. Science Studies in the [former] German Democratic Republic. Partly translated from German by W.R. Woodward. 1991
ISBN 0-7923-1286-4

135. P. Zambelli: *The Speculum Astronomiae and Its Enigma*. Astrology, Theology and Science in Albertus Magnus and His Contemporaries. 1992
ISBN 0-7923-1380-1

136. P. Petitjean, C. Jami and A.M. Moulin (eds.): *Science and Empires*. Historical Studies about Scientific Development and European Expansion.
ISBN 0-7923-1518-9

137. W.A. Wallace: *Galileo's Logic of Discovery and Proof*. The Background, Content, and Use of His Appropriated Treatises on Aristotle's *Posterior Analytics*. 1992
ISBN 0-7923-1577-4

138. W.A. Wallace: *Galileo's Logical Treatises*. A Translation, with Notes and Commentary, of His Appropriated Latin Questions on Aristotle's *Posterior Analytics*. 1992
ISBN 0-7923-1578-2.
Set (137 + 138) ISBN 0-7923-1579-0

Boston Studies in the Philosophy of Science

Boston Studies in the Philosophy of Science

157. I. Szumilewicz-Lachman (ed.): *Zygmunt Zawirski: His Life and Work.* With Selected Writings on Time, Logic and the Methodology of Science. Translations by Feliks Lachman. Ed. by R.S. Cohen, with the assistance of B. Bergo. 1994   ISBN 0-7923-2566-4

158. S.N. Haq: *Names, Natures and Things.* The Alchemist Jabir ibn Ḥayyān and His *Kitāb al-Aḥjār* (Book of Stones). 1994   ISBN 0-7923-2587-7

159. P. Plaass: *Kant's Theory of Natural Science.* Translation, Analytic Introduction and Commentary by Alfred E. and Maria G. Miller. 1994
ISBN 0-7923-2750-0

160. J. Misiek (ed.): *The Problem of Rationality in Science and its Philosophy.* On Popper vs. Polanyi. The Polish Conferences 1988–89. 1995
ISBN 0-7923-2925-2

161. I.C. Jarvie and N. Laor (eds.): *Critical Rationalism, Metaphysics and Science.* Essays for Joseph Agassi, Volume I. 1995   ISBN 0-7923-2960-0

162. I.C. Jarvie and N. Laor (eds.): *Critical Rationalism, the Social Sciences and the Humanities.* Essays for Joseph Agassi, Volume II. 1995   ISBN 0-7923-2961-9
Set (161–162) ISBN 0-7923-2962-7

163. K. Gavroglu, J. Stachel and M.W. Wartofsky (eds.): *Physics, Philosophy, and the Scientific Community.* Essays in the Philosophy and History of the Natural Sciences and Mathematics. In Honor of Robert S. Cohen. 1995
ISBN 0-7923-2988-0

164. K. Gavroglu, J. Stachel and M.W. Wartofsky (eds.): *Science, Politics and Social Practice.* Essays on Marxism and Science, Philosophy of Culture and the Social Sciences. In Honor of Robert S. Cohen. 1995   ISBN 0-7923-2989-9

165. K. Gavroglu, J. Stachel and M.W. Wartofsky (eds.): *Science, Mind and Art.* Essays on Science and the Humanistic Understanding in Art, Epistemology, Religion and Ethics. Essays in Honor of Robert S. Cohen. 1995
ISBN 0-7923-2990-2
Set (163–165) ISBN 0-7923-2991-0

166. K.H. Wolff: *Transformation in the Writing.* A Case of Surrender-and-Catch. 1995   ISBN 0-7923-3178-8

167. A.J. Kox and D.M. Siegel (eds.): *No Truth Except in the Details.* Essays in Honor of Martin J. Klein. 1995   ISBN 0-7923-3195-8

168. J. Blackmore: *Ludwig Boltzmann, His Later Life and Philosophy, 1900–1906.* Book One: A Documentary History. 1995   ISBN 0-7923-3231-8

169. R.S. Cohen, R. Hilpinen and Q. Renzong (eds.): *Realism and Anti-Realism in the Philosophy of Science.* Beijing International Conference, 1992. 1995
ISBN 0-7923-3233-4

170. I. Kuçuradi and R.S. Cohen (eds.): *The Concept of Knowledge.* The Ankara Seminar. 1995   ISBN 0-7923-3241-5

Boston Studies in the Philosophy of Science

Also of interest:

R.S. Cohen and M.W. Wartofsky (eds.): *A Portrait of Twenty-Five Years Boston Colloquia for the Philosophy of Science, 1960-1985*. 1985   ISBN Pb 90-277-1971-3

Previous volumes are still available.

KLUWER ACADEMIC PUBLISHERS – DORDRECHT / BOSTON / LONDON